T0213017

TAXONOMY AND PLANT CONSERVATION

The Cornerstone of the Conservation and the Sustainable Use of Plants

This book illustrates the key role played by taxonomy in the conservation and sustainable utilisation of plant biodiversity. Divided into four parts, the book opens with an overview of the place of taxonomy in science and in implementing the Convention on Biological Diversity. Part II outlines what taxonomy is, how it is done, its theoretical basis and why the results of taxonomy are a measure of diversity. The third part explains how taxonomy is used to establish priorities and to identify the necessary conservation action. The concluding part illustrates taxonomy in the practice and measurement of effective conservation action. The book contains authoritative contributions by taxonomists and users of taxonomy who have spent their working lives addressing these issues. These contributions together demonstrate the crucial importance of supporting the Global Taxonomy Initiative and the importance of taxonomy in implementing the targets in the Global Strategy for Plant Conservation.

The volume is a tribute to Professor Vernon Heywood who has done so much to highlight the importance of sound scholarship, training and collaboration for plant conservation. The chapters draw on and develop the unique and significant contribution that he has made to the field, resulting in a comprehensive overview of its present status, suitable for advanced students, researchers and conservation professionals.

ETELKA LEADLAY is Head of Research and Membership Services with Botanic Gardens Conservation International, whose mission is to encourage botanic gardens to work together for plant conservation and sustainable development.

STEPHEN JURY is Herbarium Curator for the University of Reading and Principal Research Fellow with responsibilities including herbarium curation, research and teaching.

TAXONOMY AND PLANT CONSERVATION

The Cornerstone of the Conservation and the Sustainable Use of Plants

Edited by

ETELKA LEADLAY
STEPHEN JURY

CAMBRIDGE
UNIVERSITY PRESS

CAMBRIDGE UNIVERSITY PRESS
Cambridge, New York, Melbourne, Madrid, Cape Town, Singapore,
São Paulo, Delhi, Dubai, Tokyo

Cambridge University Press
The Edinburgh Building, Cambridge CB2 8RU, UK

Published in the United States of America by Cambridge University Press, New York

www.cambridge.org
Information on this title: www.cambridge.org/9780521607209

First published 2006

A catalogue record for this publication is available from the British Library

Library of Congress Cataloguing in Publication data
Taxonomy and plant conservation : the cornerstone of the conservation and the sustainable use of
plants / Etelka Leadlay & Stephen Jury (ed.).
p.   cm.
Includes bibliographical references and index.
ISBN 0-521-84506-8 (hardback : alk. paper) – ISBN 0-521-60720-5 (pbk. : alk. paper)
1. Plants – Classification.   2. Plant conservation.   I. Leadlay, Etelka, 1947–   II. Jury, Stephen L.   III. Title.
QK95.T373   2005   580′.12 – dc22   2005006465

ISBN 978-0-521-84506-9 Hardback
ISBN 978-0-521-60720-9 Paperback

Transferred to digital printing 2009

This book is a tribute to the work of Vernon Hilton Heywood, following the occasion of his 75th birthday, 24th December 2002. Professor V. H. Heywood (born 1927) has made a unique contribution to plant taxonomy and the conservation of plants. This collection of papers builds on Professor Heywood's seminal work, *Principles of Angiosperm Taxonomy* (Davis & Heywood, 1965) to his present work with the DIVERSITAS Programme. Professor Heywood was an inspiring teacher encompassing undergraduate, M.Sc. and Ph.D. students as well as post-doctoral research fellows from many countries. He has formally supervised, or helped to supervise over 60 doctoral theses. He has left a wonderful trained legacy. His alumni have been given a deep respect for rigorous science and thought, plus the fundamental importance of communication and cooperation between all sectors and at all levels. Papers were invited not only from Professor Heywood's colleagues and former students but other international experts in the field.

Professor Crane's paper under the title of 'Science and the future of plant diversity' was given at Reading University on 16th December, 2002 to celebrate Professor Heywood's actual 75th birthday.

Contents

Notes on contributors

John R. Akeroyd, Lawn Cottage, West Tisbury, Salisbury, Wiltshire SP3 6SG, UK.

John Akeroyd is a freelance writer and botanist, and an Associate Editor of *Plant Talk*, which he founded with Hugh Synge in 1995. He has a Ph.D. in plant gene-cology and was principal researcher for the revision of the first volume of *Flora Europaea* (1993). Author of *Seeds of Destruction: Non-native Wildflower Seed and British Floral Biodiversity* (1994) for Plantlife, he has promoted the use of local or native seed in habitat restoration.

Stephen Blackmore, Regius Keeper, Royal Botanic Garden Edinburgh, Royal Botanic Garden Edinburgh, 20a Inverleith Row, Edinburgh EH3 5LR, UK.

Stephen Blackmore is Regius Keeper of the Royal Botanic Garden Edinburgh. He has held visiting professorships at the universities of Edinburgh, Glasgow and Reading, and at the Kunming Institute of Botany; he has written over 100 papers on taxonomy and palynology. Since working at the Royal Society Aldabra Research Station in the 1970s he has promoted the importance of research stations in long-term biodiversity research and has been involved in establishing the Las Cuevas Research Station in Belize and the Jade Dragon Field Station near Lijiang in China. He was one of the contributors to the Gran Canaria Declaration and is a member of the international coordinating group for the Global Plant Conservation Strategy.

David Bramwell, Director, Jardín Botánico Viera y Clavijo, El Palmeral 15, Tafira Alta, Las Palmas de Gran Canaria, Spain.

David Bramwell is Director of the Jardín Botánico Viera y Clavijo, Gran Canaria. David has written or co-authored 10 books and more than 110 papers on plant taxonomy, biogeography, conservation and the modern role of botanical gardens. He is an authority on the plants of the Canary Islands and island plants in general,

and is one of the founders of the Gran Canaria Group, which carried forward
the initiative to prepare the Global Strategy for Plant Conservation on behalf of
the Convention on Biological Diversity (CBD) Secretariat. He is also a member
of the World Conservation Union's Plant Conservation Committee and is on the
International Advisory Council of Botanic Gardens Conservation International. He
has been awarded an MBE.

R. K. Brummitt, The Herbarium, Royal Botanic Gardens, Kew, Richmond, Surrey, TW9 3AE, UK.

R. K. (Dick) Brummitt gained his Ph.D. at Liverpool University in 1963 under the
supervision of V. H. Heywood. After obtaining his Ph.D. he has worked as a higher
plant taxonomist in the Herbarium at the Royal Botanic Gardens, Kew and as a
retired Research Associate since 1999. His work and field experience have covered
many parts of the world, but his main interests have been in tropical Africa. At
the Nomenclature Sessions of the 1969 International Botanical Congress he was
elected to the Committee for Spermatophyta, and in 1975 he was appointed as its
Secretary, a position he has held ever since. In 1995 he was appointed Convenor of
the Species Plantarum Project set up to compile a new Flora of the world, and is
currently its Secretary.

Santiago Castroviejo, Real Jardín Botánico, Consejo Superior Investigaciones
Científicas, Plaza de Murillo, 2, 28014 Madrid, Spain.

Santiago Castroviejo is a Research Professor of the Consejo Superior de Investigaciones Científicas (CSIC) at the Real Jardín Botánico, Madrid, Spain. Professor
Castroviejo was Director, Real Jardín Botánico, Madrid from 1984 to 1994. He is
the General Co-ordinator of the *Flora Iberica* Project (Flora of the vascular plants
of the Iberian Peninsula and Balearic Islands) and a member of the Steering Committee of the Species Plantarum Project. He has published numerous papers and
books on his main research interests of plant taxonomy, ecology and chorology in
the Mediterranean and neotropics.

Peter R. Crane, Director, Royal Botanic Gardens, Kew, Richmond, Surrey, TW9
3AB, UK.

Peter Crane has been the Director of The Royal Botanic Gardens, Kew, UK since
1999. He was elected a Fellow of the Royal Society in 1988 and currently serves
on their Council. He is also a Foreign Associate of the US National Academy
of Sciences, a Foreign Member of The Royal Swedish Academy of Sciences and
a Member of the German Academy Leopoldina. Professor Crane joined the Field
Museum in Chicago in 1982 as Assistant Curator in the Department of Geology, and

from 1992 to 1999 served as Director with overall responsibility for the Museum's scientific programmes. Professor Crane's main research interests integrate studies of living and fossil plants to understand large-scale patterns and processes of plant evolution and increasingly he is also engaged in a variety of initiatives focused on the conservation of plant diversity. He is the author of more than 100 scientific publications, including several books on plant evolution. He was knighted in 2004.

Alastair Culham, School of Plant Sciences, The University of Reading, Whiteknights, PO Box 221, Reading RG6 6AS, UK.

Alastair Culham is lecturer in Plant Systematics in the School of Plant Sciences at The University of Reading, UK with a special focus on the use of molecular phylogenies in the study of plant diversification. He has a long-term interest in evolutionary radiations, in particular those in physical and habitat islands. Research projects have included the diversification of *Pelargonium* in the Cape region of South Africa, the evolution of *Echium* on the Canary Islands, the phylogenetics of the Ranunculaceae, genetic diversity and conservation – especially in *Cyclamen* – and a monographic study of the sundew family Droseraceae. Currently he is working on a major project to integrate taxonomic information to better understand patterns of plant diversification. He has published more than 50 scientific papers and many popular articles.

J. Cullen, Stanley Smith (UK) Horticultural Trust, PO Box 365, Cambridge CB2 1HR, UK.

James Cullen is Director of the Stanley Smith (UK) Horticultural Trust. He trained at the University of Liverpool but worked for a considerable period in Edinburgh, first in the University's Department of Botany (working on P. H. Davis's *Flora of Turkey*), later at the Royal Botanic Garden Edinburgh where, as Assistant Keeper, he edited *The European Garden Flora*. His main interest is in the maintenance and development of scientific collections of living plants.

Kate Davis, Conventions and Policy Section Royal Botanic Gardens, Kew, Richmond, Surrey, TW9 3AB, UK.

Kate Davis is CBD Implementation Officer, Conventions and Policy Section, Royal Botanic Gardens, Kew, working to facilitate communication between scientists and policymakers. She is a co-author with China Williams of *The* CBD *for Botanists*, a plain-language guide to the CBD and its practical implementation. Kate is a UK delegate to the meetings of the CBD. Kate has undergraduate and post-graduate qualifications in zoology and extensive experience in conservation, education and museum/herbarium curation.

John Dransfield, The Herbarium, Royal Botanic Gardens, Kew, Richmond, Surrey, TW9 3AE, UK.

John Dransfield has worked in the Herbarium of the Royal Botanic Gardens, Kew, UK since 1975 and currently leads Kew's palm research team. John Dransfield is widely acknowledged as the world expert on the biology and systematics of the palm family. He is the author of more than 200 scientific papers, 8 books and numerous floristic treatments and technical reports. His name is attached to 287 taxonomic names, the majority of which are palms, but he has also made an exceptional contribution to palm biology and tropical botany as a whole.

Doug Evans, European Topic Centre on Nature Protection and Biodiversity (European Environment Agency), Muséum National d'Histoire Naturelle, 57 rue Cuvier, 75231, Paris Cedex 05, France.

Doug Evans joined Scottish Natural Heritage in 1993 and has been on secondment to the European Topic Centre on Nature Protection and Biodiversity (Paris) since 1999. He has been involved in implementing the EU Habitats Directive since 1994, at first in the Scottish Highlands and Islands and more recently at the EU level, giving scientific advice and assistance to Environmental Director General of the European Commission, particularly for issues relating to plants and habitats. He is a graduate of Stirling and Aberdeen universities, where he studied plant ecology and has also worked for the UK Institute of Terrestrial Ecology and the French l'Institut National de la Recherche Agronomique.

Renée J. Grayer, Royal Botanic Gardens, Kew, Richmond, Surrey, TW9 3AB, UK.

Renée Grayer has been employed as a senior scientific officer at the Royal Botanic Gardens, Kew, UK since 1994, where she does phytochemical research in the Biological Interactions Section. Previously she worked at The universities of Reading, UK and Leiden, the Netherlands. Her current research covers the chemosystematics of the plant family Lamiaceae and the isolation of bioactive plant constituents. Dr Grayer is the author or co-author of 66 research or review articles and 14 books or book chapters. She was also a major contributor to *The Phytochemical Dictionary* and *The Handbook of Flavonoids* edited by J. B. Harborne and H. Baxter.

Chris J. Humphries, Department of Botany, The Natural History Museum, Cromwell Road, London SW7 5BD, UK.

Chris Humphries has been working as a biologist since 1966 and from 1972 as a taxonomist in the Department of Botany at the Natural History Museum, London. Chris has had three sabbatical trips – two to the University of Melbourne during

1979–80 and 1986 and six months at the Wissenschaftskolleg zu Berlin in 1994. Professor Humphries' areas of interest include the systematics of angiosperms and historical biogeography, both in theory and practice and he has done some of his best work on conservation using complementarity techniques. Professor Humphries' main publications can be seen at http://www.chrishumphries.com/cjh/publications.html. His awards include the Bicentenary Medal of the Linnean Society (1980), the Gold Medal of the Linnean Society (2001), and an Honorary Fellow of the American Association for the Advancement of Science (2002).

Christopher B. Johnson, Department of Natural Products and Biotechnology, Mediterranean Agronomic Institute, PO Box 85, 73100 Chania, Greece.

Christopher Johnson is a Senior Research Associate in the Department of Natural Products and Biotechnology, Mediterranean Agronomic Institute at Chania, Greece. Formerly, he was a lecturer in Plant Physiology and Biochemistry at The University of Reading, UK. He specialised extensively in the environmental and nutritional control of plant development, especially developmental effects of light; most recently he has concentrated on environmental and nutritional factors affecting secondary-product formation, particularly essential oils, in aromatic plants.

Stephen L. Jury, School of Plant Sciences, The University of Reading, Whiteknights, PO Box 221, Reading RG6 6AS, UK.

Stephen Jury is Herbarium Curator and Principal Research Fellow in the School of Plant Sciences at The University of Reading, UK. He completed a thesis on the Umbelliferae, supervised by Vernon Heywood in the 1970s. His teaching and research interests cover systematics, European and Mediterranean (especially Morocco) floristics, cultivated-plant nomenclature and conservation issues. At present he continues to coordinate and develop Euro+Med PlantBase, an Internet database providing plant information for Europe and the Mediterranean area.

Etelka Leadlay, Botanic Gardens Conservation International, Descanso House, 199 Kew Road, Richmond, Surrey, TW9 3BW, UK.

Etelka Leadlay is Head of Research and Membership Services for Botanic Gardens Conservation International (BGCI), UK. She has worked for BGCI since 1987 and has a Ph.D. in the systematics of the genus *Coincya* (Cruciferae) from The University of Reading, UK supervised by Vernon Heywood in the 1970s.

H. Noel McGough, Conventions and Policy Section, Royal Botanic Gardens, Kew, Richmond, Surrey, TW9 3AB, UK.

Noel McGough is Head of the Conventions and Policy Section at the Royal Botanic Gardens, Kew, coordinating Kew's role of UK CITES Scientific Authority for Plants and Kew's policy and implementation of its commitments under the Biodiversity Convention. Noel joined Kew in 1988 after working as an ecologist at the Wildlife Service Ireland, where he co-authored the *Irish Plant Red Book*. Noel is Co-Chair of the CITES Nomenclature Committee and a member of EU CITES Scientific Review Group and UK delegations to the CITES Conference of the Parties, Technical Committees and EU negotiations. Noel has published a range of papers and books on CITES and plant-trade-related matters, most recently as lead author on two CITES reference works in French, Spanish and English: *CITES and Plants* and also *CITES and Succulents*.

David S. Paterson, Royal Botanic Garden Edinburgh, 20a Inverleith Row, Edinburgh EH3 5LR, UK.

David Paterson is Deputy Director of Horticulture at the Royal Botanic Garden Edinburgh and Honorary Director of Horticulture at the Kunming Institute of Botany. He holds an Honours Diploma in Horticulture and was awarded a Master of Arboriculture degree by the Royal Forestry Society. He has participated in numerous field trips to the mountainous regions of southwest China. He has facilitated plant repatriation programmes in Sichuan and Guizhou Provinces and has managed a number of capacity-building projects in China. He proposed the development of the Jade Dragon Field Station near Lijiang in China and is now Project Manager for the facility. He is senior consultant to the Lijiang Botanic Garden project that is being developed adjacent to the field station.

Alan Paton, Conventions and Policy Section, Royal Botanic Gardens, Kew, Richmond, Surrey, TW9 3AB, UK.

Alan Paton is Assistant Keeper of the Herbarium, Royal Botanic Gardens, Kew, responsible for the collection, and management and research into the major dicotyledon plant families within the Herbarium. He is an author of 70 scientific papers with an interest and extensive experience in making information held within the Herbarium collections more accessible and relevant to the implementation of the CBD. Alan is currently a member of the Coordinating Mechanism of the Global Taxonomy Initiative and involved in the facilitation of Target 1 of the Global Strategy for Plant Conservation: a widely accessible list of known plant species.

Laura J. Pleasants, Directorate, Royal Botanic Gardens, Kew, Richmond, Surrey, TW9 3AB, UK.

Laura Pleasants is working at the Royal Botanic Gardens, Kew, as the Research Assistant to the Director, Professor Sir Peter Crane. Laura has a degree in Biology from the University of Southampton and an M.Sc. in Plant Diversity (Taxonomy and Evolution) from The University of Reading.

Ghillean T. Prance, c/o School of Plant Sciences, The University of Reading, Whiteknights, PO Box 221, Reading RG6 6AS, UK.

Ghillean Prance is Scientific Director of the Eden Project in Cornwall and Visiting Professor at The University of Reading. He was Director of the Royal Botanic Gardens, Kew from 1988 to 1999. His exploration of Amazonia included 15 expeditions in which he collected more than 350 new plant species. He is author of 19 books and has published 475 scientific and general papers in taxonomy, ethnobotany, economic botany, conservation and ecology. He has received numerous awards and honours – including election as a Fellow of the Royal Society – was knighted in 1995 and received the International Cosmos Prize in 1994.

T. C. G. Rich, Department of Biodiversity and Systematic Biology, National Museums and Galleries of Wales, Cathays Park, Cardiff CF10 3NP, UK.

Tim Rich holds an Ecology B.Sc. from Lancaster University and a Plant Physiology Ph.D. from Leicester University. He is currently Head of Vascular Plants at the Department of Biodiversity and Systematic Biology, National Museums and Galleries of Wales. His interests span taxonomy (e.g. *Sorbus*, Brassicaceae), rare-plant conservation, vegetation, plant physiology, environmental impact assessment and monitoring, with special reference to the flora of the British Isles. He has written numerous papers and books including the Botanical Society of the British Isles' *Crucifers of Great Britain and Ireland* Handbook and Plant Crib (1998).

Dominique Richard, European Topic Centre on Nature Protection and Biodiversity (European Environment Agency), Muséum National d'Histoire Naturelle, 57 rue Cuvier, 75231, Paris Cedex 05, France.

Dominique Richard is a graduate of Grenoble University, France where she studied Applied Ecology. She started her professional career in the French Alps within an agricultural State Department dealing with protected areas. She joined the National Museum of Natural History (Paris) in 1989 where she led a national project on sites of ecological interest on behalf of the French Ministry of the Environment until she was appointed in 1994 as Deputy Manager of the European Topic Centre on Nature Protection and Biodiversity (Paris), a branch of the European Environment Agency. In this position she has been acting at the interface between science, policy and communication, being responsible for reporting on biodiversity states and trends at European level in support of different policy processes, as well as for capacity building for biodiversity monitoring.

Melpomeni Skoula, Department of Natural Products and Biotechnology, Mediterranean Agronomic Institute, PO Box 85, 73100 Chania, Greece.

Melpomeni Skoula is a Senior Research Associate at the Department of Natural Products and Biotechnology, Mediterranean Agronomic Institute of Chania, Greece. Her work has concentrated on the biology and taxonomy of aromatic plants as well as ethnobotany and sustainable use of biodiversity. Recently her research has been centred on the phytochemistry and genetics of essential oils and flavonoids.

Paul P. Smith, Royal Botanic Gardens, Kew, Seed Conservation Department, Wakehurst Place, Ardingly, West Sussex, RH17 6TN, UK.

Paul Smith is International Coordinator Millennium Seed Bank Project (Southern Africa and Madagascar), Royal Botanic Gardens, Wakehurst Place, UK. Paul is a specialist in ecology and plant diversity in southern, central and eastern Africa. He has extensive experience in ecological surveying, botanical inventory, vegetation mapping and monitoring, including the use of geographic information systems and remote-sensing techniques.

Tod F. Stuessy, Department of Higher Plant Systematics and Evolution, Institute of Botany, University of Vienna, Rennweg 14, A-1030, Vienna, Austria.

Tod Stuessy is a Professor in the Department of Higher Plant Systematics and Evolution, Institute of Botany, University of Vienna, Austria. He has also worked in South America and Asia and been a productive and influential botanist with particular impact in plant systematics. His contributions include monographic studies in the Asteraceae, with over 200 publications including 9 edited or single-authored books. The highlight of the latter was his 1990 book, *Plant Taxonomy: the Systematic Evaluation of Comparative Data.* He has been President of the American Society of Plant Taxonomists and has held offices in many other societies.

S. Max Walters, c/o Cambridge University Botanic Garden, Cory Lodge, Bateman Street, Cambridge CB2 1JF, UK.

Max Walters has followed a distinguished career as researcher, teacher, *Flora Europaea* editor, popular writer and persistent advocate of close botanical and social links with continental Europe. Dr Walters was a Lecturer and Curator of the Herbarium at Cambridge University from 1949 until his appointment as the Director of the Cambridge University Botanic Garden in 1973; he has continued working in research, writing and advocacy in his retirement.

China Williams, Conventions and Policy Section, Royal Botanic Gardens, Kew, Richmond, Surrey, TW9 3AB, UK.

China Williams is CBD Education Officer, Conventions and Policy Section, Royal Botanic Gardens, Kew working to facilitate communication between scientists and policymakers, and to ensure that Kew's scientific work is in line with national legislation implementing the CBD. China develops teaching modules on the CBD and its implementation for Kew staff, students and international partners. She has a degree in History and is a Barrister at Law. China is co-author with Kate Davis of *The CBD for Botanists*, a plain-language guide to the CBD and its practical implementation.

Julia Willison, Botanic Gardens Conservation International, Descanso House, 199 Kew Road, Richmond, Surrey, TW9 3BW, UK.

Julia Willison is Head of Education at Botanic Gardens Conservation International (BGCI). Since joining BGCI, Julia has created an education network of over 400 botanic gardens. She co-edits the BGCI education review, *Roots,* and is responsible for publishing educational policy documents such as *Environmental Education in Botanic Gardens: Guidelines for Developing Individual Strategies* and the *Education for Sustainable Development: Guidelines for Action in Botanic Gardens*. Julia has organised international education congresses in the Netherlands, United States, India and Sydney, and has run training courses and projects in many other countries. She was also instrumental in establishing the International Diploma Course in Botanic Garden Education. Julia is committed to the idea that environmental education is crucial for a sustainable future.

Peter Wyse Jackson, Botanic Gardens Conservation International, Descanso House, 199 Kew Road, Richmond, Surrey, TW9 3BW, UK.

Peter Wyse Jackson has been Secretary General of Botanic Gardens Conservation International since 1994, joining in 1987 when it was known as IUCN Botanic Gardens Conservation Secretariat. He is author of over 250 scientific papers and articles on plant conservation, plant taxonomy, gardening and horticulture, Irish floristics and systematics, botanic garden development and management, and the conservation of endangered island floras; and is author, co-author or editor of 9 books. He is an adviser to botanic gardens in more than 30 countries and a leading contributor to the development, coordination and implementation of the Global Strategy for Plant Conservation adopted by the CBD in 2002. He is also Chairman (from February 2004) of the Global Partnership for Plant Conservation.

Preface

The objective of the book is to demonstrate the critical importance of taxonomy for the conservation and sustainable use of plant biodiversity.

All conventions, initiatives, strategies and programmes for plant conservation highlight the need for taxonomy. These initiatives assume that everyone understands why taxonomy is important but this is not the case and the work of conservation biologists, biodiversity management agencies and users of biodiversity is often compromised through failure to ensure an adequate taxonomic basis. To some extent taxonomy is the victim of its own success; quite simply, scientists and the public take it for granted and think 'it has been done'. This view leads some scientists to think that taxonomy has no value and is an unnecessary complication. Scientists are not always aware of its rigorous and painstaking base and do not fully understand that taxonomy is a continuous process of incorporating new information. Furthermore, taxonomists appear to work in isolation and are criticised for not collaborating more closely with conservation agencies or making their taxonomic work and expertise easily available for conservationists. This book attempts to bridge the gap between taxonomists and conservation practitioners.

Plants are universally recognised as a vital part of the world's biological diversity and an essential resource for the planet. At present we do not have a complete inventory of the plants of the world, but estimates are in the order of 250 000 to 400 000 species. Of particular concern is the fact that many of these species are in danger of extinction, threatened by habitat transformation, over-exploitation, alien invasive species, pollution and climate change. According to the 1997 *IUCN Red List of Threatened Plants* nearly 34 000 plant species, or 12.5% of the world's vascular flora, are threatened with extinction (Walter & Gillett, 1998). However, the reduction in abundance and range of many numerous and widespread species is also an expression of overall biodiversity loss. The disappearance of such vital and large amounts of biodiversity sets one of the greatest challenges for the world community: to halt the destruction of the plant diversity that is so essential to meet the present and future needs of humankind.

It was growing concern over the effects of biodiversity loss on progress towards sustainable development that led to the establishment of the Convention on Biological Diversity (CBD) in 1992 (UNEP, 1992). The identification of the 'taxonomic impediment' on our ability to manage and use our biological diversity led to the Global Taxonomy Initiative (GTI) to increase capacity in taxonomy (UNEP, 2002a). Concern that insufficient resources were being directed towards plant conservation led to the Global Strategy for Plant Conservation (GSPC) which has set targets that are to be met by 2010 with the object of halting 'the current and continuing loss of plant diversity' (UNEP, 2002b). Furthermore, it is thought that biodiversity loss, together with other forms of environmental degradation, has the potential to undermine progress towards the achievement of the Millennium Development Goals, to be achieved by 2015 (www.undp.org/mdg); in particular Goal 1 to eradicate poverty and hunger, Goals 4 and 5 to improve health and Goal 7 to ensure environmental sustainability. Biodiversity contributes to poverty reduction in five key areas: food security, health improvement, income generation, reduced vulnerability to unpredictable events (e.g. access to food and environmental risks) and ecosystem services (e.g. generation of water, prevention of erosion) (Koziell & McNeill, 2002). The CBD and thematic programmes of the CBD (GTI and GSPC) and the Millennium Development Goals show that it is more important than ever to identify and conserve biodiversity for the sustainable development of the planet.

This book addresses this issue by describing and illustrating the importance of taxonomy in conservation. The Introduction (Part I) provides an overview of the place of taxonomy in science and in implementing the CBD (UNEP, 1992); the introduction also outlines areas of taxonomy that will be of particular importance in the future. Part II describes taxonomy and the work of taxonomists, and shows how a taxonomist makes decisions and why their outputs are valuable. Part III shows how taxonomy is essential in measuring and analysing plant diversity to establish priorities and develop plans for conservation. Part IV demonstrates how taxonomy is used in the practice and measurement of effective conservation action. These chapters cover the problems of: island floras, critical vascular plants and reintroduction of plants into the wild; the sustainable development and use of plants; *ex situ* and *in situ* approaches; legislation and the importance of conservation networks.

This book endeavours to show that sound taxonomy underpins conservation and that there is an urgent need for taxonomists to work in partnership with the managers of diversity for the conservation and the sustainable use of plants.

References

UNEP (1992). *Convention on Biological Diversity (CBD): Text and Annexes*. Geneva, Switzerland: CBD Interim Secretariat. www.biodiv.org.

(2002a). *Global Taxonomy Initiative (GTI)*. Decision VI/8, UNEP/CBD/COP/6/20. Montreal, Canada: CBD Secretariat. www.biodiv.org/programmes/cross-cutting/taxonomy/default.asp.

(2002b). *Global Strategy for Plant Conservation (GSPC)*. Decision VI/9, UNEP/CBD/COP/6/20. Montreal, Canada: CBD Secretariat. (Available in hard copy in English, Spanish and Chinese from the Secretariat.) www.biodiv.org/programmes/cross-cutting/plant/.

Koziell, I. & McNeill, C. I. (2002). *Building on Hidden Opportunities to Achieve the Millennium Goals: Poverty Reduction Through Conservation and Sustainable Use of Biodiversity*. New York: The Equator Initiative (UNDP).

Walter, K. S. & Gillett, H. J. eds. (1998). *1997 IUCN Red List of Threatened Plants*. Cambridge, UK: World Conservation Monitoring Centre and Gland, Switzerland: IUCN.

Acknowledgements

We are very grateful to everyone who has given advice on this book. We would like to thank Barbara Pickersgill, Chris Humphries, Hugh Synge and Alastair Culham for help in developing the scope of the book; and all those who have read chapters and given their comments and help, in particular Dick Brummitt and David Moore. We would also like to thank family and friends who have been supportive, especially John Davey. Above all, we would like to thank our authors who have been so willing to write these excellent chapters to illustrate the importance of taxonomy.

Part I

Introduction

1

Taxonomy and the future of plant diversity science

Peter R. Crane and Laura J. Pleasants

> taxonomy . . . the lonely voice speaking on behalf of an interest in diversity
> *V. Heywood, 1967*

In the last 50 years, science has experienced a remarkable period of rapid and irreversible growth. There are more scientists alive today than all those that have lived in previous generations. More money is now being spent on science than has ever been the case in the past; and new knowledge is being created at an unprecedented rate. However, over these five decades, science has not only grown, it has also changed. It has become increasingly fragmented into ever-narrower specialisms. It has become increasingly reliant on sophisticated instrumentation. It has become highly internationalised. It has also become ever-more dependent on the coordinated activities of multiple practitioners working together in teams to make real progress in many areas.

All these factors, together with many other changes in society, mean that science is now more central in the lives and thoughts of people and governments than it has ever been in the past. But as science's prominence has grown, public demands on science and scientists have also increased. Governments and tax payers want to know what they are getting for their money. Many areas of society are also beginning to question why they need such knowledge in the first place, and to what uses that knowledge will be put. The science of plant diversity has not been immune from these changes or these pressures; and among the world's leading practitioners few have seen these decadal trends so clearly, and anticipated their outcomes so effectively, as Vernon Heywood.

As observers of just part of Vernon Heywood's career, it seems to us that much of what he has accomplished – and much of his success – has been based on four key 'principles' that have allowed him to anticipate, or stay abreast of, broader changes in science: the importance of synthesis, to make sense of knowledge that is increasingly information-rich, but also increasingly fragmented; the necessity of broad collaboration to complete significant tasks; the imperative of taking an international perspective to frame and accomplish important goals; and the need, at all times, to continue to build worldwide capacity in plant diversity science. These principles have been embedded in Vernon Heywood's career for decades but they remain important and will become even more relevant in the future. The enormous potential for scientific advancement in the field of plant diversity will not be fulfilled without such an approach, nor will we be able to respond effectively to increasing societal demands.

Forty years ago, Vernon Heywood – together with his colleague Peter Davis – provided a classic synthetic perspective on plant taxonomy that summarised progress in the first half of the twentieth century (Davis & Heywood, 1963). *Principles of Angiosperm Taxonomy* is still essential reading for any serious student of plant diversity science. It starts from the perspective that '[taxonomy] extends beyond the frontiers of biology' and that 'there is surely no aspect of biology more deeply intertwined in man's history, economy, literature, aesthetics and folklore' (Davis & Heywood, 1963). Davis and Heywood emphasised that taxonomy, the science of naming and identifying species,[1] is the foundation of all biology. They also articulated the view that taxonomy is fundamental to communication and research in all areas of plant science including ecology. Today, we also understand that taxonomy has a central role to play in the contemporary environmental arena. Taxonomy is inextricably linked to the study of biodiversity, defined as variation among organisms and the ecological complexes of which they are part, and hence to issues of conservation and sustainability. Even in the face of rapid expansion in technology, an estimated 40% of the global economy is still based on biological products and processes (WEHAB, 2002). In this context, as Davis and Heywood emphasised more than 40 years ago, taxonomy remains of central importance, but it will only flourish if it continues to evolve and rise to the challenge presented by its new context.

The origins of plant diversity science

The origins of taxonomy are rooted in informal folk taxonomies, as well as the works of early biologists and herbalists who were concerned mainly with ethnobotanically important species (Davis & Heywood, 1963). Later – through the 'Enlightenment' – the exploration, description and naming of the exuberant diversity of organisms came to be seen both as intellectually and economically important. At the close of the seventeenth century, de Tournefort stabilised what in hindsight became the

---

[1] The terms 'taxonomy' and 'systematics' are used today more or less synonymously. However, in the narrow sense, taxonomy refers to classification and all that that entails (e.g. nomenclature and types). Systematics is the broader science of biological diversity that seeks to describe not only the kinds of diversity of organisms, but also the relationships among them (Simpson, 1961).

concept of the genus. He recognised 698 genera (de Tournefort, 1700). By the mid eighteenth century, Linnaeus, the founder of the binomial system of naming species, recognised 1313 genera (Jarvis *et al.*, 1993). Today we recognise about 14 000 genera (Brummitt, 2001).

In 2003, we celebrated the 250th anniversary of the publication of Linnaeus' *Species Plantarum* (1753). In this landmark attempt to document all the world's known species, Linnaeus produced one of the most enduring syntheses in the history of botany; and, like many works of synthesis, it provided the foundation for subsequent rapid scientific advances. Linnaeus' work was both a summary of existing knowledge and a catalyst for future research. But Linnaeus was also unaware of the scale of the task before him. He declared confidently: 'That the number of plants in the whole world is much less than is commonly believed I ascertained by fairly safe calculation, in as much as it hardly reaches 10,000.' (Linnaeus, 1753). Today, some estimates suggest that there may be about 420 000 described plant species (Govaerts, 2001; Bramwell, 2002).

Throughout the lifetime of Linnaeus, and the rest of the eighteenth century, European plant hunters explored the newly accessible interiors of North America, they first encountered Australia, and they also began to penetrate parts of Africa and Asia. They were motivated both by a desire for scientific discovery and the potential economic rewards from horticultural novelties or other plants of practical value. In the nineteenth century these explorations continued, especially in western North America, South America, Africa and over larger parts of Asia. Vast collections of herbarium specimens and living plants were accumulated, and the transfer of plants around the world took place with extraordinary speed.

Until the mid nineteenth century, most of the science of biological diversity focused on documenting the variety of life. But attention then turned to understanding how this diversity may have arisen. Darwin first presented his ideas about evolution by natural selection in a joint paper with Alfred Wallace published in 1858 (Darwin & Wallace, 1858), and this was quickly followed by the publication of *The Origin of Species* (Darwin, 1859). Supplemented by an increased understanding of genetics during the twentieth century, this theory has been refined and has come to be widely accepted as the explanation for how biological diversity has been generated. However, as Davis and Heywood (1963) pointed out, this huge theoretical advance brought with it no new approaches or techniques to assist practising taxonomists in their work.

Plant diversity science, 1960–2000

In their review of developments in taxonomy up to the mid twentieth century, *Principles of Angiosperm Taxonomy*, Davis and Heywood (1963) drew together the theory and methods underpinning the classification of plants. Davis and Heywood (1963) brought plant taxonomy up to date, made its principles and methods more accessible, and sought to place them in the context of contemporary science. They noted that the 1940s and 1950s brought taxonomy a wealth of new techniques and theories in fields such as cytology, ecology, genetics and evolution. As early as the 1960s they pointed out how the face of taxonomy might change in the future from being purely 'intuitive' or 'taught by imitation' to greater maturity as a repeatable, and even experimental, science based on a multidisciplinary approach that draws evidence not only from morphology, but also from a myriad of other sources.

After the publication of *Principles of Angiosperm Taxonomy*, during the 1960s, practitioners in taxonomy experimented with a wide range of ever more sophisticated techniques. The widespread use of scanning and transmission electron microscopy allowed plant and pollen surfaces to be examined in detail for the first time. Advances in gas–liquid chromatography and mass spectrometry enabled phytochemistry to be studied in unprecedented detail (Harborne, 1970). As a result a great new range of potential taxonomic characters were uncovered.

In addition, in the late 1960s (e.g. Heywood, 1968), the first inklings of new quantitative approaches – as exemplified in Sneath and Sokal's *Numerical Taxonomy* (1973) – were beginning to permeate botanical consciousness. To quote Heywood (1968, p. 7): 'The last preserve of the taxonomist – the taxonomic eye, the so-called intuition – is being probed and analysed into its components with a view to machine copying.' Numerical taxonomy introduced new ideas, new analytical approaches and fresh debates. It subsequently exploded as the aspirations of its theory were gradually matched by DNA technology and the increased sophistication and capacity of computers to process and analyse large data sets.

Over the last 40 years, advances in phylogenetic theory (e.g. Hennig, 1966), continued increases in computational power and an improved understanding of DNA and the genome, have together fuelled a new revolution in taxonomy. Taxonomy has developed to levels of sophistication that were simply unimaginable at the beginning of Vernon Heywood's career. Perhaps most significantly, the invention of the polymerase chain reaction (PCR) (Saiki *et al.*, 1988) allowed plentiful amounts of DNA to be obtained from small amounts of plant material (including herbarium specimens). This has permitted the development of 'high-throughput' methods and, together with increasingly sophisticated computer analyses, has revolutionised our ability to analyse the relationships between different groups of plants. To a very large extent (although not entirely), subjectivity has been reduced. Different regions of the genome can now be examined to reveal different aspects of evolutionary history, the combined use of morphological and molecular data – when analysed in computationally intensive ways – now results in explicit, repeatable, testable and robust phylogenies. Before the current molecular age, Davis and

Heywood (1963) regarded classifications based on phylogeny as completely unrealistic in groups with an inadequate fossil record. The prevailing view was that classifications should be based on overall resemblances, 'which may then be interpreted in phylogenetic terms'. Today's phylogenies are still models that need to be interpreted, but they are models that rest on a more sophisticated theoretical and empirical base. Phylogenetics has breathed new life into taxonomy. It has also stimulated botanists to expand their horizons and their methods to encompass a world-view that taxonomy too can be 'big science'.

Plant diversity science in the twenty-first century

Even though the scientific study of plant diversity began well before words such as evolution, genetics, ecology or DNA were ever used in a scientific context it is perhaps surprising that at the beginning of the twenty-first century, after more than 300 years of effort, the basic process of discovering plant diversity is still far from complete. We now find ourselves in the extraordinary situation where some branches of taxonomy are highly developed, while others have clear deficiencies in even the most basic information. Perhaps most worryingly of all, we remain uncertain about what it is that we actually do know, largely because much of the practice of systematics – which is classically a decentralised scientific activity – is surprisingly unsystematic.

It is also clear that we still have much to learn about even the basic units of botanical diversity. Entries in IPNI (the International Plant Names Index) (2004) reveal that over 2000 new species are described each year; a large proportion of which come from the tropics, the most botanically diverse, unique and undercollected areas of the world. Here, the practical difficulties of exploratory fieldwork are often exacerbated by the fact that many of the new species now being discovered exist only in small populations with very narrow distributions (Prance *et al.*, 2000). For example, the Wollemi Pine (*Wollemia nobilis*) – a new genus in a family in which two genera were known previously – was discovered in 1994 in a canyon of the Wollemi National Park about 150 km northwest of Sydney, Australia. This remarkable living fossil was found growing in the bottom of a canyon where the microenvironment was moist and protected from bush fires; today it persists as a population of only 43 individuals (Botanic Gardens Trust, 2004). Similarly, the recent rediscovery of *Takhtajania perrieri* in Madagascar, 85 years after its original discovery (Schatz & Lowry, 1998), exemplifies the difficulties of locating small populations in remote areas. In Brazil alone, one of the world's centres of plant diversity (Davis *et al.*, 1997) which contains significant hotspots of biodiversity (Myers *et al.*, 2000), 39 000 new species names have been described from type specimens and listed in *Index Kewensis* since 1898.

It is also very obvious that there is currently insufficient taxonomic capacity to keep up with the rate of discovery of new species. An increase in taxonomic capacity is needed even to process the outputs of current fieldwork. Building up human capacity, requiring a major increase in training programmes for taxonomists throughout the world, needs to remain a high priority. Vernon Heywood expended enormous effort in this area throughout his career and this important work must continue.

It is also crucial to recognise that even in countries where capacity for plant diversity research has been increasing (e.g. Brazil), a persistent major impediment to taxonomic work is that the key literature and specimens, which are needed to understand both new and previously described species, are not easily available to the in-country scientists who need them to do their work. Efforts devoted to capacity building should therefore include digitising and increasing online availability of scientifically important historical information, including images of type specimens and primary literature. Much of this material is physically located in the 'North', but is needed in the 'South'. New initiatives to make such information electronically available point the way for other efforts around the world (e.g. the Australian Virtual Herbarium (2004), the International Plant Names Index (2004), digitisation of the Missouri Botanical Garden's nomenclatural database and associated images VAST (VAScular Tropicos) (2004), the Herbarium of the New York Botanical Garden (2004) and the electronic Plant Information Centre (ePIC) of the Royal Botanic Gardens, Kew (2004)).

Much of the current interest in taxonomy, especially from governments, is driven by the relevance to plant conservation of the data contained in such electronic resources. The link between plant diversity science and conservation biology is also encouraging the development of new scientific approaches. Computer map-based geographic information systems (GIS) are used increasingly to understand environmental information and to manage and process spatial data (Burrough & McDonnell, 1998). These approaches are now being used increasingly to integrate biological and ecological information. At the same time, population genetics is also becoming increasingly important to conservation biology. Genetic diversity within species (Guarino *et al.*, 1999) is an often-neglected aspect of biodiversity. Before the advent of DNA-based techniques, morphometric and protein-based analyses were commonly used to assess levels of genetic diversity at the population level; now, however, DNA sequencing, amplified fragment length polymorphism (AFLPs, a PCR-based method) and plastid microsatellite-based techniques can all be used to identify isolated lineages, determine genetic diversity within and among populations, and hence guide conservation science. The data provided by such techniques are increasingly important in guiding management decisions focused on small populations of key species.

The impact of the Convention on Biological Diversity (CBD)

Increased concern about the conservation and sustainable use of plant diversity has been driven by significant changes in the political and social spheres over the last 30 years. In turn, this has changed the context in which plant diversity science is pursued. During the 1960s, concern about the impacts of humans on the environment increased dramatically, and culminated in the 1992 United Nations Conference on Environmental Development (UNCED) in Rio de Janeiro, more commonly referred to as the 'Earth Summit'. At Rio, world governments sought to find ways to halt the destruction of natural resources by committing to the CBD, as well as the associated need for national strategies of sustainable development. For plant taxonomy, adoption of the CBD was a key event. The Convention established three main goals: the conservation of biological diversity, the sustainable use of its components, and the fair and equitable sharing of the benefits from the use of genetic resources (UNEP, 1992). The CBD recognised that biodiversity is a natural resource that needs to be managed in a sustainable way. It also recognised that sustainable use of biodiversity, and the opportunity to benefit from biodiversity in an economic as well as an aesthetic way, creates incentives for conservation, especially among the poorer countries of the world.

The Sixth Conference of the Parties to the CBD (Decision VI/8) recognised 'taxonomy to be a priority in implementing the Convention on Biological Diversity' and endorsed the Global Taxonomy Initiative (GTI) (UNEP, 2002a). The framework of the GTI aims to support maintenance of reference collections and taxonomic-capacity building, to improve accessibility of taxonomic data and to generate taxonomic information to underpin decision making concerning species conservation and sustainable development. As no formal knowledge assessment was carried out before the negotiation of the CBD, the UN Environment Program (UNEP) commissioned the *Global Biodiversity Assessment*, edited by Heywood and Watson (1995). This effort was a global synthesis of the state of the global biodiversity knowledge created from the peer-reviewed contributions of more than 1500 experts from the world of academia and international organisations, from developing and developed countries alike.

The *Global Biodiversity Assessment* undertook a key scientific analysis of the issues surrounding biodiversity under five main headings: (i) agriculture, fisheries and the over-harvesting of resources; (ii) habitat destruction, conservation, fragmentation and degradation; (iii) introduction of exotic/invasive organisms and diseases; (iv) pollution of soil, water and the atmosphere; and (v) global change. The ideas that the book contained helped to lay the foundation for the 2002 World Summit on Sustainable Development (WSSD, 2002).

The Hague and Johannesburg 2002

During the preparations for the WSSD in Johannesburg in late August and early September 2002, biodiversity was identified as one of five areas in which action must be taken to help alleviate poverty and to achieve the broad aim of sustainable development. Discussions at the WSSD emphasised the five crucially interrelated areas of Water, Energy, Health (and the Environment), Agriculture and Biodiversity (and Ecosystems Management). This was the first crystallisation of an increasingly influential international view that highlighted the importance of biodiversity in the context of poverty alleviation, and the importance of biodiversity in preserving the integrity of other vital resources.

Also in Johannesburg the world committed itself to a goal of '[achieving] by 2010 a significant reduction in the current rate of loss of biological diversity' – an ambitious target. However, the WSSD did not even begin to develop specific mechanisms to measure how much biodiversity we have, or the rate at which it is being lost. As a contribution on how to approach such a challenge, the Royal Society has published a framework for measuring biodiversity. This makes explicit existing best practice for carrying out long-term monitoring, so that we can apply consistent sampling methods for the collection of new biodiversity data, and address the gaps in our current knowledge (Royal Society Working Group, 2003). This has now been followed by an international meeting and workshop that drew together governments, non-governmental organisations and scientists to help develop proactive approaches to global-biodiversity measurement.

In the context of the post-Rio development of the CBD in 2002, we also saw for the first time clear recognition by the global community that a more comprehensive, synthetic and cohesive understanding of the plant world must be an international priority. In April 2002, the Sixth Conference of the Parties to the CBD in the Hague adopted a Global Strategy for Plant Conservation (GSPC) (UNEP, 2002b). This strategy outlines 16 clear, time-limited targets towards improving our understanding of, and conserving, the world's valuable plant resources (Table 1.1). The first and most fundamental of these sets down the challenge to produce, by 2010, a taxonomically standardised world checklist of plant species, as a first step towards completing a World Flora. Completing even this first step is a massive task that cannot be accomplished without synthesis, collaboration, an international perspective and expansion of the worldwide capacity in plant diversity science.

Science, conservation and sustainability

In spite of the remarkable theoretical and technological developments over the past few decades, the science of taxonomy is sometimes still criticised for being too inward looking. Increasingly, however, in light of the CBD process and the Rio and Johannesburg summits, taxonomists are now considering the wider conservation and sustainability management issues to which their work relates. The GSPC is the most obvious manifestation of the trend that recognises that taxonomy is the foundation on which wider issues and decisions regarding the future of plant diversity must be based. The formulation of the GSPC – largely by taxonomists and botanic garden specialists – and its acceptance by the international community, was a landmark accomplishment.

Anticipating these developments, in recent years there has been a significant increase in the extent to which conservation-orientated initiatives are being undertaken by traditional taxonomic organisations. At one extreme, new scientific research has been devoted to methods of prioritising conservation efforts more effectively. For example, the relative phylogenetic distinctiveness of taxa with respect to patterns of phylogenetic branching can now be approached objectively. The ongoing Natural History Museum-based 'Worldmap' project uses this approach to link problems in biogeography, biodiversity assessment and conservation needs at a variety of spatial scales (Williams, 1996). At the other extreme, traditional methods of *ex situ* conservation are being applied and developed in new ways and on an expanded scale. *Ex situ* conservation represents the mainstay of botanic gardens, and is a key way to educate the public and raise awareness of conservation and sustainability issues. The GSPC explicitly acknowledges the key role that *ex situ* conservation has to play in the conservation of species and genetic diversity. Botanic Gardens Conservation International (BGCI) has estimated that botanic gardens in the European Union alone represent living collections of up to 50 000 plant species, or almost 20% of the world's known flora (Cheney *et al.*, 2000). It is also significant that *ex situ* collections comprise not just growing plants, but also seed banks, and cell and tissue culture collections.

The Millennium Seed Bank (MSB), based at Wakehurst Place, UK is the largest *ex situ* conservation project focusing on the preservation of wild plant species (Smith *et al.*, 1998). All overseas projects undertaken by the MSB are carried out in collaboration with local partners and are regulated by formal Access and Benefit-Sharing Agreements, as recommended by the CBD. In some cases benefit-sharing constitutes a financial agreement, but it can also involve capacity building and a great variety of other initiatives, including training. For example, every year the MSB runs a two-week course in seed-conservation methods and in-country courses are also arranged with individual countries. Such efforts are also given increasing priority on an international scale. For example, at the Rio Summit in 1992, the UK Government introduced the 'Darwin Initiative for the Survival of Species' scheme which brings together individuals from biodiverse, but economically poor, countries and UK centres of expertise. Through grants provided by this scheme, the United Kingdom has been able to welcome many hundreds of individuals from developing countries and to participate in joint conservation projects, many of which include a significant training component.

Table 1.1. *The 5 main objectives and 16 outcome-oriented targets of the Global Strategy for Plant Conservation (UNEP, 2002b)*

Ultimate aim: to halt the current and continuing loss of plant diversity

Objectives	Sub-objectives	Targets to be achieved by 2010
(a) Understanding and documenting plant diversity	• Document the plant diversity of the world • Monitor the status and trends in global plant diversity and identify species, communities, habitats and ecosystems at risk, including consideration of 'red lists' • Develop an integrated, distributed, interactive information system to make information on plant diversity manageable and accessible • Promote research on genetic diversity, systematics, taxonomy, ecology and conservation biology of plants, plant communities, habitats and ecosystems, and on social, cultural and economic factors that impact biodiversity	(i) A widely accessible working list of known plant species, as a step towards a complete world flora; (ii) A preliminary assessment of the conservation status of all known plant species, at national, regional and international levels; (iii) Development of models with protocols for plant conservation and sustainable use, based on research and practical experiences;
(b) Conserving plant diversity	• Improve long term conservation, management and restoration of plant diversity, plant communities, habitats and ecosystems, *in situ* and, where necessary *ex situ*, preferably in the country of origin. Special attention will be paid to the world's important areas of plant diversity and to the conservation of plant species of direct importance to human societies	(iv) At least 10 per cent of each of the world's ecological regions conserved; (v) Protection of 50 per cent of the most important areas for plant diversity assured; (vi) At least 30 per cent of production lands managed consistent with the conservation of plant diversity; (vii) 60 per cent of the world's threatened species conserved in *situ*; (viii) 60 per cent of threatened plant species in accessible *ex situ* collections, preferably in the country of origin, and 10 per cent of them included in recovery and restoration programmes;

    (ix) 70 per cent of the genetic diversity of crops and other major socio-economically valuable plant species conserved, and associated indigenous and local knowledge maintained;

    (x) Management plans in place for at least 100 major alien species that threaten plants, plant communities and associated habitats and ecosystems;

(c) Using plant diversity sustainably

    (xi) No species of wild flora endangered by international trade;

    (xii) 30 per cent of plant-based products derived from sources that are sustainably managed;

    (xiii) The decline of plant resources, and associated indigenous and local knowledge, innovations and practices that support sustainable livelihoods, local food security and health care, halted;

- Strengthen measures to control unsustainable utilisation of plant resources
- Support the development of livelihoods based on sustainable use of plants, and promote the fair and equitable sharing of benefits arising from the use of plant diversity

(d) Promoting education and awareness about plant diversity

    (xiv) The importance of plant diversity and the need for its conservation incorporated into communication, educational and public-awareness programmes;

- Articulate and emphasise the importance of plant diversity, the goods and services that it provides, and the need for its conservation and sustainable use, in order to mobilise popular and political support for its conservation and sustainable use

(e) Building capacity for the conservation of plant diversity

    (xv) The number of trained people working with appropriate facilities in plant conservation increased, according to national needs, to achieve the targets of this Strategy;

    (xvi) Networks for plant conservation activities established or strengthened at national, regional and international levels

- Enhance the human resources, physical and technological infrastructure necessary, and necessary financial support for plant conservation
- Link and integrate actors to maximise action and potential synergies in support of plant conservation

Taxonomists are also making new efforts to deploy massive amounts of data in herbarium collections to guide conservation action. For example, the MSB project uses herbarium data to produce conservation assessments and seed collection guides using GIS and remote-sensing technology. The field notes accompanying herbarium specimens are being databased to pinpoint the localities of populations of fruiting species at different times of the year (S. Balding, personal communication, 2002), leading to more productive fieldwork. More efficient fieldwork also helps to speed up basic plant-inventory work.

Population dynamics and genetic processes are also increasingly the focus of research by taxonomists and are crucial to devising specific intensive strategies for species conservation. For example, a recent species-recovery programme for the Lady's Slipper Orchid (*Cypripedium calceolus*), which was jointly undertaken by Kew and English Nature, utilised primers designed to amplify two repetitive regions of the plastid genome. These were used to screen *C. calceolus* samples from the United Kingdom and Europe to identify different genotypes. This information was then used to make suggestions for the targeting of specific populations of *C. calceolus* for seed storage and seedling production for reintroduction purposes in the United Kingdom (Fay & Cowan, 2001). It is true that, in some circumstances, genetic studies may be too expensive or too labour intensive to provide timely information for practical management (Heywood & Iriondo, 2003). However, in other situations they are invaluable for guiding conservation and restoration activities.

Conclusions

In the last 50 years, taxonomy has grown and developed as a science, in ways that were unimaginable when Vernon Heywood was at the beginning of his career. Taxonomy is now stronger as a science than it has ever been. It is also widely recognised as being of great societal relevance. The science of plant taxonomy must continue to develop, but the ongoing intelligent and systematic global synthesis of knowledge of plant diversity must be an equally important focus of future efforts for those organisations that hold globally, regionally and locally important collections and that specialise in plant diversity. Such efforts must be continued in imaginative ways with the fundamental task of building physical and human capacity in those parts of the world in which plant diversity is especially high.

In the twenty-first century, it will not be just the quality of the science that counts; there is also the political and ethical imperative to connect taxonomy to the central concerns of contemporary society. The world around taxonomy and taxonomists is continuing to change. This is no time to be complacent. Taxonomy and taxonomists have undergone a remarkable few decades of rapid evolution, but if they do not continue to evolve then they, like the species they are studying, will gradually become extinct. On the one hand, taxonomy must continue to enhance its status as a modern scientific discipline, to retain credibility with its peers and to provide a sound foundation for intellectual development. On the other hand, it must also find and exploit new ways to emphasise its broader societal relevance. For example, the provision of species-conservation assessments should become a routine part of all taxonomic treatments. New and more appropriate tools must be developed; tools that need to build on, and extend, traditional Floras to provide practical ways to help establish and monitor protected areas *in situ*. Through developing a greater understanding of plant diversity and the processes by which it is sustained, plant diversity science needs to contribute increasingly to the conservation decisions of land managers working in the field.

Securing the future of plant diversity will be difficult to achieve without international collaboration. There remain a variety of local, national and regional impediments that make such global collaboration difficult; however, we are much better placed than previous generations. Electronic dissemination of information can help to catalyse, support and accelerate collaborative efforts in ways that were previously unimaginable. Furthermore, Vernon Heywood recognised very early in his career that international collaborative work is the only way to tackle the large-scale challenges in plant diversity science. *Flora Europaea* (Tutin *et al.*, 1964–80), for example, would not have been completed without a massive, brilliantly coordinated and tenaciously executed international effort. Similar approaches will be needed in the future if we are to draw effectively and efficiently on the expertise of scientists from all around the world, and to take an international view of biodiversity. The CBD has been useful in helping to initiate activities within national boundaries. But there is also an urgent need to establish global databases and overviews. Collaboration among scientific and conservation organisations around the world is crucial if this need is to be met.

April 2003 marked the 50th anniversary of the elucidation of the structure of DNA by James Watson and Francis Crick. In 50 years, DNA-based science has moved with extraordinary speed to give us complete genome sequences for several plant species. The challenge for taxonomy is to formulate similarly bold and aggressive objectives, and then to complete them through greater coordination and cooperation among conservation groups, academic scientists, and governmental and intergovernmental agencies (Crane, 2003).

In the 'North', our privileged, but often exploitative, colonial past gives us special responsibilities. Collectors from the eighteenth, nineteenth and early twentieth centuries have endowed us with unrivalled collections of global plant diversity. We are well-equipped with research facilities. Through the efforts of Vernon Heywood and others, in training the next generation of taxonomists, we also have human resources who are well-qualified in the techniques of plant taxonomy. All these capabilities now need to be brought to bear on the crucial issues of sustainability and plant conservation through international programmes of synthesis and research that are combined imaginatively with education, training and capacity building. The GSPC has presented us with both a challenge and an unrivalled opportunity to show what we can do. We now need to work with our partners around the world to help deliver against the targets with a shared sense of purpose that transcends national boundaries.

References

Australian Virtual Herbarium (2004).
    http://www.chah.gov.au/avh/.
Botanic Gardens Trust (2004).
    http://www.rbgsyd.gov.au/information_about_plants_wollemi_pine.
Bramwell, D. (2002). How many plant species are there? *Plant Talk*, **28**, 32–3.
Brummitt, R. K. (2001). *Vascular Plant Families and Genera*. Kew, UK: The Royal
    Botanic Gardens.
Burrough, P. A. & McDonnell, R. A. (1998). *Principles of Geographical Information
    Systems*. Oxford, UK: Oxford University Press.
Cheney, J., Navarro, J. N. & Wyse-Jackson, P. (2000). *Action Plan for Botanic Gardens in
    the European Union*. Meise, Belgium: Ministry for SMEs and Agriculture,
    Directorate of Research and Development, National Botanic Garden of Belgium.
Crane, P. R. (2003). Plundering the planet. *New Scientist*, **178**(2396), 25.
Darwin, C. (1859). *On the Origin of Species by Means of Natural Selection or the
    Preservation of Favoured Races in the Struggle for Life*. London: John Murray.
Darwin, C. & Wallace, A. (1858). On the tendency of species to form varieties; and on the
    perpetuation of varieties and species by natural means of selection. *Proceedings of
    the Linnean Society*, **3**, 45–62.
Davis, P. H. & Heywood, V. H. (1963). *Principles of Angiosperm Taxonomy*. Edinburgh
    and London: Oliver and Boyd.
Davis, S. D., Heywood, V. H., Herrera-MacBryde, O., Villa-Lobos, J. & Hamilton, A. C.
    (1997). *Centres of Plant Diversity: a Guide and Strategy for their Conservation*,
    vol. 3, *The Americas*. Cambridge, UK: IUCN and WWF.
Electronic Plant Information Centre (ePIC) (2004).
    http://www.kew.org/epic/index.htm.
Fay, M. & Cowan, R. S. (2001). Plastid microsatellites in *Cypripedium calceolus*.
    (Orchidaceae): genetic fingerprints from herbarium specimens. *Lindleyana*, **16**, 151–6.
Govaerts, R. (2001). How many species of seed plants are there? *Taxon*, **50**(4), 1085–90.
Guarino, L., Maxted, N. & Sawkins, M. (1999). Analysis of georeferenced data and the
    conservation and use of plant genetic resources. In *Linking Genetic Resources and
    Geography: Emerging Strategies for Conserving and Using Crop Biodiversity*, eds.
    S. L. Greene & L. Guarino. Crop Science Society of America Special Publication 27.
    Madison, WI: American Society of Agronomy Inc.
Harborne, J. B. (ed.) (1970). *Phytochemical Phylogeny*. London and New York: Academic
    Press.
Hennig, W. (1966). *Phylogenetic Systematics*. Urbana, IL: University of Illinois Press.
Herbarium of the New York Botanical Garden (2004).
    http://scisun.nybg.org:8890/searchdb/owa/wwwspecimen.searchform.
Heywood, V. H. (1967). *Plant Taxonomy*. London: Edward Arnold.
    (1968). *Modern Methods in Plant Taxonomy*. London and New York: Academic Press.
Heywood, V. H. & Iriondo, J. M. (2003). Plant conservation: old problems, new
    perspectives. *Biological Conservation*, **113**, 321–55.

Heywood, V. H. & Watson, R. T. (eds.) (1995). *Global Biodiversity Assessment*. United Nations Environment Programme. Cambridge, UK: Cambridge University Press.

IPNI (International Plant Names Index) (2004). http://www.ipni.org.

Jarvis, C. E., Barrie, F. R., Allan, D. M. & Reveal, J. L. (1993). A list of Linnaean generic names and their types. *Regnum Vegetabile*, **127**, 1–100.

Linnaeus, C. (1753). *Species plantarum, exhibentes plantas rite cognitas, ad genera relatas, cum differentiis specificis, nominibus trivialibus, synonymis selectis, locis natalibus, secundum systema sexuale digestas*. Holmiae: L. Salvii.

Missouri Botanical Garden (2004). VAST (VAScular Tropicos). http://mobot.mobot.org/W3T/Search/vast.html.

Myers, N., Mittermeier, R. A., Mittermeier, C. G., da Fonseca, G. A. B. & Kent, J. (2000). Biodiversity hotspots for conservation priorities. *Nature*, **403**, 853–8.

Prance, G. T., Beentje, H., Dransfield, J. & Johns, R. (2000). The tropical flora remains undercollected. *Annals of the Missouri Botanical Garden*, **87**, 67–71.

Royal Society Working Group (2003). *Measuring Biodiversity for Conservation*. Policy Document 11/03. London: The Royal Society.

Saiki, R. K., Gelfand, D. H., Stoffel, S. *et al.* (1988). Primer-directed enzymatic amplification of DNA with a thermostable DNA polymerase. *Science*, **239**, 487–97.

Schatz, G. E. & Lowry P. P. II (1998). *Takhtajania perrieri* rediscovered. *Nature*, **391**, 133–4.

Simpson, G. G. (1961). *Principles of Animal Taxonomy*. New York: Columbia University Press.

Smith, R. D., Linington, S. H. & Wechsberg, G. E. (1998). The Millennium Seed Bank, the Convention on Biological Diversity and the dry tropics. In *Plants for Food and Medicine*, eds. H. D. V. Prendergast, D. R. Etkin & P. J. Houghton. Kew, UK: The Royal Botanic Gardens.

Sneath, P. H. A. & Sokal, R. R. (1973). *Numerical Taxonomy: the Principles and Practice of Numerical Classification*. San Francisco, CA: W. H. Freeman.

Tournefort, J. P. de (1700). *Institutiones rei herbariae, editio altera*. Paris: Typographia regia.

Tutin, T. G., Heywood, V. H., Burges, N. A. *et al.* (1964–80). *Flora Europaea*, vols. 1 to 5. Cambridge, UK: Cambridge University Press.

UNEP (1992). *Convention on Biological Diversity (CBD): Text and Annexes*. Geneva, Switzerland: CBD Interim Secretariat. www.biodiv.org.

(2002a). *Global Taxonomy Initiative (GTI)*. Decision VI/8, UNEP/CBD/COP/6/20. Montreal, Canada: CBD Secretariat. www.biodiv.org/programmes/cross-cutting/taxonomy/default.asp.

(2002b). *Global Strategy for Plant Conservation (GSPC)*. Decision VI/9, UNEP/CBD/COP/6/20. Montreal, Canada: CBD Secretariat. (Available in hard copy in English, Spanish and Chinese from the Secretariat.) www.biodiv.org/programmes/cross-cutting/plant/.

WEHAB Working Group (2002). *A Framework for Action on Biodiversity and Ecosystem Management*. www.johannesburgsummit.org/html/documents/summit_docs/wehab_papers/wehab_ biodiversity.pdf.

Williams, P. H. (1996). Worldmap: software and help document 4.2. Distributed privately. http://www.nhm.ac.uk/science/projects/worldmap/.

WSSD (World Summit on Sustainable Development) (2004). www.johanesburgsummitt.org.

2

Taxonomy in the implementation of the Convention on Biological Diversity

Alan Paton, China Williams and Kate Davis

This chapter gives an overview of the role of taxonomy in the Convention on Biological Diversity (CBD). It outlines how taxonomy underpins the successful implementation of the Convention, and how the implementation is currently being severely held back by the shortage of taxonomic knowledge and expertise, particularly in biodiversity-rich developing countries. The way that the two new initiatives of the CBD, the Global Taxonomy Initiative (GTI) and the Global Strategy for Plant Conservation (GSPC), tackle this problem is examined. The need to focus and coordinate existing taxonomic skills and activities is demonstrated, as is the importance of adequate funding being made available for taxonomic-capacity building. In conclusion, the need for taxonomists to work more closely with other sectors to support the taxonomy necessary to implement the CBD is stressed.

The importance of taxonomy within the CBD

The CBD represents a commitment by the nations of the world to conserve biological diversity, to use biological resources sustainably and to share the benefits arising from that use fairly and equitably (Article 1, Objectives) (UNEP, 1992). It currently has 188 Parties, more countries than any other international convention. The CBD is a framework treaty in that it does not lay down a set list of activities or closely defined obligations that Parties to the Convention must follow. Its provisions, which are laid out as a series of Articles, are expressed as overall goals. At the CBD's heart is the recognition that countries have sovereign rights over their biological diversity. Therefore, decision making is placed at the national level: individual countries interpret the provisions of the CBD according to their own national priorities, and implement them through the development of national strategies, plans or programmes. Parties to the CBD meet every two years at the Conference of the Parties (or 'COP') to coordinate and guide this work, to adopt further work programmes and take decisions to steer development. The

COP is provided with guidance by the Subsidiary Body on Scientific Technical and Technological Advice (or 'SBSTTA') (Glowka *et al.*, 1994; Synge, 1995–6; Williams *et al.*, 2003).

While the Articles of the CBD set out the framework for the conservation of biodiversity across the globe, taxonomy supplies the indispensable baseline information needed to achieve the objectives of the CBD. Simply put, if we do not know what biodiversity we have and cannot identify what we find, how can we manage, monitor, utilise and conserve it? Taxonomy is integral both to the functioning of the CBD and to the measurement of its success. In the original 1992 Convention text, Article 7 provided the most explicit reference to taxonomy, calling for Parties to 'Identify components of biological diversity important for its conservation and sustainable use'. However, over time there has been a growing acknowledgement of taxonomy's role by the Parties to the CBD, and this has led to further guidance and decisions focusing on the importance of a coordinated global initiative promoting taxonomy (see COP Decisions II/8, III/10, IV/1, VI/8, VII/9, SBSTTA recommendation II/2 (CBD website) and the January 1998 Darwin Declaration (Environment Australia, 1998). Table 2.1 illustrates how taxonomic knowledge underpins many of the articles and work programmes of the CBD.

Despite the obvious importance of taxonomy in implementing the CBD, there are still yawning gaps in taxonomic knowledge across many groups of organisms. At present, 1.7 million species have been described, but there may be anything between 4 and 15 million species that remain undescribed (UNEP, 2003a). The deficiency of baseline knowledge is compounded by the lack of taxonomic expertise necessary to document and identify the components of biodiversity. In addition, many countries lack the infrastructure in terms of collections, technological support and actual buildings capable of housing collections permanently. These shortages are most severe in the biodiversity-rich developing countries, and this lack of taxonomic expertise and infrastructure has been termed the 'taxonomic impediment' to the successful implementation of the Convention (Environment Australia, 1998).

The importance of addressing the taxonomic impediment has now been formally recognised with the adoption of the GTI as a cross-cutting programme by the Sixth Conference of the Parties in April 2002 (UNEP, 2002b); GTI web pages. The overall objectives of the GTI are to assess taxonomic needs and capacities, strengthen networks and infrastructures, facilitate systems for access to information and include taxonomic components where needed in the thematic work programmes and cross-cutting areas of the CBD. These are very broad aims, but the more specific operational objectives set out a useful structure for Parties and regional groups to follow (for examples, see UNEP (2002b) and UNEP (2003a), Appendix 3). A gap analysis of the CBD's existing work plans with respect to their taxonomic components will give greater focus to the GTI's programme of work (Decision VII/9).

Table 2.1. *Taxonomy and the implementation of the CBD's articles and work programmes*

The CBD: Articles and work programmes	Importance of taxonomy for successful implementation
Develop National Biodiversity Strategies and Action Plans (Article 6)	Taxonomy provides vital accurate and up-to-date national information
Identification and monitoring (Article 7)	Taxonomy is required to identify and monitor components of plant diversity to ensure conservation and sustainable use
In situ conservation (Article 8)	Taxonomy is required to identify components of ecosystems and protected areas and assists identification of appropriate *in situ* conservation areas
Ex situ conservation (Article 9)	Taxonomy is required for identification and curation of *ex situ* collections, and assists targeting of acquisitions for *ex situ* conservation
Sustainable use (Article 10)	Taxonomy is required to identify resources, to develop protocols for sustainable use and to design measures to minimise adverse impacts on biodiversity
Research and training (Article 12)	Training in taxonomy is central to identification, conservation and sustainable use of biological diversity. Taxonomic expertise is necessary to promote and encourage research
Access to genetic resources and benefit-sharing (Articles 15 and 19; Decision VI/24)	National inventories required to facilitate access to and use of genetic resources, to ensure fair and equitable sharing of benefits. Elements of an access and benefit-sharing regime set out in the Bonn Guidelines (UNEP, 2002a) rely on accurate identification of genetic resources
Technology transfer, information exchange, scientific and technical cooperation (Articles 16, 17 and 18)	Taxonomic expertise, data and capacity necessary at national, regional and global levels.
Global Taxonomy Initiative (Decisions VI/8 and VII/9)	GTI formally recognises taxonomic impediment and suggests objectives and a programme of work
Global Strategy for Plant Conservation (Decisions VI/9 and VII/10)	Taxonomy is needed to understand and document plant diversity. Targets 1 and 2 on the global checklist of all known plant species and conservation assessments rely on taxonomy
Alien species that threaten ecosystems, habitats or species (Article 8(h) and Decision VI/23)	Guiding Principles (Decision VI/23) emphasise the need to identify and monitor invasive species

Table 2.1. (*cont.*)

The CBD: Articles and work programmes	Importance of taxonomy for successful implementation
Traditional knowledge (Article 8(j))	Taxonomy is required to identify resources and monitor links between Western and traditional classification systems. There is a need to involve local and indigenous people in taxonomic identification
Thematic work programmes (forests, inland waters, dry and subhumid lands, islands, agricultural, marine and coastal, mountains)	Taxonomic expertise and infrastructure are necessary in all these areas, e.g. to establish baselines and monitor progress
Ecosystem approach	Ecosystem assessment requires taxonomic information for reporting on patterns of ecosystem diversity

Above all, the GTI has provided a much-needed prompt to governments and policy makers. It reminds them that accurate taxonomy is necessary for decision making, and that they should promote and facilitate taxonomic research. For example the Bonn Guidelines (UNEP, 2002a) specifically state: 'Taxonomic research, as specified by the Global Taxonomy Initiative, should not be prevented, and providers should facilitate acquisition of material for systematic use and users should make available all information associated with the specimens thus obtained.' In an international arena where developing national access legislation could potentially restrict taxonomic research for conservation, this is useful support. However, simple recognition that there is a taxonomic impediment is not enough. Effective action requires greater focus of existing taxonomic expertise and activities, and the provision of adequate and accessible funding.

Focusing on implementation of the Convention

The first step in this focusing process is for countries to undertake a taxonomic needs assessment (called for by COP in Decisions IV/1/D, V/9 and VI/8). Secondly, each Party has been asked to nominate a National Focal Point for the GTI to coordinate national strategies. However, the second wave of National Reports of the CBD indicate that only four countries have been able to complete a needs assessment (Klopper *et al.*, 2002; UNEP, 2003a), and many countries have not yet identified their GTI National Focal Points. Where focal points have been identified, the nominated person may have no links to the taxonomic community or related activities in that country. It is clear that in order to move the process forward, taxonomists need to take a much more proactive role in the GTI. For instance, taxonomic communities within countries have a role to play in nominating and supporting their GTI focal points, in building stronger national networks, and shaping and influencing the development of National Biodiversity Strategies and Action Plans so that such strategies fully address the taxonomic impediment.

Although implementation of the GTI has really only just begun, regional meetings have already been held in Central America, Africa and east Asia. These meetings have highlighted some priority issues, most notably the urgent need for capacity building at the national level (Klopper *et al.* 2001; UNEP, 2001, 2003b (Appendix 3)). In addition, its initial operations have been greatly helped by a number of bodies with a proven record of successful work in the area, and these pre-existing structures are being used by the GTI to support its work, provide useful models and contribute expertise.

For example, the Mexican governmental research organisation CONABIO (Comisión nacional para conocimiento y uso de la biodiversidad, or National Commission for the Knowledge and Use of Biodiversity) was established in 1992 to promote and coordinate activities within Mexico related to the study and use of the country's biological resources. CONABIO's overall objective is to conserve Mexico's ecosystems and produce criteria for their sustainable management (CONABIO website). It has been remarkably successful in focusing skills and harnessing knowledge on a national level. In addition, CONABIO provides financial support for projects, and gives scientific and technical advice to both the public and private sectors. Information and vegetation maps produced by CONABIO have been used by (among others) the Ministry of the Environment to plan reforestation programmes. The National Information System on Biodiversity (SNIB) being developed by CONABIO brings together a wide range of information including data from biological inventories, researchers, non-governmental organisations (NGOs) and other institutions. SNIB is designed to serve the entire biological community – including taxonomists, conservationists, biogeographers, ecologists and others. However, there are too-few examples of successful models like CONABIO.

Countries can also benefit from a regional approach. For example, the global network BioNET INTERNATIONAL is dedicated to taxonomic-capacity building on a regional level to support sustainable development (BioNET INTERNATIONAL website). BioNET largely operates through sub-regional networks of institutions from developing countries (called 'loops'), and these networks have been identified as appropriate structures through which much of the GTI can be effectively implemented (UNEP, 2002b). Currently BioNET actively supports the GTI's programme of work, liaising closely with GTI National Focal Points and helping to identify priority actions. So far, BioNET has helped to build nine regional taxonomic capacity-building networks with others under development, each with a regional and a national coordinator. In many cases these individuals have gone on to become GTI National Focal Points.

The Southern African Botanical Diversity Network (SABONET) is another example of regional cooperation. SABONET is a capacity-building network of southern-African herbaria and botanic gardens with the objective: 'to develop a strong core of professional botanists, taxonomists, horticulturists and plant diversity specialists within the ten countries of southern Africa, competent to inventory, monitor, evaluate and conserve the botanical diversity of the region in the face of specific development challenges, and to respond to the technical and scientific needs of the Convention on Biological Diversity' (SABONET website). Again, it pre-dates the GTI and provides a useful model for the GTI to learn from and build upon. As well as the development of taxonomic products, SABONET's successes include all the benefits of good networking (support, technology transfer, contacts, etc.), permanent employment of many of the taxonomists trained and increased awareness of the need for producing data to support conservation (Timberlake & Paton, 2001; Golding, 2002). In its latter stages, SABONET conducted user needs assessments and these are important in defining some of the taxonomic barriers to implementation of the Convention (Steenkamp & Smith, 2002). It is only by increasing communication between taxonomists and conservationists that these taxonomic barriers can be identified and overcome (Golding & Timberlake, 2003; Lowry & Smith, 2003).

The 'taxonomic impediment' cannot be overcome by working only at national and regional levels. Much of the world's existing taxonomic expertise and collection base is still housed in the northern hemisphere and making this information accessible to all is a major challenge. The Global Biodiversity Information Facility (GBIF) aims to make the world's primary data on biodiversity freely and universally available via the internet (GBIF, 2004). Most institutions throughout the world have their own programmes to make their data more widely accessible. Missouri Botanical Garden's VAST (VAScular Tropicos) is an example of this (Missouri Botanical Garden website). With so much information to gather and make accessible, it is crucial to identify priorities.

An important development which helps provide a focus for activities is the targets of the GSPC (UNEP, 2002c; GSPC web pages). The GSPC was, like the GTI, adopted as a cross-cutting programme by the Sixth Conference of the Parties. Its ultimate objective is to 'halt the current and continuing loss of plant diversity'. The novel development of the GSPC is that it has 16 outcome-orientated targets to be achieved by 2010; the first internationally agreed targets in biodiversity conservation ever adopted. The deadline of 2010 is consistent with that of the World Summit on Sustainable Development (WSSD), 'to reduce the loss of biodiversity by 2010' (WSSD, 2002), as well as that of the CBD's strategic plan (Decision VI/26). The agreed, clear targets of the GSPC facilitate partnerships at national, regional and global levels, between taxonomists and other sectors working for conservation and sustainable use of plants, to achieve common goals. Although the GTI ensures the need for taxonomy remains on the political agenda, without targets and without explicit connection to agendas in conservation and sustainable use, activity is likely to remain diffuse and uncoordinated. The first two targets of the GSPC emphasise the need for taxonomic baselines and several other targets rely on taxonomic input. Target 1 calls for a widely accessible working list of known plant species, as a step towards a complete world flora. This target will be integrated into the GTI's work programme, following Decision VII/10. Target 2 calls for a preliminary assessment of the conservation status of all known plant species – at national, regional and international levels.

Funding taxonomy to implement the CBD

As discussed above, activities are under way at a range of levels. Although some funding has been directed towards taxonomy under the CBD (UNEP 2003a), it has been limited. Clearly, funding sources need to be more secure for genuine success in overcoming the taxonomic impediment. Neither the GTI nor GSPC can supply funds directly, but they do both provide a framework for negotiating and programming international financial support for taxonomy. The CBD has identified the Global Environment Facility (GEF) as its interim financial mechanism and thus the source of financial support for the GTI, GSPC and other activities under the Convention (GEF website). Access to grants from the GEF starts with a collaborative project proposal backed by the relevant GEF and CBD National Focal Points. These project submissions are administered through one of the GEF's implementing agencies: United Nations Environment Programme (UNEP), United Nations Development Programme (UNDP) or the World Bank (GEF website). The Conference of Parties of the CBD has given over 100 recommendations to the GEF as to where funds should be directed. Therefore, not only does taxonomy have to compete for funding with all other areas of the CBD, but also with recommendations of all other environmentally related conventions, such as the Convention on Climate Change. With such a high number of recommendations it is difficult to prioritise funding towards taxonomy. The GEF itself cannot make taxonomy a priority; this remains the responsibility of each Party. The submission of GEF projects through National Focal Points helps to ensure that projects meet national priorities, but can be a barrier for larger regional or global projects. Regional projects need to ensure that the relevant authorities in each country have been fully consulted and are willing to support the proposal. Although some projects, such as SABONET, have been successful in obtaining funding through the GEF, it is a lengthy procedure. Existing networks can help facilitate project development, but the time and bureaucracy involved has probably meant that a lot of good projects – such as the sample framework projects outlined by Diversitas (UNEP, 1999) – have never passed the initial planning stages.

In order to access GEF funds for taxonomy, taxonomists must be proactive in identifying taxonomic needs and projects that help implement the Convention. However, GEF funding for taxonomy is required most in countries with low taxonomic capacity and with overstretched taxonomists who have the least resources to prepare lengthy proposals and convince governments of the need. Although the GEF does fund 'enabling activities', which could be used to support taxonomy within country, governments still have to make the case that such funds should be used for taxonomy as opposed to any other priority relevant to the Convention. Thus, although the GEF may be willing to fund taxonomy, countries may lack the resources to articulate this need and thus may not receive funding for this purpose.

There are clearly other means of funding taxonomy, and other institutions that are already carrying out taxonomic work. Within the GTI an attempt has been made to identify participants who are willing to support areas of the GTI work plan (UNEP, 2003c). This is a useful first step, but if the taxonomic impediment is really to be overcome then additional resources will be required, and access to existing ones simplified.

Conclusions: overcoming the taxonomic impediment

As we have seen, the taxonomic impediment remains a significant barrier to the implementation of the CBD. More needs to be done to focus taxonomic skills on the problems of implementation, and more funding is needed.

Action has begun with taxonomists themselves. Although supportive of conservation projects, they may sometimes lack knowledge of what information the conservation project needs and how best it could be delivered (Golding & Timberlake, 2003; Lowry & Smith, 2003). This is in part due to an academic hierarchy that still does not adequately value or respect the production of tools such as a user-friendly field guide or simple provision of taxonomic data in usable forms such as distribution maps. On the other hand, conservationists sometimes see taxonomy as irrelevant, or at least find it difficult to see how it can be made to be relevant. However, there has been some movement in this area. Taxonomists are beginning to shift their work to produce more user-friendly identification guides without losing scientific rigour (Lowry & Smith, 2003). In addition, they are starting to include more conservation-linked information in their more traditional work (e.g. Red List status included in monographs) and developing better ways to use data for the purposes of conservation (Willis *et al.*, 2003). However, levels of communication between taxonomists and conservationists will have to increase in order to identify areas of synergy and outputs that facilitate implementation of the Convention.

In order to address the shortfalls in funding, taxonomists will have to become more involved in the political process by creating stronger links with the GTI, GEF and CBD National Focal Points. This is not always easy, but the alternative is to miss out on CBD funding for taxonomy. Networks such as BioNET have an important role in disseminating case studies and best practice so that those in countries with little taxonomic capacity can learn from the experiences of others and better articulate their needs. Taxonomists have to recognise that within the CBD and the funding remit of the GEF, taxonomy is seen as a means towards implementation – serving conservation and sustainable use – rather than a pure baseline science. In a world where biodiversity is disappearing faster than we can describe it, the days of taxonomy for taxonomy's sake are numbered (Heywood, 2001).

Although increased funding for taxonomy through the GEF and via organisations such as the GBIF will go some way towards overcoming some of the taxonomic impediment, funding from other sources is vital. Clear agreed goals, such as those of the GSPC, will help facilitate partnerships, coordinate existing resources and identify new resources. However, increased communication between taxonomists, business and other sectors of society will still be required to increase funding significantly as the GEF is unlikely to be able or willing to provide all the resources necessary. Several large companies have recently devoted significant funds to biodiversity and this is a hugely important development. For example, the banking corporation HSBC has formed a US$50-million partnership over five years to fund conservation projects around the world (HSBC Investing in Nature Partnership, 2003). There are obvious concerns about close links between conservation work and funding by commercial companies. However, although there is clearly still much work to be done in developing appropriate agreements that can be good models for commercial partnerships, and in protecting intellectual property rights and the rights of local communities and indigenous peoples, the involvement of the commercial sector in conservation is important.

Taxonomists will have to look beyond the traditional sources of funding for their work and need to persuade all sectors of society, including the business sector, of the value of biodiversity and the sustainable use of its components. Without funding from new sources, the taxonomy necessary to support implementation will be under-resourced and the goals of the CBD will remain an unfulfilled dream.

Acknowledgements

Our thanks to Matthew Mustard and Lucy Ellerbeck for comments and assistance with the manuscript.

References

BioNet INTERNATIONAL. The Global Network for Taxonomy.
www.bionet-intl.org (accessed 14 May 2004).

CONABIO (Comisión nacional para conocimiento y uso de la biodiversidad).
www.conabio.gob.mx (accessed 14 May 2004).

Environment Australia (1998). *The Darwin Declaration.* Australian Biological Resources
Study. Canberra: Environment Australia.
www.biodiv.org/programmes/cross-cutting/taxonomy/darwin-declaration.asp
(accessed 14 May 2004).

GBIF (Global Biodiversity Information Facility).
www.gbif.org (accessed 14 May 2004).

GEF (Global Environment Facility).
www.gefweb.org (accessed 14 May 2004).

Glowka, L., Burhenne-Guilmin, F. & Synge, H. (1994). *A Guide to the Convention on
Biological Diversity.* Cambridge, UK: Gland, Switzerland and IUCN.
http://www/iucn.org/themes/law/index.html (accessed 14 May 2004).

Golding, J. S. (2002). *Southern African Plant Red Data Lists.* Southern African Botanical
Diversity Network Report Series 14. Pretoria, South Africa: SABONET.

Golding, J. S. & Timberlake, J. (2003). How taxonomists can bridge the gap between
taxonomy and conservation science. *Conservation Biology*, **17**, 1177–8.

GSPC (Global Strategy for Plant Conservation).
www.biodiv.org/programmes/cross-cutting/plant/ (accessed 14 May 2004).

GTI (Global Taxonomy Initiative).
www.biodiv.org/programmes/cross-cutting/taxonomy/default.asp.

Heywood, V. (2001). Floristics and monography – an uncertain future? *Taxon*, **50**, 361–81.

HSBC Investing in Nature Partnership (2003).
www.hsbc.com.au/information/community/env˙nature.html (accessed 14 May 2004).

Klopper, R. R., Smith, G. F. & Chikuni, A. C. (eds.) (2001). *The Global Taxonomy
Initiative: Documenting the Biodiversity of Africa.* Proceedings of a workshop held at
the Kirstenbosch National Botanical Garden, Cape Town, South Africa, 27 February
to 1 March 2001. *Strelitzia*, **12**. Pretoria, South Africa: NBI, pp. 1–202.

(2002). The Global Taxonomy Initiative in Africa. *Taxon*, **51**, 159–65.

Lowry, P. P., II & Smith, P. P. (2003). Closing the gulf between botanists and
conservationists. *Conservation Biology*, **17**, 1–2.

Missouri Botanical Garden. VAST (VAScular Tropicos).
http://mobot.mobot.org/W3T/Search/vast.html (accessed 14 May 2004).

SABONET (Southern African Botanical Diversity Network).
www.sabonet.org (accessed 14 May 2004).

Steenkamp, Y. & Smith, G. F. (2002). *Addressing the Needs of Users of Botanical
Information.* Southern African Botanical Diversity Network Report Series 15.
Pretoria, South Africa: SABONET.

Synge, H. (1995–6). The Biodiversity Convention explained. *Plant Talk*, **1**, 14–5; **2**, 22–3;
3, 26–8; **4**, 26–7; **5**, 29–32; **6**, 26–7.
http://www.plant-talk.org/Pages/cbdconts.html (accessed 14 May 2004).

Timberlake, J. & Paton, A. (2001). SABONET Mid-Term Review. *SABONET News*, **6**(1), 5–13.

UNEP (1992). *Convention on Biological Diversity (CBD): Text and Annexes.* Geneva, Switzerland: CBD Interim Secretariat. www.biodiv.org.

(1999). *The Global Taxonomy Initiative: Shortening the Distance Between Discovery and Delivery, Report of a Meeting Held at the Linnean Society, London, UK on 10–11 September 1998*, submitted by DIVERSITAS. UNEP/CBD/SBSTTA/4/Inf.1. Montreal, Canada: CBD Secretariat.

(2001). *Global Taxonomy Initiative, Progress Report on the Global Taxonomy Initiative.* UNEP/CBD/SBSTTA/6/INF/4/Add.1. Montreal, Canada: CBD Secretariat.

(2002a). *Bonn Guidelines on Access to Genetic Resources and Fair and Equitable Sharing of the Benefits Arising out of their Utilization.* Decision VI/24, UNEP/CBD/COP/6/20. Montreal, Canada: CBD Secretariat. (Available in hard copy, in all UN languages, from the Secretariat.)

(2002b). *Global Taxonomy Initiative (GTI).* Decision VI/8, UNEP/CBD/COP/6/20. Montreal, Canada: CBD Secretariat. www.biodiv.org/programmes/cross-cutting/taxonomy/default.asp.

(2002c). *Global Strategy for Plant Conservation (GSPC).* Decision VI/9, UNEP/CBD/COP/6/20. Montreal, Canada: CBD Secretariat. (Available in hard copy in English, Spanish and Chinese from the Secretariat.) www.biodiv.org/programmes/cross-cutting/plant/.

(2003a). *Draft Guide to the Global Taxonomy Initiative.* UNEP/CBD/SBSTTA/9/INF/30. Montreal, Canada: CBD Secretariat.

(2003b). *Preliminary Report of First Global Taxonomy Initiative Workshop in Asia.* UNEP/CBD/SBSTTA/9/INF/17. Montreal, Canada: CBD Secretariat.

(2003c). *Global Taxonomy Initiative: Progress and implementation of the Programme of Work.* UNEP/CBD/SBSTTA/9/INF/16. Montreal, Canada: CBD Secretariat.

Williams, C., Davis, K. & Cheyne, P. (2003). *The CBD for Botanists: An Introduction to the Convention on Biological Diversity for People Working with Botanical Collections.* Kew, UK: The Royal Botanic Gardens.

Willis, F., Moat, J. & Paton, A. (2003). Defining a role for herbarium data in Red List assessments: a case study of *Plectranthus* from East and Southern Tropical Africa. *Biodiversity and Conservation*, **12**, 1537–52.

WSSD (World Summit on Sustainable Development) (2002). Plan of Implementation. www.iisd.ca/wssd/portal.html (accessed 14 May 2004).

Part II

The practice of taxonomy

3

Principles and practice of plant taxonomy

Tod F. Stuessy

Introduction

Taxonomy, one of the cornerstones of biology, allows us to deal in an organized way with the living world. At this point in human history, we are concerned about inventorying all known life forms, and at the same time worrying about their loss through human destabilization of the environment. Within these concerns, taxonomy commands central stage. In this age of genetic engineering, one of the crucial questions to ask is: To what extent will genetically modified organisms pose threats to existing biodiversity and functioning of ecosystems? To assess these threats, taxonomy provides a logical, scientific, legal and practicable framework. We continue to be interested in the origins of biodiversity, especially the origins of our own human species. The critical evaluation of what species really are (or should be) and how they are formed can only be addressed within a broad taxonomic context. Taxonomy celebrates diversity, specifically biodiversity, but also extends to human cultural diversity and in this way contributes to social progress. Taxonomy is fundamental to all aspects of the human experience.

Taxonomy may be defined as the study of the principles and methods of classification. Classification is the grouping and ranking of organisms based on similarities and/or differences. Taxonomy, therefore, is concerned with classification, its concepts and its practical execution. Allied to this is nomenclature, which seeks to provide a stable system of naming for all life forms. Identification should not be confused with classification. The former is the referral of an individual organism into an already classified group (or taxon). The relationships between these concepts and terms are illustrated in Figure 3.1.

Associated with taxonomy are investigations on dimensions of the evolutionary process, through which all life forms have developed. Studies may focus on details of population-level phenomena, or micro-evolutionary processes, which involve revealing the origin of variation, its distribution within and among populations,

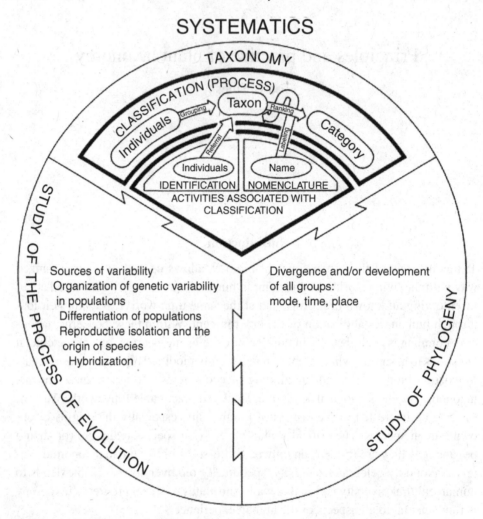

Figure 3.1 Diagram showing relationships of concepts within systematics. From Stuessy (1979).

speciation and subsequent reticulate patterns through hybridization. Other studies investigate longer-term evolutionary divergence, or phylogeny, in which is sought the time, place, and specific origins of different taxa. Much time and effort are presently being invested in reconstructing the entire Tree of Life (Maddison, 2003). Encompassing all these activities, whereby taxonomy is arguably fundamental, is systematics. Simply defined, systematics is the study of diversity, or more elaborately, the study of the kinds and diversity of organisms and their many interrelationships (Simpson, 1961). Systematics is the umbrella under which we investigate diversity, and taxonomy provides the tools for its classification.

Human need to classify: brief history of plant taxonomy

In view of its central role in biology and human affairs, it should come as no surprise that taxonomy has a long history. The first recorded documents in the Western world on plants and their relationships come from the Ancient Greeks. Theophrastus (*c.* 300 BC) wrote several manuscripts dealing with plants, which have been published later as books. His most remarkable, *Enquiry into Plants*, carefully sets down many morphological aspects, including how they grow and how they reproduce (but without a clear understanding of plant sexuality). He observes (in the translation by Hort (1916), p. 27) that 'exact classification [is] impracticable', predicting a battleground of philosophical conflicts that became reality more than 2000 years later. The practical Romans contributed little to understanding plant relationships, being content to compile existing information, such as the work done by Pliny the Elder (Gaius Plinius Secundus, AD 23–79) in his colossal *Natural History*. This work still makes interesting reading, with the botanical portion taking up 62 pages, plus fascinating cogent arguments for an Earth that is round! (see pp. 173 and 373 in the translation by Rackham (1938)).

The fall of Rome and the chaos of the Middle Ages finally led to the Renaissance and a renewed examination of plants for their structures and value for human welfare. Due to poor sanitary conditions, increased crowding in rapidly growing cities and ignorance of causes of disease, human suffering rose to appalling levels during the fifteenth to seventeenth centuries. Physicians turned to plants and their known efficacy (due to the diverse biochemistry of their secondary products) for preventing and ameliorating the effects of disease. The Age of the Herbalists (*c.* 1460–1660), therefore, was a time of looking at plants in their native habitats, of describing them from nature, of naming them for more efficient communication and of attempting simple systems of classification.

These positive developments led, during and soon after the Age of the Herbalists, to the early botanical classifiers who were interested not so much in the medicinal import of the plants but rather how they could best be classified in their own right. Workers such as Andreas Caesalpino from Italy, Gaspar Bauhin from Switzerland, John Ray from England and Josef Pitton de Tournefort from France are representative. The objective was to reveal God's system of classification that was used at the Creation. The great Swedish botanist, the 'father of plant taxonomy', Carl Linnaeus (1707–78), extended this perspective and believed (at least early in his career) that the number of sexual structures in a plant – the stamens and carpels – was the signature that God had given for their proper classification. This system, however useful it was for bringing comprehensive order to what had previously been a smorgasboard of efforts, was clearly artificial; that is, based on only a few characters.

The information retrieval capacity of the system was limited to the features used to establish it.

Awareness of the limitations of an artificial system of plant classification led to developments of a more natural system. Antoine Lauren de Jussieu, working in the Royal Botanical Garden in Paris, and stimulated by his uncle Bernard's earlier attempts to organize plants in the Petit Trianon in Versailles, published *Genera Plantarum* (1789), the first major and successful effort to show plant relationships based on many different morphological features. Adanson (1763) from Paris, earlier had also attempted such a system based on many selected characteristics, but his system failed due to radical new generic concepts in many plant groups and eccentric nomenclature. The natural system, because it was based on many features, contained a much higher level of information. Many new classifications ensued from this point, one of the most elaborate and best known being from the father and son Candolle team in Geneva, Switzerland, embodied in their *Prodromus*, the last World Flora at the specific level (excluding monocots) and published in 17 volumes (A. P. de Candolle, 1824–38; A. L. P. P. de Candolle, 1844–73).

All these positive developments, however, occurred in the absence of evolutionary theory. That is, classifications were being refined to reflect more information about plants, but the reasons for the observed relationships were still unclear. It was Darwin with his *Origin of Species* (1859) who provided strong and convincing arguments for the reality of change through time based on natural selection as the mechanism. From this time forward, classifications were viewed in a different light. Although the features of the plants being used in classification did not change, the philosophical perspective of the meaning of the classification changed completely. The hierarchical structure was no longer viewed as simply the work of the Creator, but rather as a result of the Earthly process of organic evolution. Classifications were organized along phylogenetic principles or ideas as to what might be primitive, what derived and which groups might have evolved from others. This approach spawned numerous new systems of plant classification: beginning with that of Engler and Prantl (1887–1915); and continuing with Bessey (1915) and Hutchinson (1926, 1934); and more recently Takhtajan (1969, 1997), Cronquist (1968, 1981, 1988) and the Angiosperm Phylogeny Group (APG) (1998, 2003), to name just a few.

The point of illustrating the history of plant taxonomy in the above paragraphs is to show that people have always needed to classify the living world. We classify, in fact, all aspects of our existence, including inanimate objects (e.g. furniture, automobiles, styles of clothing, types of architecture), philosophical concepts (e.g. theological viewpoints, schools of ethical thought, etc.) and even theories (e.g. origin of the universe). This need to classify may relate to our use of language and its logical structure, it may reflect our general insecurity about the life-experience and our desire to control it better or it could possibly reflect how life itself is really organized.

The importance of plant taxonomy in human affairs

For whatever fundamental reasons, biological classifications provide many positive benefits for humans in our dealing with the living world. Classification refers to the process as well as products from the process (the classifications themselves). The first benefit of classification, mentioned above, is storage and retrieval of information. Biologists make many observations and measurements, and this information must be placed somewhere for use by us and future generations; classifications make these data accessible. The second benefit of a classification is that it allows us to predict attributes of organisms not yet observed or measured. If we find, for example, a secondary plant product from some plant species that shows activity in inhibiting growth of cancerous cells, we would wish to investigate related species (or genera) to discover similar compounds with perhaps even greater potency. The classification enables us to know where to sample next for greater efficiency and success. Without this predictive framework, we would be reduced to tedious and expensive sampling across many taxa each time such a need arose. The third benefit accruing from classifications is that they stimulate investigations on evolutionary and biogeographic patterns and processes. The structure of relationships within hierarchical classification reveals close relatives that have developed through time by evolutionary processes. To study these processes, therefore, one must examine these close relatives. It is no coincidence that intensive evolutionary studies are undertaken in groups for which a predictive classification already exists. Likewise, one cannot explain geographic distributions of taxa that are not already well classified.

The recent use of DNA data for reconstruction of phylogenies of plants has ushered in a new era of quantitative assessment of relationships. Sequences in regions of nuclear or organelle (chloroplasts or mitochondria) DNA that presumably are not under intense selection, and hence accumulate mutations in a pattern reflecting common ancestry and phylogenetic history, are compared and phylogenetic trees built. Statistical tests are used to determine the robustness of these branching patterns. Such diagrams, called cladograms or phylograms (if the length of the branches is also indicated), show hierarchical patterns of relationship and are useful for constructing classifications. The large amounts of data obtainable, the simplicity of characters (the site on the double helix) and their states (the four different base pairs), the quantitative approach to classification (more objective and repeatable) and the statistical methods to test the veracity of the relationships seen, have all led to a higher acceptance and appreciation of phylogenetic trees then ever before. This is true not only in plant systematics but in all areas of comparative biology (e.g. ecology, developmental biology and genetics; Harvey *et al.*, 1996).

Principles of plant taxonomy

Articulating the principles of any field is bound to delight some and disappoint others. To my knowledge, there has never been a listing of principles of plant taxonomy, perhaps due to its multifaceted nature. At the risk of appearing presumptive, I list and discuss below those points that, in my opinion, define and distinguish our field:

1. All known life has originated on Earth during the past 3.5 billion years.
2. Due to evolutionary processes (speciation, hybridization, extinction, etc.), life forms show natural patterns of relationship to one another.
3. Using selected features of organisms (characters and states), we determine patterns of relationship that we assume reflect these evolutionary processes.
4. Humans need hierarchical systems of information storage and retrieval to live and survive, including dealing with the living world.
5. The assessed patterns of organismal relationship are used to construct hierarchical classifications of coordinate and subordinate groups that are information-rich and have high predictive efficacy; these are the taxonomic hypotheses that change with new information and new modes of analysis.
6. Names are assigned to the classified groups to facilitate communication about them.

These six principles are general ones, applicable to organismic taxonomy (i.e. all life forms). For plant taxonomy one only has to substitute 'plant' or 'green plant' etc. at the appropriate place instead of 'organism'.

We assume, based on present knowledge, that life originated on Earth. New data, however, raise doubts about whether this is the only place in the universe where life occurred. Recent investigations on a meteorite from Mars have suggested microstructures and compounds that might be organic in nature (perhaps bacterial; McKay *et al.*, 1996), even including magnetite structures that resemble modern magnetite-chain-forming bacteria (Friedmann *et al.*, 2001; Thomas-Keprta *et al.*, 2001) (these conclusions are controversial, however, e.g. Barber and Scott (2002)). More recent interpretations suggest that water was formerly present on Mars, some billions of years earlier (e.g. Head *et al.*, 1999). Furthermore, it is estimated that the surface temperature of a meteorite ejected from the surface of Mars and arriving into Earth's atmosphere would be no higher that 40 °C internally (Weiss *et al.*, 2000), a temperature easily tolerated by many modern species of bacteria. In any event, even if life evolved in parallel on Mars, or if life on Earth came from that already developed earlier on Mars, it would presumably still be monophyletic in our world; that is, having evolved from a single origin. Hence the other five principles would still apply. It is not impossible, although unlikely, that Earth was 'salted' several times from early bacteria from Mars. Depending upon how different these introductions were from each other, life on Earth might still be

Grouping

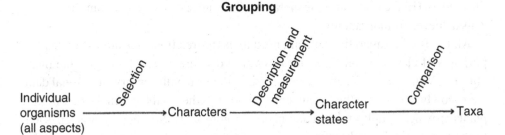

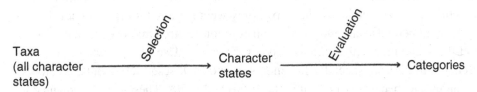

Ranking

Figure 3.2 Flow diagram showing specific steps involved with grouping and ranking, the two basic operations of biological classification. From Stuessy (1990).

regarded as monophyletic, or alternatively as polyphyletic if different arrivals were divergent – perhaps helping to explain the remarkable differences in RNA/DNA sequences among bacteria and other micro-organisms (instead of, or in addition to, relying on the huge amounts of evolutionary time available to explain the divergence).

Practice of plant taxonomy

To appreciate better the principles of plant taxonomy, we need to look closely at its practice. All classification involves operations similar to those shown in Figure 3.2. The crucial point is selecting from all available features of organisms those aspects to be compared for purposes of making groups and their ranking. A character of an organism is a feature that we judge useful in classification (i.e. in making comparisons to establish the degree of relationship) and a character state is one particular aspect of this character. For example, color of petals might be a character and character states might be yellow, white and pink. One cannot compare characters for establishing relationships; rather we compare character states. Through this process, groups are formed, and these groups are called taxa. These can be classified informally into coordinate and subordinate units. For accurate and efficient communication, however, they must be placed into accepted categories in the modern (Linnean) taxonomic hierarchy (i.e. into species, genera, families, etc.). This achieves formal ranking.

All kinds of comparative data are used to assess relationships among plants for purposes of classification. The oldest types of data are external features, the morphology. With aid of microscopes invented in the sixteenth century, additional data were slowly obtained in the seventeenth century; studies of the internal structures of plants provided details of their anatomy and embryology. Stronger microscopes led to the use of cytological features, especially chromosomes, which became established in plant taxonomy in the early decades of the twentieth century. Research on the nature of species in the 1930s through to the 1960s led to artificial hybridizations among taxa to determine their reproductive isolation and genetic similarities/differences, called cytogenetics. At this same time, the study of pollen-grain structure and its potential in plant taxonomy were explored, furthered enormously by the commercial availability of the transmission electron microscope in the 1950s and the scanning electron microscope in the 1970s. Use of secondary metabolic products in plants, especially flavonoids, mono- and sesquiterpenoids, alkaloids, betalains, etc. became important in the 1960s and 1970s. These were supplanted to a large extent by the use of isozymes in population-level studies during the 1980s. In the last decade of the twentieth century, however, DNA data – either through sequences or fingerprints – have provided enormous new and stimulating data in plant taxonomy. One might argue that this progression from the external to internal characters was nothing more than obtaining more detailed structural data, or 'deep morphology' (Stuessy, 2003). There is now such an array of data for assessing relationships that the challenge is to understand which is best to solve a particular systematic question (i.e. at what level of the hierarchy, in which type of reproductive system, within what general pattern of genetic divergence, etc.).

Although advances in numbers and types of characters have greatly enhanced our ability to characterize and evaluate structures of plants for purposes of classification, the importance of learning about plants in their natural habitats must be stressed. The reasons for this are several. First, characters and states vary in natural populations due to genetic differences among individuals and due to plasticity. Mutation, recombination, migration and extinction all affect the genetic composition of natural populations. Furthermore, because plants have no mobility as adults, they possess patterns of development that add new, and sometimes different, modules depending upon immediate ecological conditions. This is plasticity. Smaller sun versus larger shade leaves, or highly dissected submerged versus lobed emergent leaves in aquatic plants, are two familiar examples. Fieldwork helps reveal the nature of characters and states, and whether they will be suitable for classification. In addition, we can begin to suggest hypotheses for the adaptive nature of many of these characters and states, which aids our understanding of their ecological roles within the population, their relevance to isolating mechanisms and speciation, and their significance within the entire ecosystem (pollination syndromes, predator–prey relationships, reproductive behaviors, symbiotic relationships, etc.).

Within the general process of classification of grouping and ranking, and based on whatever types of data are available, three different schools of classification have developed over the past 40 years: phenetics, cladistics and phyletics (quantitative evolutionary classification). In general, the principal themes have been to: (i) determine more carefully what we really do in classification and (ii) produce classifications more quantitatively (i.e. more objectively and reproducibly, or more 'scientifically'). Phenetics at the end of the 1950s and into the 1960s was the first attempt to place biological classification on a more explicit footing. To do so, it openly rejected evolutionary interpretations, at least in the process itself, as being too complex and mired in circular reasoning (e.g. Sokal & Sneath, 1963). It treated characters and states as particles of information to be assessed mathematically for the purposes of grouping and ranking. All characters and states were treated as equal (usually) and overall similarity was emphasized. It stressed the importance of the 'basic data matrix' of all life – the matrix of characters and states that must be used in classification. It is not coincidental that the development of phenetics was correlated with the availability of computing machines, without which it would have been impossible to compare and evaluate large quantities of data. Cladistics, developing in the 1970s, viewed the quantitative aspects of phenetics favorably, but advocated the reinstatement of evolutionary dimensions into classification (e.g. Hennig, 1966; Wiley, 1981). This approach focused on determining branching patterns of evolution (representing phylogeny), which could be determined mathematically. Particular characters and states were selected that were believed to have maximum phylogenetic value. This approach has intuitive appeal because diversity has resulted from evolution, and one would, therefore, presume that classification should reflect this history. Cladistics, then, restored evolutionary interpretations back into classification, but it also maintained quantitative comparisons. Phyletics (or quantitative evolutionary classification) (Stuessy, 1987, 1997) builds upon cladistics by also emphasizing evolutionarily significant characters and states and quantitative assessment in producing branching patterns (the phylogenetic diagrams). The difference is that phyletics also takes into account the degree of divergence within lineages whereas cladistics uses only the branching pattern. With the onslaught of new DNA data, phenetic, cladistic and phyletic portrayals of relationships are all now being used routinely, depending upon the type of data being used, the nature of the plant group involved and the preferences of the individual taxonomist. Cladistics was truly dominant during the 1990s, but more phenetic algorithms and phyletic portrayals (phylograms) of relationship are being used as we enter the twenty-first century. This emphasizes the dynamic and progressive nature of biological classification.

Relationship of taxonomy to conservation of biodiversity

The relationship of taxonomy to understanding biodiversity is in many ways obvious. First, if we do not know what diversity exists on Earth, it is clearly impossible to conserve it. We need an inventory of life forms to appreciate and evaluate our biological heritage. This means knowing their existence, having them described and named (for efficient communication), understanding their evolutionary relationships (through classification) and studying their biology to understand the limits of their survival. This inventory would include learning about geographic distribution, ecological tolerances, reproductive behavior (e.g. pollination, breeding and dispersal systems), secondary products, genetics and development.

Secondly, we need life forms classified in formal hierarchies to facilitate communication and establishment of conservation priorities. Without a hierarchical arrangement of information about life, we are unable to compare regions of the world for conservation import. Although there is nothing magic about the categories we call species, genus and family, these three levels are fundamental for evaluating biodiversity in the context of conservation. In recent years, there has been much interest in establishing priorities for biosphere reserves, national parks, etc. (e.g. Campbell & Hammond, 1989; Myers *et al.*, 2000), and decisions to protect areas often depend upon awareness of high levels of taxic diversity. Although the employment of these three levels in different plant groups is not always consistent, even among taxonomists, it does provide a reasonable yardstick by which to assess levels of biodiversity. As an aside, the recent call for the possible use of an alternative system of nomenclature – the PhyloCode (Cantino & de Queiroz, 2000) – which would abandon all formal ranks above the species, seems ill-advised. This new nomenclature seeks compatibility with results from cladistic analysis, focusing on circumscriptions and naming of unranked clades. In my opinion, this approach would seriously hinder communication about the living world at a time when conservation is most needed.

Thirdly, we need more taxonomists to help do the basic work. This lack of trained personnel has been labelled the 'taxonomic impediment' (Environment Australia, 1998), and it is not trivial. Because of the enormous challenge of cataloging life, we need skilled personnel to accomplish the task. The best solution would be to have many more professional taxonomists worldwide, especially in megadiversity countries (e.g. Brazil, Ecuador, etc.). Another practical and less expensive solution is the formation of a guild of parataxonomists who can collect and identify plants in their region in collaboration with professionals (Basset *et al.*, 2000).

There are two additional initiatives that should be encouraged. The first is the more extensive use of the World Wide Web for dissemination of taxonomic information. Several authors (e.g. Godfray, 2002) have advocated that the systematics community should learn how to post more information on the World Wide Web, not only information that already exists in hard-copy, but also new information as it is being accumulated. The Web is also well-suited to color graphics, of obvious importance for portraying biodiversity. Interactive identification keys are also available now on the Web, and this should be expanded. One might also imagine that the Web could be used for original descriptions of new taxa, serving simultaneously as an automatic registration mechanism (the present rate is about 2500 new plant species per year; from *Index Kewensis*, Prance, 2005). The problem here is data permanence. Because the literature of systematics endures for hundreds of years, it is essential that original descriptions last for many generations. The solution will be electronic archiving, but no one agency or institution is yet responsible.

The second new initiative is the use of DNA data directly to assay levels of biodiversity. With procaryotes, analysis of DNA from deep-sea samples finds sequences that are 30% different from any known organisms (Fuhrman & Campbell, 1998). The vast bacterial and protist world has only begun to be inventoried. The same can be said for soil organisms. It has been advocated that the quickest way to determine levels of biodiversity is through DNA analyses of specific DNA gene sequences (such as SSU (small subunit) ribosomal RNA; Blaxter, 2003; Tautz *et al.*, 2003). Advocates point to continuing habitat destruction and the need for innovative and more rapid approaches to determining biodiversity; the traditional methods (describing, comparing, etc.) being viewed as too slow – better reserved for future efforts after natural areas have been set aside. Although a DNA inventory would not tell us much about the organisms themselves, if done on a worldwide scale it would reveal concentrations of genetic diversity. Such information, if used carefully, could be extremely helpful in conservation initiatives.

Conclusion

If one steps back and looks at the numerous challenges for conserving the living world, especially in the present political climate of global political and economic instability, it is easy to become despondent. The task is both urgent and immense. A small comfort is that in the course of human history, large impacts have often been accomplished through small but persistent efforts by dedicated (and often charismatic) individuals. The systematic biology community is fortunate to have individuals such as Thomas Lovejoy, Peter Raven, E. O. Wilson and others, who have been successful in emphasizing conservation problems with the general public. Vernon Heywood has also been most effective in stressing conservation in the context of plant taxonomy – the need for taxonomists to reach out from their own small, professional communities into society at large. As professional taxonomists, we must bring as much expertise to bear on these problems as quickly as possible.

Taxonomy stands ever ready to help by providing a framework into which we can organize information. However fast we inventory the planet, the taxonomic structure absorbs new data and adjusts accordingly, providing new and more robust predictions about the living world. Taxonomy is fundamental for conservation and all other human activities.

References

Adanson, M. (1763). *Familles des Plantes*. Paris: Vincent.

Angiosperm Phylogeny Group (APG) (1998). An ordinal classification for the families of flowering plants. *Annals of the Missouri Botanical Garden*, **85**, 531–53.

(2003). An update of the Angiosperm Phylogeny Group classification for the orders and families of flowering plants: APG II. *Botanical Journal of the Linnean Society*, **141**, 399–436.

Barber, D. J. & Scott, E. R. D. (2002). Origin of supposedly biogenic magnetite in Martian meteorite Allan Hills 84001. *Proceedings of the National Academy of Sciences of the USA*, **99**, 6556–61.

Basset, Y., Novotny, V., Miller, S. E. & Pyle, R. (2000). Quantifying biodiversity: experience with parataxonomists and digital photography in Papua New Guinea and Guyana. *BioScience*, **50**, 899–908.

Bessey, C. E. (1915). The phylogenetic taxonomy of flowering plants. *Annals of the Missouri Botanical Garden*, **2**, 109–64.

Blaxter, M. (2003). Counting angels with DNA. *Nature*, **421**, 122–4.

Campbell, D. G. & Hammond, H. D. (eds.) (1989). *Floristic Inventory of Tropical Countries: the Status of Plant Systematics, Collections, and Vegetation, plus Recommendations for the Future*. New York: New York Botanical Garden.

Candolle, A. L. P. P. de (ed.) (1844–73). *Prodromus systematis naturalis regni vegetabilis*, vols. 8 to 17. Paris: Fortin, Masson.

Candolle, A. P. de (ed.) (1824–38). *Prodromus systematis naturalis regni vegetabilis*, vols. 1 to 7. Paris: Treuttel and Würtz.

Cantino, P. D. & de Queiroz, K. (2000) PhyloCode: a Phylogenetic Code of Biological Nomenclature.
http://www.ohiou.edu/phylocode

Cronquist, A. (1968). *The Evolution and Classification of Flowering Plants*. Boston, MA: Houghton Mifflin.

(1981). *An Integrated System of Classification of Flowering Plants*. New York: Columbia University Press.

(1988). *The Evolution and Classification of Flowering Plants*, 2nd edn. New York: New York Botanical Garden.

Darwin, C. (1859). *On the Origin of Species by Means of Natural Selection or the Preservation of Favoured Races in the Struggle for Life*. London: John Murray.

Engler, A. & Prantl. K. (eds.) (1887–1915). *Die natürlichen Pflanzenfamilien*. Leipzig, Germany: Wilhelm Engelmann.

Environment Australia (1998). *The Darwin Declaration*. Australian Biological Resources Study. Canberra, Australia: Environment Australia.
www.biodiv.org/programmes/cross-cutting/taxonomy/darwin-declaration.asp (accessed 2004).

Friedmann, E. I., Wierzchos, J., Ascaso, C. & Winklhofer, M. (2001). Chains of magnetite crystals in the meteorite ALH84001: evidence of biological origin. *Proceedings of the National Academy of Sciences of the USA*, **98**, 2176–81.

Fuhrman, J. A. & Campbell, L. (1998). Microbial microdiversity. *Nature*, **393**, 410–11.

Godfray, H. C. J. (2002). Challenges for taxonomy: the discipline will have to reinvent itself if it is to survive and flourish. *Nature*, **417**, 17–19.

Harvey, P. H., Leigh Brown, A. J., Smith, J. M. & Nee, S. (eds.) (1996). *New Uses for New Phylogenies*. Oxford, UK: Oxford University Press.

Head, J. W., III, Hiesinger, H., Ivanov, M. A. *et al.* (1999). Possible ancient oceans on Mars: evidence from Mars orbiter laser altimeter data. *Science*, **286**, 2134–7.

Hennig, W. (1966). *Phylogenetic Systematics*. Trans. D. D. Davis & R. Zangerl. Urbana, IL: University of Illinois Press.

Hort, A. (1916). *Theophrastus, Enquiry into Plants*, vol. 1. Cambridge, MA: Harvard University Press.

Hutchinson, J. (1926). *The Families of Flowering Plants I. Dicotyledons. Arranged According to a New System Based on Their Probable Phylogeny*. London: Macmillan.
(1934). *The Families of Flowering Plants II. Monocotyledons. Arranged According to a New System Based on Their Probable Phylogeny*. London: Macmillan.

Jussieu, A. L. de. (1789). *Genera plantarum*. Paris: Viduam Herrissant and Theophilum Barrois.

Maddison, D. R. (ed.) (2003). The Tree of Life Web Project. http://www.tol.web.org.

McKay, D. S., Gibson, E. K., Jr, Thomas-Keprta, K. L. *et al.* (1996). Search for past life on Mars: possible relic biogenic activity in martian meteorite ALH84001. *Science*, **273**, 924–30.

Myers, N., Mittermeier, R. S., Mittermeier, C. G., da Fonseca, G. A. B. & Kent, J. (2000). Biodiversity hotspots for conservation priorities. *Nature*, **403**, 853–8.

Prance, G. T. (2005). Completing the inventory. In *Species Plantarum: 250 years*, ed. I. Hedberg, *Symbolae Botanicae Upsalienses*, **33**, 207–19.

Rackham, H. (1938). *Pliny, Natural History*, Preface and Books 1–2. Cambridge, MA: Harvard University Press.

Simpson, G. G. (1961). *Principles of Animal Taxonomy*. New York: Columbia University Press.

Sokal, R. R. & Sneath, P. H. A. (1963). *Principles of Numerical Taxonomy*. San Francisco, CA: Freeman.

Stuessy, T. F. (1979). Ultrastructural data for the practicing plant systematist. *American Zoologist*, **19**, 621–36.
(1987). Explicit approaches for evolutionary classification. *Systematic Botany*, **12**, 251–62.
(1990). *Plant Taxonomy: the Systematic Evaluation of Comparative Data*. New York: Columbia University Press.
(1997). Classification: more than just branching patterns of evolution. *Aliso*, **15**, 113–24.
(2003). Morphological data in plant systematics. In *Deep Morphology: Toward a Renaissance of Morphology in Plant Systematics*, eds. T. F. Stuessy, V. Mayer & E. Hörandl. Ruggell, Liechtenstein: Gantner Verlag, pp. 299–315.

Takhtajan, A. (1969). *Flowering Plants: Origin and Dispersal*. Trans. C. Jeffrey. Edinburgh: Oliver and Boyd.
(1997). *Diversity and Classification of Flowering Plants*. New York: Columbia University Press.

Tautz, D., Arctander, P., Minelli, A., Thomas, R. H. & Vogler, A. P. (2003). A plea for DNA taxonomy. *Trends in Ecology and Evolution*, **18**, 70–4.

Thomas-Keprta, K. L., Clemett, S. J., Bazylinski, D. A. *et al.* (2001). Truncated hexa-octahedral magnetite crystals in ALH84001: presumptive biosignatures. *Proceedings of the National Academy of Sciences of the USA*, **98**, 2164–9.

Weiss, B. P., Kirschvink, J. L., Baudenbacher, F. J. *et al.* (2000). A low temperature transfer of ALH84001 from Mars to Earth. *Science*, **290**, 791–5.

Wiley, E. O. (1981). *Phylogenetics: the Theory and Practice of Phylogenetic Systematics*. New York: John Wiley.

4

Flowering-plant families: how many do we need?

J. Cullen and S. Max Walters

By the classification of any series of objects, is meant the actual, or ideal, arrangement together of those which are like and the separation of those which are unlike; the purpose of this arrangement being to facilitate the operations of the mind in clearly conceiving and retaining in the memory, the characters of the objects in question. Thus, there may be as many classifications of any series of natural, or of other, bodies, as they have properties or relations to one another, or to other things; or, again, as there are modes in which they may be regarded by the mind.

T. H. Huxley, 1869

It is with most sytematicians in relation to their systems as with a man who has built a vast palace and himself occupies a barn close by: they do not themselves live in their vast systematic structures.

S. A. Kierkegaard, 1846

It is an enormous pleasure for us to participate in this volume in honour of Vernon Heywood's 75th birthday. We have known Vernon for around 50 years – as student, teacher, colleague and friend – and we are pleased to pay tribute to his great contributions to plant taxonomy and plant conservation, and to his enormous effect as a teacher of these two subjects.

In a recent paper (Heywood, 2001), Vernon drew attention to the importance of taxonomy to science and to society in general. In the paper he presented 'a new paradigm for taxonomy', as follows:

It [taxonomy] must be:

* socially responsive to the needs of society;
* efficient and consistent in its methods and procedures;
* both scientifically sound and practicable;
* cooperative, and work closely with other disciplines, both pure and applied;
* outward-looking;

- accessible;
- able to produce outputs that are suitable for various consumers;
- capable of the best practice in informatics.

We would like to take this opportunity to extend the discussion of some of these points in relation to the relatively recent sharp increase in the number of flowering-plant families proposed by taxonomists with an interest in phylogeny – whether traditional, cladistic or molecular.

It is notorious that the results of taxonomic activity – presented as Floras, mono-graphs, revisions or 'phylogenies' – are usually written by taxonomists with other taxonomists in mind, but that these results are used by a much wider community: biologists of many different kinds, gardeners, landscape architects, agriculturists, agronomists, conservationists and land managers, etc. Until very recently, many taxonomists have been essentially dismissive of these consumer groups, despite the frequently made claims that taxonomy is the underpinning of biology, or at least of communication within it, and that classifications must serve a wide variety of purposes. This dismissive attitude is very much to the fore in connection with the proliferation of plant families we are currently enduring.

Most non-taxonomic users of plant classification require two things of it: (i) that it should provide an efficient reference system and means for plant identification, i.e. accurate and speedy access to all the available information about particular plants and reliable predictive properties (based on the taxonomic hierarchy); and (ii) that it should also provide a reasonable level of stability, so that, in making use of the past literature, there is not too great a confusion between different studies and publications.

Such users of plant classifications tend to make use essentially of three levels in the taxonomic hierarchy: the family, the genus and the species (and all that fall within this – subspecies, varietas, forma, cultivar, etc.). The other taxonomic categories – such as orders, subfamilies, tribes, etc. – are generally of no particular interest or value. The family is the highest level so used, as its identification is the first necessary step in the total identification of a completely unknown plant; it is notable that most serious books on, for instance, plants in cultivation, list the family for each genus they include (e.g. the Royal Horticultural Society's *New Dictionary of Gardening* (Royal Horticultural Society, 1999) and *RHS Plant Finder* (Lord, 2003), etc.), even when the family level as such is not really used at all.

General considerations

The history of plant taxonomy has been well studied and recorded (e.g. Croizat, 1945; Lawrence, 1951; Walters, 1961, 1986; Davis & Heywood, 1963; Morton, 1981). Its progress has involved broadly five phases as given below.

1. Artificial systems, culminating in that used in Linnaeus' *Species Plantarum* of 1753.
2. From 1770 to 1880: systems thought to be more and more natural (see Stevens (1994) for much detailed information on the early stages of this phase).
3. 1880–1950: phylogenetic systems.
4. 1950–80: phenetic studies (eschewing phylogenetic speculation and returning to a more 'natural' approach, although widening this by attempting to make use of, or at least to consider, all available information and sometimes using numerical methods).
5. Finally, 1980 to present: a resurgence of interest in phylogenetic ideas, based – at least in part – on cladistics and the use of molecular information.

In this long history, from the pre-Linnean period to the present, three aspects have been contending for predominance: the analytic, that is the need to provide for an accurate and clear method of identification and reference; the synthetic, that is the need to provide groupings of units on the basis of their similarities and differences into higher and higher categories so that reliable inferences (predictions) may be made about the characteristics and properties of the various units; and the phylogenetic, which seeks to demonstrate evolutionary relationships. At different periods, different aspects have been in the ascendant, at least as far as taxonomic theory is concerned. Most of the practical taxonomic results produced have been in the form of Floras, monographs and revisions, which – on the whole – are concerned only with the first two aspects above. Running alongside this whole process is the need for a stable nomenclatural system to facilitate communication.

Artificial classifications, designed as they are for easy and accurate identification, clearly are obviously designed with the user in mind. Natural or phenetic classifications, which combine the identificatory and the predictive aspects, are also intensely practical as they have to make use of information of various kinds in ways that can be sensibly used. This practicality necessitates compromises, and the realisation that at least the higher units into which the basic units are grouped are mental constructs, the results of the effects of the natural world on our minds, developed so that we can handle as efficiently as possible all the information provided by that natural world. Gilmour (1936/1989) puts this very elegantly: 'It [the taxonomic process] is a tool by the aid of which the human mind can deal effectively with the almost infinite variety of the universe. It is not something inherent in the universe, but is, as it were, a conceptual order imposed on it by man for his own purposes.' Thus an element of convenience is built in to the idea of these higher groups from the beginning.

The phylogenetic approach is different from many points of view. In particular, its practitioners seem to regard the higher groupings in classifications as real, fixed and discoverable, whose evolutionary relations can be – at least in principle – known and used to form classifications that are 'true', without regard to how practical they are in use. These topics are dealt with in several very influential papers by Gilmour (1936/1989, 1937, 1940) and a further paper by Heywood (1989).

Hayata (1921) realised that this intellectual split was important, and proposed that classifications should be produced by 'natural' methods, but that, for other kinds of studies – notably the phylogenetic – the classification could be put into a 'dynamic' condition. This suggestion clarified the division in uses between types of output, and allowed for the existence of phylogenetic views without these causing alterations to the natural system. Unfortunately, Hayata's ideas were forgotten; perhaps their time has now come.

The family: history and development

Our concern here is with the one particular supraspecific level, the family. This was first put forward as a category serving to define groups of genera by Pierre Magnol (Magnol, 1689). Earlier workers had, in fact, grouped genera in various ways before this (especially Ray's groupings of monocotyledons and dicotyledons), and some of these groupings, deriving ultimately from European folk taxonomy (Walters, 1986), persist in our classifications today.

Linnaeus was very aware that his sexual system, while ideal for identification (causing the grouping of plants that had the same numbers of stamens and carpels but were otherwise very unlike), was very artificial and he sought more natural groupings of genera. In his *Systema Naturae* (1756), he produced groupings of genera which he called '*ordines naturales*' (natural orders) rather than families. In the early years of the nineteenth century many purportedly natural systems were proposed (see Lindley (1853), for a reasonably detailed review of these). In most of these systems the suprageneric groupings were called 'natural orders', following Linnaeus, but some French authors, notably Adanson (1763) continued to use the term 'family'.

As early as 1815, Mirbel (Mirbel, 1815) had noticed that there were two kinds of family (or natural order): the 'group' family ('*famille en groupe*'), in which all the members were broadly similar to one central type, showing a syndrome of similar morphological features – e.g. Cruciferae, Umbelliferae, Compositae, etc.; and those, 'chain' families ('*familles par enchainement*'), which were more diverse – the genera forming links in a chain such that genus B is similar to genus A, genus C to genus B, genus D to genus C, genus E to genus D, but there is no obvious close similarity between genus E and genus A, e.g. Rosaceae, Ranunculaceae, etc. It is notable that the eight families generally recognised today whose names do not end in the characteristic 'aceae' (Cruciferae, Guttiferae, Leguminosae, Umbelliferae, Labiatae, Compositae, Palmae and Gramineae) are all of the 'group' type, and all come through from pre-Linnaean times, indeed from European folk taxonomy (Walters, 1986). Their individual names carry this history, whereas their anodyne alternatives (Brassicaceae, Clusiaceae, Fabaceae, Apiaceae, Lamiaceae,

Asteraceae, Arecaceae and Poaceae) do not. The use of these alternative names elevates consistency over content, and should in our view be discouraged. The distinction between the two types of family recognised by Mirbel was not widely taken up (but see Baillon, 1871, English translation, vol. 1; Croizat, 1962; Walters, 1961); it needs further consideration.

Throughout the rest of the nineteenth century, numerous taxonomic systems were put forward, culminating in Bentham and Hooker's *Genera plantarum* (1862–83) and the inception of Engler and Prantl's *Die natürlichen Pflanzenfamilien* (1887–1915) and *Das Pflanzenreich* (1900–). Most of these systems used the term '*ordines naturales*' for their suprageneric groups, but around 1900 – and especially in the years leading up to the International Botanical Congress of 1905 held in Vienna – the term 'family' became more and more widely used, largely because zoologists had begun to use the term 'order' (*Ordo*) to define groupings higher than the immediately suprageneric, and botanists wanted to follow suit (Barnhart, 1895). Hence, 'natural order' became a somewhat confusing term, and the Vienna Code clearly preferred the term 'family' (Briquet, 1904; see also Briquet, 1905), with natural order as a more or less downgraded synonym (persisting, of course, in the 'order beds' of many British botanic gardens).

The influence of Darwin's theories as to the existence and mechanism for evolution produced little effect on botanical classifications until the early years of the twentieth century. Then, numerous systematists attempted to produce classifications that reflected the course of evolution (e.g. Bessey, 1915; Hutchinson, 1959, etc.). These varied considerably among themselves, in both the taxa recognised and in the linear order in which they were presented. Interestingly they fall into two camps: those beginning with the buttercups/magnolias, following Candolle (1819) who had clearly stated that the order he had chosen was purely for convenience; and those following the Engler school and beginning with the catkin-bearing plants. Continuing plant exploration and research in many fields (morphology, anatomy, chemistry, etc.) failed to produce any reconciliation between these two camps, and by the late 1950s most taxonomists had become disillusioned with the phylogenetic idea, and attention became concentrated on general-purpose or phenetic classifications, designed for use and taking into consideration as many features of the plants as possible (Gilmour, 1940). The arrival of computers in the 1970s allowed for some mathematical methods to be used in dealing with the large amounts of information available.

During this period, a few taxonomists – notably Arthur Cronquist, Robert Thorne, Armen Takhtajan and Rolf Dahlgren – had continued with phylogenetic methods, and such approaches received a boost when cladistic methods (begun in the 1950s by the zoologist Willi Hennig) became more well known, discussed and used.

The rise of cladistics

Cladistics is an overtly phylogenetic methodology that produces, via computer programmes (i.e. 'black boxes' to most of us), cladograms – diagrams showing the supposed evolutionary trajectory of lineages, indicating branching and the relationships of the various branches, but not including a time scale. Currently, studies making use of these techniques appear to be in the ascendant in plant taxonomy, as a rapid survey of recent taxonomic journals shows quite clearly, and these studies have accelerated the increase in the number of new families already effected by pre-cladistic phylogenetic studies. The subject is pursued and discussed with semi-religious zeal by many of its practitioners, and Jim Endersby has produced an interesting and amusing survey of this and its results in both the botanical/taxonomic and the public arena (Endersby, 2001).

This is not the place for a detailed critique of cladistics; neither are we in any way competent to provide it, having been working as general-purpose taxonomists for many years, with, over the last 20 years, a particular interest in plants in cultivation. However, we do feel competent to comment quite freely on the results produced, and this has necessitated our acquisition of at least some general knowledge of the methodology. For this purpose we have used the textbook *Plant Systematics: a Phylogenetic Approach* (Judd *et al.*, 1998), which provides a particularly clear and often candid description of the methods and procedures involved, and a detailed classification at the family level, with descriptions and observations. We have also supplemented this with a zoologist's view, as represented by the early chapters of Janet Moore's *An Introduction to the Invertebrates* (2001).

Our reading of these sources has given rise to the following observations. Cladistics is a methodology: the results of the application of this methodology to a particular group are as they are. It requires an act of considerable faith to see these as representing either a classification or as representing the course of evolution. This is largely because the methodology rests on three a-priori axioms: bifurcation of lineages, parsimony and monophyly.

The third of these, monophyly (the name indicating a certain circularity in reasoning), is the principle that any classificatory group defined should be 'monophyletic', by which is meant not what the ordinary taxonomist understands by that term, but that it should be traceable back to a single branch of the cladogram (i.e. a methodological definition whose relation to the older definition is not in any way clear, although the two meanings are often conflated in practice). This principle has been much discussed over the last few years and its value as a classificatory principle has been shown to be zero in a series of papers by Brummitt (1996, 1997, 2002) and Brummitt and Sosef (1998); we can add nothing to these discussions, except to say that this 'principle' has resulted in the mechanical production of several new families because the older, larger families cannot be shown to fulfil its (discredited) criterion.

The second axiom, parsimony, has been much discussed over the last ten years. It is the view that the evolution of a group proceeds by the minimum number of possible steps; Occam's razor (which states that 'one should not increase, beyond what is necessary, the number of entities required to explain anything', i.e. that one should not make more assumptions than the minimum needed) is often trundled out in support of this totally counter-intuitive idea (at least to anyone with a good knowledge of the details of plant and animal classification). It (or something like it) is, however, absolutely necessary to cladistic methodology, as the computer programmes involved produce innumerable potential cladograms for any particular group, and there has to be some way of reducing these to a manageable number. Even with the application of parsimony there may be several to many equally eligible cladograms, and even more dubious statistical methods have to be used to choose among these. Moore (2001, p. 15) puts all this very neatly: 'The worst problem is that usually many possible cladograms can be drawn and to pick the correct one (*sic!*) the principle of "parsimony" is invoked. Parsimony in this context means the selection of the cladogram with the smallest number of evolutionary steps . . . The difficulty of choosing between what may be large numbers of equally "parsimonious" cladograms is a further problem.'

Finally, the whole of cladistic thought is shot through with the first of the axioms: that almost every evolutionary event is a bifurcation of a lineage into two. This implies that each group has one, and one only, other group to which it is most closely related – the sister-group as the jargon has it. Furthermore, in order to give an evolutionary vector to a cladogram, it is necessary to relate it to a (the) sister-group to the whole taxonomic unit under investigation; this sister-group is known as the outgroup. Again, Moore (2001, p. 15) describes the problem very neatly: 'Rooting the cladogram (i.e. defining the primitive condition) is often very difficult. The root is compared with an "outgroup", chosen as being related to the group but not part of it. Choice of the outgroup is subjective and difficult. It has reasonably been claimed that whenever outgroup analysis can be applied unambiguously it is not needed, and whenever it is needed it cannot be applied unambiguously.'

In fact, in flowering plants, where the families and higher groups are generally defined by different combinations of the same, relatively few, characters, bifurcation seems unlikely to be the rule. Vicariance biogeography (Croizat, 1958; Nelson & Rosen, 1981) and punctuated equilibria (Gould & Eldredge, 1977) both suggest that the progress of evolution consists of long periods when little is happening, punctuated by 'explosions' of taxa more or less all at once (and certainly not analysable into a series of compressed bifurcations). One advantage of avoiding bifurcation is that the 'problems' of parallelism and convergence, which are seen as considerable by cladists, become part of the normal recombination process, and almost entirely cease to be problematic (see, for example, Croizat (1962)).

Molecular taxonomy

The availability of molecular data for many plants (actually a relatively small number of the total number of species has been examined, few more than once) over the last ten years or so has brought an enormous amount of new material for consideration by taxonomists. Because it is at the molecular level, there is a tendency in some circles to claim for it a greater 'reality' than information available at the morphological or anatomical levels, and to claim that classifications produced by it should be more reliable than those traditionally produced (similarly for phylogenies). The problem is that the amount of data produced is so voluminous that simple inspection of it does not help to produce a classification; because of this, workers in the molecular field have turned to cladistics as a method of dealing with the new information and incorporating it into classifications based on more traditional methods. This means that molecular studies suffer from all the drawbacks involved with cladistics, which have been lightly touched on above; further, additional problems come to light. So Moore (2001, p. 19) lists five advantages of molecular data (the equivalence of the data, the enormous size of the data set, the possibility of statistical analysis of cladograms to avoid the 'parsimony' problems, the possibility of tracing relationships far back in time when change in a particular molecule (e.g. genes coding for ribosomal RNA) occurred, and the avoidance of non-heritable variation). As opposed to these, she lists four disadvantages, which seem to outweigh the advantages. These deserve verbatim quotation:

1. The underlying assumption for most methods is that change in a gene will depend only on the mutation rate and the time elapsed, i.e. that an unvarying 'molecular clock' is ticking at a regular rate. However, the clock is known to be variable under certain conditions, and the whole idea of functionally neutral changes in genes is controversial. Some branches of the evolutionary tree are known to evolve very fast: should we compensate by a subjective decision to omit such species (or groups of species) from our calculations?
2. There is no record of past changes in characters. This is a serious disadvantage, as there are only four possible nucleotides for any site in the DNA molecule. If there have been changes from one nucleotide to another and back again, such 'multiple hits' cannot be detected.
3. There is no recognisable intermediate condition between characters and, worse, no primitive condition for a given site can be recognised.
4. Functional correlates of character change can very seldom be traced.

In the light of all these criticisms – of molecular studies themselves and of the cladistics that are apparently needed to produce anything like classifications out of them – it seems safest to assume that the results of all these activities, while interesting and thought-provoking in themselves, and potentially capable of considerable improvement with further research and refinement of techniques, should not be taken as the last word in classification, at the family or any other level. Indeed, it would probably be best if the people carrying out such studies desisted from proposing taxonomic conclusions that appear to affect the classifications we all use for reference and prediction (they can, perhaps, make use of other systems, such as PhyloCode – see Heywood (2001)). Of course, it has always been the case in the history of taxonomy that anyone can propose any change to the existing classifications that they choose (and some of them have been remarkably stupid, such as the sinking of all cruciferous genera into one by Krause (1902)), but that the general taxonomic consensus will pick them up if they are useful and discard them if not – and this is generally true: patience will solve all these problems. However, the polemic and vast claims about 'reality' and 'truth' surrounding cladistic and molecular taxonomy need clearing away, so that a realistic assessment can be made. For the moment, the results of phylogenetic work, cladistics and molecular taxonomy can be considered as interesting and suggestive, but not final, at least as far as reference classification is concerned.

In a remarkably candid statement, the authors of the textbook we have been using as a source (Judd *et al.*, 1998, p. 21) say that the cladograms, classifications and phylogenies produced by cladistics (whether or not including molecular data) are 'simply models or hypotheses, best guesses about the history of a group of plants. It follows that some guesses might be better, or at least more convincing, than others'. We agree entirely with this statement which reinforces, from within as it were, the contention that such classifications are methodological artefacts and have no special status *vis à vis* 'truth' or 'reality'. From this it follows that taxonomists interested in the main business of taxonomy, i.e. reference and prediction, must treat the vast increase in families that this kind of work has produced with considerable suspicion. Figure 4.1 shows the increase in family numbers over a set of classifications arranged in time sequence, from Adanson (1763) to the present.

Table 4.1 gives some details of the current situation with regard to the families. A rapid search of recent literature has yielded a total of 662 families overall, and there are probably more lurking in the farther recesses. The table shows how these families relate in five recent systems or arrangements of genera. These are:

1. Melchior (1964). The oldest of the systems, but one widely used in Europe (see *Flora Europaea* (Tutin *et al.*, 1964–80), *The European Garden Flora* (The European Garden Flora Editorial Committee, 1984–2000) and many Floras produced since 1964). It has the great advantage that the family placement of almost every genus is determinable, either

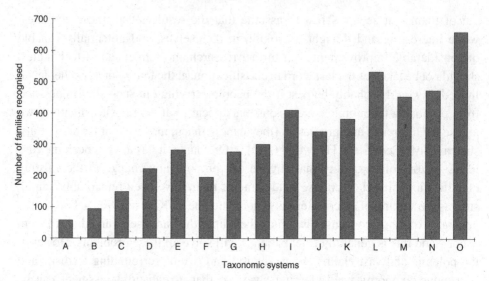

Figure 4.1 Numbers of families recognised in selected taxonomic systems. A, Adanson (1763); B, Jussieu (1789); C, Candolle (1819); D, Bartling (1830); E, Lindley (1853); F, Bentham and Hooker (1862–83); G, Dalla Torre and Harms (1900–7); H, Bessey (1915); I, Hutchinson (1959); J, Melchior (1964); K, Cronquist (1981); L, Takhtajan (1987); M, Brummitt (1992); N, Angiosperm Phylogeny Group (1998); O, total number of families found in recent publications.

from the volume itself or from the latest edition of *Die natürlichen Pflanzenfamilien* (Engler & Prantl, 1887–1915).

2. Takhtajan (1987), slightly modified in Takhtajan (1997). This system is used because of its inclusion in Brummitt (1992), see below, which allows for the placing of almost every genus. The largest of the systems, with 533 families included. Takhtajan has stated that he prefers many small families, because these are more homogeneous and easier to define and identify: there is something in this, but the large number tends to obfuscation when using previous literature.

3. Cronquist (1981), with a few alterations from Cronquist (1988). This system includes fewer families than most of the others, largely because of the amalgamation of some of the larger families. The placement of most genera in this system can be derived from Mabberley (1997).

4. Brummitt (1992). This most useful book lists all the genera in use in the herbarium at Kew, with indication of the families to which they belong. Compiled with the assistance of many experts, the book also indicates how the various families are placed in several other systems (it is not a system in itself, in that the families and genera are listed alphabetically rather than systematically).

5. The Angiosperm Phylogeny Group (APG) (1998). This paper is a summation of molecular and cladistic work up to 1998, and its combination with more traditional approaches. It includes many new small families, but also combines together many other, traditional

groups (e.g. all of the traditional Malvales are placed in the Malvaceae), and the contents of some families are heavily rearranged (e.g. Labiatae and Verbenaceae; cf. Cantino *et al.* (1992)).

In Table 4.1 below, the 662 family names we have found are included in the first column (headed 'Names'). In this column, names in square brackets are *not* included in any of the five systems mentioned above, but occur in other books and papers. The next five columns indicate how the five systems above have dealt with the group in the 'Names' column. A question mark (?) next to the family name indicates an assumption by the authors. Family names that are found in all five systems are shaded. It must be remembered that this simply means that there is a family of that name in each system: the contents of the family may well vary from system to system. Families thought to be monogeneric are boxed. This information is not always easy to obtain for all of the systems, and it must be remembered that a family can be monogeneric in one system but not in others; e.g. the Akaniaceae, which is monogeneric in Melchior and Brummitt (definitely), and in Cronquist and Takhtajan (probably), but is not monogeneric (owing to the inclusion of the Bretschneideriaceae), in APG. In the column headed 'Melchior' any name that is a subfamily, tribe or subtribe in this system is so indicated, after the name of the family in which it is included.

The conclusions that can be drawn from Table 4.1 are many. First, although 662 families are to be found here, no one system includes them all. Takhtajan's system, with 533 families, is the most divided of those we have used, but, if we add to the Melchior total (344) the total of names that are subfamilies, tribes or subtribes in this system (166), we arrive at a total of 510; a figure not too far from the Takhtajan total. This figure shows how many of the 'newer' families are merely the old subfamilies, tribes, etc., in disguise, and this 'rank inflation' seems to be a widespread phenomenon. There is much to be said for keeping these groups at the infrafamilial level, as they then do not cause confusion for the ordinary user.

There is also, in general, a tendency to increase the number of monogeneric families; this, again, is generally not helpful, as many of these can be, and have been, treated at infrafamilial levels where they are available for those who want them and effectively hidden for those who do not need them.

Some further conclusions can be gathered from taking the Melchior system (the earliest, and smallest in terms of number of families) as a basis, ignoring those families that are not in any of the five systems (i.e. those with names in square brackets in the 'Names' column), and following through what has happened to these Melchior families in any or all of the new systems. This means counting the number of times any particular name occurs in the column headed 'Melchior', which will indicate how much 'noise' this particular Melchior family creates in terms of its division. Table 4.2 (summarized in Table 4.3 and Figure 4.2) shows the results of this calculation.

As regards the splitting of families, the main conclusions to be drawn from this are as follows: there are 3 families (Liliaceae, Saxifragaceae and Cornaceae) (Tables 4.2 and 4.3, and Figure 4.2, I to III) that produce the maximum number of splits. Together, they contain 312 genera and 4795 species; thus they are 0.9% of the families, 2.6% of the genera and 2.1% of the species – a very small proportion to cause so much trouble. All authors agree that these families should be split, but there is no general agreement as to *how* this is to be done for the best result. Of the splits, in the Liliaceae, 21 families (50%) already exist in Melchior at infrafamilial rank and 12 (28%) are generally thought to be monogeneric. For the Saxifragaceae, 20 families (71%) exist at subfamilial level and 11 (39%) are considered to be monogeneric in at least some systems. For the Cornaceae, 6 families (60%) exist at subfamilial level and 8 (80%) are monogeneric (at least in some systems).

There is a group of families that is much less split, those with between 3 and 7 entries (Tables 4.2 and 4.3, and Figure 4.2, IV to VIII): 43 families, 2480 genera and 48 971 species; thus 12.5% of the families, 20.6% of the genera and 22.4% of the species.

Finally, there is a group of 298 families that are not split, or scarcely split at all, and remain much the same throughout all 5 systems. These contain 9279 genera and 166 530 species, and include all the completely monotypic families (mono-generic + monospecific) as well as some of the largest (and oldest in terms of their recognition): Compositae, Orchidaceae, Gramineae, Cruciferae and Umbel-liferae. These represent 86.6% of the families, 76.8% of the genera and 75.6% of the species.

These figures all confirm existing trends: that much of the problem is caused by a small number of families (genera, species), rank inflation and the increase in number of very small, mainly monogeneric, families (Tables 4.2 and 4.3, and Figure 4.2).

When we turn to the reverse problem, that of the amalgamation of already existing families, the situation is more complex. The main problem areas are the *Malvales*; the enlarged Plantaginaceae (as by APG); the potential joining of the Araliaceae with Umbelliferae, and Apocynaceae with Asclepiadaceae; and, more recently and later than any of the five systems under consideration, the merging under the name Salicaceae, of Flacourtiaceae and Salicaceae (Chase *et al.*, 2002). Also relevant is the change in the content of the Labiatae/Verbenaceae mentioned above.

All of these groups are known to have been difficult to discriminate precisely since the days of Lindley (1853) and Bentham and Hooker (1862–83); however, from the purely classificatory point of view, any advantage in their combination is difficult, if not impossible, to discern. For instance, the precise discrimination

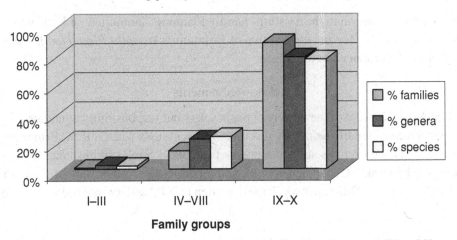

Figure 4.2 Family groups based on subsequent splits. Family groups IX and X, which represent most of the families, are scarcely split at all (see Tables 4.2 and 4.3)

of the Papaveraceae and Fumariaceae, is not difficult and a conventional, though less-precise, discrimination is perfectly easy to use, causing no problems whatsoever. The putting together of these families has become accepted because they are really rather small. This is not the case with the families mentioned above, where the numbers of genera and species are significant.

We are anxious not to seem to be discouraging research, or the inclusion of new research results into taxonomic systems. We are happy to accept families such as the Hydrangeaceae as a family separate from the Saxifragaceae where it was formerly included, and to think of it as more closely similar (the difficulties in the use of the term 'related' are discussed by Gilmour (1936/1989, 1937, 1940)) to the Caprifoliaceae. The problem is that each of us can have our own particular favourite families and there is no general agreement as to what is the best solution at any particular time.

We recommend that the users of taxonomy should take a very cautious approach to the new families (whether splits or mergers) as a whole, not accepting changes until and unless their usefulness has been clearly demonstrated. To the cladists and phylogenists who will, of course, continue with their work, we recommend again a cautious approach, preferably one that does not cause confusion in the reference and prediction system: hence we recommend the use of subfamilies and tribes where possible (particularly if they already exist) rather than new families; or perhaps the use of newer systems, such as PhyloCode, which don't have the unwelcome complications of upsetting a well-understood and effective reference and prediction system. Perhaps the families to be used could be reviewed every 10 or 12 years (say at alternate Botanical Congresses), by a group on which the users were as strongly represented as the producers. This inclusion of users might also be a useful expansion to the existing 'Family Planning Committee' at Kew, which will be responsible for any new edition of Brummitt's *Vascular Plant Families and Genera* to be produced in the future.

Acknowledgements

Although the content of the present paper is solely our responsibility as authors, we have been considerably helped by several long discussions with Dr R. K. Brummitt, who also passed on to us some comments from Vicki Funk (Smithsonian Institution, Washington, DC) and helped with finding the earliest publication of Krause's *Crucifera* (1902). This paper was largely written in 2002 and had not been amended since then.

References

Adanson, M. (1763). *Familles des Plantes*. Paris: Vincent.

Angiosperm Phylogeny Group (APG) (1998). An ordinal classification for the families of flowering plants. *Annals of the Missouri Botanical Garden*, **85**, 531–53.

Baillon, H. (1866–95). *The Natural History of Plants*, 13 vols., vols. 1 to 8 trans. M. Hartog (1871–88). London: L. Reeve.

Barnhart, J. (1895). Family nomenclature. *Bulletin of the Torrey Botanical Club*, **22**, 1–24.

Bartling, F. G. (1830). *Ordines naturales plantarum*. Göttingen, Germany: sumptibus Dieterichianis.

Bentham, G. & Hooker, J. D. (1862–83). *Genera plantarum ad exemplaria imprimis in Herbariis Kewensibus servata definita/auctoribus G. Bentham et J. D. Hooker*. London: Reeve.

Bessey, C. (1915). The phylogenetic taxonomy of flowering plants. *Annals of the Missouri Botanical Garden*, **2**, 109–64.

Briquet, J. (1904). *Texte synoptique des documents destinés à servir de base aux débats du Congrès Internationale de nomenclature botaniques de Vienne, 1905*. Berlin: R. Friedländer.

Briquet, J. (1905). Règles internationales de la Nomenclature botanique. Adoptées par le Congrès international de Botanique de Vienne 1905. In *Verhandlungen des Internationalen Botanischen Kongresses in Wien 1905*, eds. R. von Wettstein, J. Wiesner & A. Zahlbruckner. Jena, Germany: Gustav Fischer, pp. 165–261.

Brummitt, R. K. (ed.) (1992). *Vascular Plant Families and Genera*. Kew, UK: The Royal Botanic Gardens.

(1996). In defence of paraphyletic taxa. In *The Biodiversity of African Plants*, eds. L. J. G. Maesen, X. M van den Bingt & J. M. van Medenbach van Rooy. Proceedings of the XIVth AETFAT Congress, 22–27 August 1994, Wageningen. Dordrecht, the Netherlands: Kluwer Academic Press.

(1997). Taxonomy versus cladonomy: a fundamental controversy in biological systematics. *Taxon*, **46**, 723–34.

(2002). How to chop up a tree. *Taxon*, **51**, 31–41.

Brummitt, R. K. & Sosef, M. S. M. (1998). Paraphyletic taxa are inherent in Linnean classification: a reply to Freudenstein. *Taxon*, **47**, 425–9.

Candolle, A. P. de (1819). *Théorie élementaire de la botanique*, 2nd edn. Paris, Deterville.

Cantino, P. D., Harley, R. M. & Wagstaff, S. J. (1992). Genera of Labiatae: status and classification. In *Advances in Labiate Science*, eds. R. M. Harley & T. Reynolds, Kew, UK: The Royal Botanic Gardens.

Chase, M., Zmarsty, S., Lledó, M. D. *et al.* (2002). When in doubt, put it in Flacourtiaceae: a molecular phylogenetic analysis based on plastid *rbcL* DNA sequences. *Kew Bulletin*, **57**, 141–81.

Croizat, L. (1945). History and nomenclature of higher units of classification. *Bulletin of the Torrey Botanical Club*, **72**, 52–75.

(1958). *Panbiogeography*. Caracas, Venezuela: L. Croizat.

(1962). *Space, Time, Form: the Biological Synthesis*. Caracas, Venezuela: L. Croizat.

Cronquist, A. (1981). *An Integrated System of Classification of Flowering Plants*. New York: Columbia University Press.

(1988). *The Evolution and Classification of Flowering Plants*, 2nd edn. New York: Columbia University Press.

Dalla Torre, T. S. & Harms, H. A. T. (1900–7). *Genera Siphonogamarum*. Leipzig, Germany: Engelmann.

Davis, P. H. & Heywood, V. H. (1963). *Principles of Angiosperm Taxonomy*. Edinburgh and London: Oliver and Boyd.

Endersby, J. (2001). 'The realm of hard evidence': novelty, persuasion and collaboration in botanical cladistics. *Studies in History and Philosophy of Biological and Biomedical Science*, **32c**, 343–60.

Engler, A. & Prantl, K. (1887–1915). *Die natürlichen Pflanzenfamilien*. Leipzig, Germany: Engelmann.

(1900–) *Das Pflanzenreich*. Leipzig, Germany: Engelmann.

Gilmour, J. S. L. (1936/1989). Whither taxonomy? Address to Section K of the British Association for the Advancement of Science, 1936. Reprinted in *Plant Systematics and Evolution*, **167**(1–2), 98–103 (1989).

(1937). A taxonomic problem. *Nature*, **139**, 1040–2. Reprinted in *Plant Systematics and Evolution*, **167**(1–2), 103–7 (1989).

(1940). Taxonomy and philosophy. In *The New Systematics*, ed. J. Huxley. Oxford: Oxford University Press, pp. 461–74. Reprinted by the Systematics Association, London (1971).

Gould, S. J. & Eldredge, N. (1977). Punctuated equilibria: the tempo and mode of evolution reconsidered. *Paleobiology*, **3**, 115–51.

Hayata, B. (1921). The natural classification of plants according to the dynamic system. *Iconum Plantarum Formosanarum*, **10**, 75–95.

Heywood, V. H. (1989). Nature and natural classification. *Plant Systematics and Evolution*, **167**(1–2), 87–92.

(2001). Floristics and monography: an uncertain future. *Taxon*, **50**, 361–80.

Hutchinson, J. (1959). *The Families of Flowering Plants*, 2nd edn., 2 vols. Oxford, UK: Oxford University Press.

Huxley, T. H. (1869). *An Introduction to the Classification of Animals*. London: John Churchill.

Judd, W. S., Campbell, C. S., Kellogg, E. A. & Stevens, P. F. (1998). *Plant Systematics: a Phylogenetic Approach*. Sunderland, MA: Sinauer Associates Inc.

Jussieu, A. L. de (1789). *Genera plantarum*. Paris, Viduam Harrisant and Theophilum Barrois.

Kierkegaard, S. A. (1846). *Journals*. Trans. P. Rohde.

Krause, E. H. L. (1902). *J. Sturm's Flora von Deutschland*, **6**, 32–168, Stuttgart: Verlag K. Lutz.

Kubitzki, K. (ed.) (1998). *The Families and Genera of Vascular Plants*, vol. 3, *Flowering Plants, Monocotyledons, Lilianae (except Orchidaceae)*. Berlin, Heidelberg, New York: Springer Verlag.

Lawrence, G. H. M. (1951). *Taxonomy of Vascular Plants*. New York: Macmillan.

Lindley, J. (1853). *The Vegetable Kingdom*, 3rd edn. London: Bradbury & Evans.

Linnaeus, C. (1753). *Species plantarum. A facsimile of the first edition, 1753*. London: Ray Society. Originally printed Stockholm, Sweden: L. Salvic.

(1756). *Systema naturae. Facsimile 1960*. Uppsala: Bokgillet.

Lord, T. (2003). *RHS Plant Finder 2003–2004*, 16th edn. London: Dorling Kindersley.

Mabberley, D. J. (1997). *The Plant Book: a Portable Dictionary of the Vascular Plants*, 2nd edn. Cambridge, UK: Cambridge University Press.

Magnol, P. (1689). *Prodromus historiæ generalis plantarum*. Monspelij: Gabrielis & Honorati Pech Frantrum.

Melchior, H. (ed.) (1964). *Syllabus der Pflanzenfamilien*, 12th edn, vol. 2. Berlin: Gebrüder Borntraeger.

Mirbel, C. F. B. de. (1815). *Elémens de physiologie végétale et de botanique*, part 2. Paris: Magimel.

Moore, J. (2001). *An Introduction to the Invertebrates*. Cambridge, UK: Cambridge University Press.

Morton, A. G. (1981). *History of Botanical Science*. London: Academic Press.

Nelson, G. & Rosen, D. E. (eds.) (1981). *Vicariance Biogeography: a Critique*. New York: Columbia University Press.

Royal Horticultural Society (1999). *The New Royal Horticultural Society Dictionary of Gardening*. London: Macmillan.

Stevens, P. F. (1994). *The Development of Biological Systematics*. New York: Columbia University Press.

Takhtajan, A. (1987). *Systema Magnoliophytorum*. Leningrad: Komarov Botanical Institute.

(1997). *Diversity and Classification of Flowering Plants*. New York and Chichester: Columbia University Press.

The European Garden Flora Editorial Committee (eds.) (1984–2000). *The European Garden Flora*, vols. 1 to 6. Cambridge, UK: Cambridge University Press.

Tutin, T. G., Heywood, V. H., Burges, N. A. *et al.* (1964–80). *Flora Europaea*, vols. 1 to 5. Cambridge, UK: Cambridge University Press.

Walters, S. M. (1961). The shaping of angiosperm taxonomy. *New Phytologist*, **60**, 74–84.

(1986). The name of the rose: a review of ideas on the European bias in angiosperm classification. *New Phytologist*, **104**, 527–46.

Table 4.1. *The comparison of families in five recent systems or arrangements of genera (see text for explanation)*

Names	Melchior (1964)	Takhtajan (1987)	Cronquist (1988)	Brummitt (1992)	APG (1998)
Abolbodaceae	in Xyridaceae	? Xyridaceae	? Xyridaceae	in Xyridaceae	Abolbodaceae
Acanthaceae	Acanthaceae	Acanthaceae	Acanthaceae	Acanthaceae	Acanthaceae
Aceraceae	Aceraceae	Aceraceae	Aceraceae	Aceraceae	Sapindaceae
Achariaceae	Achariaceae	Achariaceae	Achariaceae	Achariaceae	Achariaceae
Achatocarpaceae	Achatocarpaceae	Achatocarpaceae	Achatocarpaceae	Achatocarpaceae	Achatocarpaceae
Acoraceae	Araceae	Araceae	Araceae	Acoraceae	Acoraceae
Actinidiaceae	Actinidiaceae	Actinidiaceae	Actinidiaceae	Actinidiaceae	Actinidiaceae
Adoxaceae	Adoxaceae	Adoxaceae	Adoxaceae	Adoxaceae	Adoxaceae
[Aegialitidaceae]	Plumbaginaceae	Plumbaginaceae	Plumbaginaceae	Plumbaginaceae	Plumbaginaceae
Aegicerataceae	Myrsinaceae	Aegicerataceae	Myrsinaceae	Myrsinaceae	Myrsinaceae
Aextoxicaceae	Aextoxicaceae	Aextoxicaceae	Aextoxicaceae	Aextoxicaceae	Aextoxicaceae
Agapanthaceae	Liliaceae	? Alliaceae	Liliaceae	in Alliaceae	Agapanthaceae
Agavaceae	Agavaceae	Agavaceae	Agavaceae	Agavaceae	Agavaceae
Agdestidaceae	Phytolaccaceae	Phytolaccaceae	Phytolaccaceae	Agdestidaceae	Phytolaccaceae
[Aitoniaceae]	Meliaceae	Meliaceae?	Meliaceae?	Meliaceae	Meliaceae?
Aizoaceae	Aizoaceae	Aizoaceae	Aizoaceae	Aizoaceae	Aizoaceae
Akaniaceae	Akaniaceae	Akaniaceae	Akaniaceae	Akaniaceae	Akaniaceae
Alangiaceae	Alangiaceae	Alangiaceae	Alangiaceae	Alangiaceae	Cornaceae
Alismataceae	Alismataceae	Alismataceae	Alismataceae	Alismataceae	Alismataceae
Alliaceae	Liliaceae	Alliaceae	Liliaceae	Alliaceae	Alliaceae
Aloaceae	Liliaceae	Asphodelaceae	Aloeaceae	Aloaceae	Asphodelaceae
Alseuosmiaceae	Caprifoliaceae	Alseuosmiaceae	Alseuosmiaceae	Alseuosmiaceae	Alseuosmiaceae
Alstroemeriaceae	Liliaceae	Alstroemeriaceae	Liliaceae	Alstroemeriaceae	Alstroemeriaceae
Altingiaceae	Hamamelidaceae	Altingiaceae	? Hamamelidaceae	Hamamelidaceae	Altingiaceae
Alzateaceae	?	Alzateaceae	?	Alzateaceae	Alzateaceae
Amaranthaceae	Amaranthaceae	Amaranthaceae	Amaranthaceae	Amaranthaceae	Amaranthaceae

(cont.)

Table 4.1. (*cont.*)

Names	Melchior (1964)	Takhtajan (1987)	Cronquist (1988)	Brummitt (1992)	APG (1998)
Amaryllidaceae	Amaryllidaceae	Amaryllidaceae	Liliaceae	Amaryllidaceae	Amaryllidaceae
Amborellaceae	Amborellaceae	Amborellaceae	Amborellaceae	Amborellaceae	Amborellaceae
Anacardiaceae	Anacardiaceae	Anacardiaceae	Anacardiaceae	Anacardiaceae	Anacardiaceae
Anarthriaceae	Restionaceae	Anarthriaceae	Restionaceae	Anarthriaceae	Anarthriaceae
Ancistrocladaceae	Ancistrocladaceae	Ancistrocladaceae	Ancistrocladaceae	Ancistrocladaceae	Ancistrocladaceae
[Androstachydaceae]	Euphorbiaceae	Euphorbiaceae	Euphorbiaceae	Euphorbiaceae	Euphorbiaceae
Anemarrhenaceae	Liliaceae	Asphodelaceae	Liliaceae	in Anthericaceae	Anemarrhenaceae
Anisophylleaceae	Rhizophoraceae	Anisophyllaeaceae	Anisophyllaeaceae	Anisophylleaceae	Anisophyllaeaceae
Annonaceae	Annonaceae	Annonaceae	Annonaceae	Annonaceae	Annonaceae
[Anomochlooaceae]	Gramineae	Gramineae	Gramineae	Gramineae	Gramineae
Anthericaceae	Liliaceae	Asphodelaceae	Liliaceae	Anthericaceae	Anthericaceae
[Antoniaceae]	Loganiaceae	Loganiaceae	Loganiaceae	Loganiaceae	Loganiaceae
Aphloiaceae	Flacourtiaceae	Aphloiaceae	? Flacourtiaceae	Flacourtiaceae	Aphloiaceae
Aphyllanthaceae	Liliaceae	Aphyhllanthaceae	Liliaceae	Aphyllanthaceae	Aphyllanthaceae
Apocynaceae	Apocynaceae	Apocynaceae	Apocynaceae	Apocynaceae	Apocynaceae
Apodanthaceae	Rafflesiaceae	Rafflesiaceae	Apodanthaceae	Rafflesiaceae	Rafflesiaceae
Aponogetonaceae	Aponogetonaceae	Aponogetonaceae	Aponogetonaceae	Aponogetonaceae	Aponogetonaceae
[Apostasiaceae]	Orchidaceae	Orchidaceae	Orchidaceae	Orchidaceae	Orchidaceae
[Aptandraceae]	Olacaceae	Olacaceae	Olacaceae	Olacaceae	Olacaceae
Aquifoliaceae	Aquifoliaceae	Aquifoliaceae	Aquifoliaceae	Aquifoliaceae	Aquifoliaceae
[Aquilariaceae]	Thymelaeaceae	Thymelaeaceae	Thymelaeaceae	Thymelaeaceae	Thymelaeaceae
Araceae	Araceae	Araceae	Araceae	Araceae	Araceae
Araliaceae	Araliaceae	Araliaceae	Araliaceae	Araliaceae	Araliaceae
Aralidiaceae	? Cornaceae	Aralidiaceae	Cornaceae	Aralidiaceae	Aralidiaceae
Argophyllaceae	Saxifragaceae	Argophyllaceae	? Grossulariaceae	Escalloniaceae	Argophyllaceae
Aristolochiaceae	Aristolochiaceae	Aristolochiaceae	Aristolochiaceae	Aristolochiaceae	Aristolochiaceae
Asclepiadaceae	Asclepiadaceae	Asclepiadaceae	Asclepiadaceae	Asclepiadaceae	Apocynaceae

Asparagaceae	Liliaceae	Asparagaceae	Liliaceae	Asparagaceae	Asparagaceae
Asphodelaceae	Liliaceae	Asphodelaceae	Liliaceae	Asphodelaceae	Asphodelaceae
Asteliaceae	Liliaceae	Asteliaceae	Liliaceae	Asteliaceae	Asteliaceae
[Asteranthaceae]	Lecythidaceae	Lecythidaceae	Lecythidaceae	Lecythidaceae	Lecythidaceae
Asteropeiaceae	Theaceae	Asteropeiaceae	Theaceae	Asteropeiaceae	Asteropeiaceae
Atherospermataceae	Monimiaceae	Atherospermataceae	Monimiaceae	in Monimiaceae	Atherspermataceae
Aucubaceae	Cornaceae	Aucubaceae	Cornaceae	Aucubaceae	Aucubaceae
Austrobaileyaceae	Austrobaileyaceae	Austrobaileyaceae	Austrobaileyaceae	Austrobaileyaceae	Austrobaileyaceae
[Averrhoaceae]	Oxalidaceae	Oxalidaceae	Oxalidaceae	Oxalidaceae	Oxalidaceae
Avicenniaceae	Verbenaceae	Verbenaceae	Avicenniaceae	Avicenniaceae	Avicenniaceae
Balanitaceae	Zygophyllaceae	Balanitaceae	Zygophyllaceae	Balanitaceae	Zygophyllaceae
Balanopaceae	Balanopaceae	Balanopaceae	Balanopaceae	Balanopaceae	Balanopaceae
Balanophoraceae	Balanophoraceae	Balanophoraceae	Balanophoraceae	Balanophoraceae	Balanophoraceae
Balsaminaceae	Balsaminaceae	Balsaminaceae	Balsaminaceae	Balsaminaceae	Balsaminaceae
[Bambusaceae]	Gramineae	Gramineae	Gramineae	Gramineae	Gramineae
Barbeiuaceae	Phytolaccaceae	Barbeiuaceae	Phytolaccaceae	Barbeiuaceae	Phytolaccaceae
Barbeyaceae	Ulmaceae	Barbeyaceae	Barbeyaceae	Barbeyaceae	Barbeyaceae
Barclayaceae	Nymphaeaceae	Barclayeaceae	Barclayaceae	Nymphaeaceae	Nymphaeaceae
[Barringtoniaceae]	Lecythidaceae	Lecythidaceae	Lecythidaceae	Lecythidaceae	Lecythidaceae
Basellaceae	Basellaceae	Basellaceae	Basellaceae	Basellaceae	Basellaceae
Bataceae	Bataceae	Bataceae	Bataceae	Bataceae	Bataceae
Baueraceae	Saxifragaceae	Baueraceae	? Cunoniaceae	Cunoniaceae	Cunoniaceae
Begoniaceae	Begoniaceae	Begoniaceae	Begoniaceae	Begoniaceae	Begoniaceae
Berberidaceae	Berberidaceae	Berberidaceae	Berberidaceae	Berberidaceae	Berberidaceae
Berberidopsidaceae	Flacourtiaceae	Berberidopsidaceae	Flacourtiaceae	in Flacourtiaceae	Berberidopsidaceae
Betulaceae	Betulaceae	Betulaceae	Betulaceae	Betulaceae	Betulaceae
Biebersteiniaceae	Geraniaceae	Biebersteiniaceae	Geraniaceae	Geraniaceae	Biebersteiniaceae
Bignoniaceae	Bignoniaceae	Bignoniaceae	Bignoniaceae	Bignoniaceae	Bignoniaceae
[Bishofiaceae]	Euphorbiaceae	Euphorbiaceae	Euphorbiaceae	Euphorbiaceae	Euphorbiaceae
Bixaceae	Bixaceae	Bixaceae	Bixaceae	Bixaceae	Bixaceae

(cont.)

Table 4.1. (cont.)

Names	Melchior (1964)	Takhtajan (1987)	Cronquist (1988)	Brummitt (1992)	APG (1998)
Blandfordiaceae	Liliaceae	Blandfordiaceae	Liliaceae	Blandfordiaceae	Blandfordiaceae
Blehniaceae	???	???		???	Blehniaceae
[Blepharocaryaceae]	Anacardiaceae	Anacardiaceae	Anacardiaceae	Anacardiaceae	Anacardiaceae
Bombacaceae	Bombacaceae	Bombacaceae	Bombacaceae	Bombacaceae	Malvaceae
Bonnetiaceae	Guttiferae	Bonnetiaceae	Guttiferae	Guttiferae	Bonnetiaceae
Boraginaceae	Boraginaceae	Boraginaceae	Boraginaceae	Boraginaceae	Boraginaceae
Boryaceae	Liliaceae	? Asphodelaceae	Liliaceae	in Anthericaceae	Boryaceae
Bretschneideriaceae	Bretschneideriaceae	Bretschneideriaceae	Bretschneideriaceae	Bretschneideriaceae	in Akaniaceae
Brexiaceae	Saxifragaceae	Brexiaceae	? Grossulariaceae	Escalloniaceae	Celastraceae
Bromeliaceae	Bromeliaceae	Bromeliaceae	Bromeliaceae	Bromeliaceae	Bromeliaceae
Brunelliaceae	Brunelliaceae	Brunelliaceae	Brunelliaceae	Brunelliaceae	Cunoniaceae
Bruniaceae	Bruniaceae	Bruniaceae	Bruniaceae	Bruniaceae	Bruniaceae
Brunoniaceae	Brunoniaceae	Brunoniaceae	Brunoniaceae	Goodeniaceae	Goodeniaceae
Buddleiaceae	Buddleiaceae	Buddlejaceae	Buddlejaceae	Buddleiaceae	Buddlejaceae
Burmanniaceae	Burmanniaceae	Burmanniaceae	Burmanniaceae	Burmanniaceae	Burmanniaceae
Burseraceae	Burseraceae	Burseraceae	Burseraceae	Burseraceae	Burseraceae
Butomaceae	Butomaceae	Butomaceae	Butomaceae	Butomaceae	Butomaceae
Buxaceae	Buxaceae	Buxaceae	Buxaceae	Buxaceae	Buxaceae
Byblidaceae	Byblidaceae	Byblidaceae	Byblidaceae	Byblidaceae	Byblidaceae
Cabombaceae	Nymphaeaceae	Cabombaceae	Cabombaceae	Cabombaceae	in Nymphaeaceae
Cactaceae	Cactaceae	Cactaceae	Cactaceae	Cactaceae	Cactaceae
Caesalpiniaceae	Leguminosae	Leguminosae	Caesalpiniaceae	Leguminosae	Leguminosae
Calectasiaceae	Xanthorrhoeaceae	Dasyopogonaceae	Xanthorrhoeaceae	Calectasiaceae	Dasypogonaceae
Callitrichaceae	Callitrichaceae	Callitrichaceae	Callitrichaceae	Callitrichaceae	Plantaginaceae
Calochortaceae	Liliaceae	Calochortaceae	Liliaceae	Liliaceae	Liliaceae

Calycanthaceae	Calycanthaceae	Calycanthaceae	Calycanthaceae	Calycanthaceae
Calyceraceae	Calyceraceae	Calyceraceae	Calyceraceae	Calyceraceae
Campanulaceae	Campanulaceae	Campanulaceae	Campanulaceae	Campanulaceae
Campynemataceae	Melianthaceae	? Melianthaceae	in Melianthaceae	Campynemataceae
Canellaceae	Canellaceae	Canellaceae	Canellaceae	Canellaceae
Cannabaceae	Cannabaceae	Cannabaceae	Cannabaceae	Cannabaceae
Cannaceae	Cannaceae	Cannaceae	Cannaceae	Cannaceae
Canotiaceae	?Celastraceae	Celastraceae	Canotiaceae	Celastraceae
Capparaceae	Capparaceae	Capparaceae	Capparaceae	Cruciferae
Caprifoliaceae	Caprifoliaceae	Caprifoliaceae	Caprifoliaceae	Caprifoliaceae
Cardiopteridaceae	Cardiopteridaceae	Cardiopteridaceae	Cardiopteridaceae	Cardiopteridaceae
Caricaceae	Caricaceae	Caricaceae	Caricaceae	Caricaceae
Carlemanniaceae	Carlemanniaceae	Caprifoliaceae	Carlemanniaceae	Carlemanniaceae
[Carpinaceae]	Betulaceae	Betulaceae	Betulaceae	Betulaceae
Carpodetaceae	Saxifragaceae	? Grossulariaceae	Escalloniaceae	Carpodetaceae
[Cartonemataceae]	Commelinaceae	Commelinaceae	Commelinaceae	Commelinaceae
Caryocaraceae	Caryocaraceae	Caryocaraceae	Caryocaraceae	Caryocaraceae
Caryophyllaceae	Caryophyllaceae	Caryophyllaceae	Caryophyllaceae	Caryophyllaceae
Casuarinaceae	Casuarinaceae	Casuarinaceae	Casuarinaceae	Casuarinaceae
Cecropiaceae	Moraceae	Cecropiaceae	Cecropiaceae	Cecropiaceae
Celastraceae	Celastraceae	Celastraceae	Celastraceae	Celastraceae
Celtidaceae	Ulmaceae	Ulmaceae	Ulmaceae	Celtidaceae
Centrolepidaceae	Centrolepidaceae	Centrolepidaceae	Centrolepidaceae	Centrolepidaceae
Cephalotaceae	Cephalotaceae	Cephalotaceae	Cephalotaceae	Cephalotaceae
Ceratophyllaceae	Ceratophyllaceae	Ceratophyllaceae	Ceratophyllaceae	Ceratophyllaceae
Cercidiphyllaceae	Cercidiphyllaceae	Cercidiphyllaceae	Cercidiphyllaceae	Cercidiphyllaceae
Chenopodiaceae	Chenopodiaceae	Chenopodiaceae	Chenopodiaceae	Amaranthaceae
[Chloanthaceae]	Verbenaceae?	Verbenaceae?	Verbenaceae	Verbenaceae?
Chloranthaceae	Chloranthaceae	Chloranthaceae	Chloranthaceae	Chloranthaceae
Chrysobalanaceae	Chrysobalanaceae	Chrysobalanaceae	Chrysobalanaceae	Chrysobalanaceae

(cont.)

Table 4.1. (*cont.*)

Names	Melchior (1964)	Takhtajan (1987)	Cronquist (1988)	Brummitt (1992)	APG (1998)
Circeastraceae	Ranunculaceae	Circeasteraceae	Circeastraceae	Circeastraceae	Circeastraceae
Cistaceae	Cistaceae	Cistaceae	Cistaceae	Cistaceae	Cistaceae
[Cleomaceae]	Capparaceae	Capparaceae	Capparaceae	Capparaceae	Capparaceae
Clethraceae	Clethraceae	Clethraceae	Clethraceae	Clethraceae	Clethraceae
Cneoraceae	Cneoraceae	Cneoraceae	Cneoraceae	Cneoraceae	Rutaceae
Cobaeaceae	Polemoniaceae	Cobaeaceae	Polemoniaceae	Cobaeaceae	Polemoniaceae
Cochlospermaceae	Cochlospermaceae	Cochlospermaceae	Bixaceae	Cochlospermaceae	Cochlospermaceae
Colchicaceae	Liliaceae	Melanthiaceae	Liliaceae	Colchicaceae	Colchicaceae
Columelliaceae	Columelliaceae	Columelliaceae	Columelliaceae	Columelliaceae	Columelliaceae
Combretaceae	Combretaceae	Combretaceae	Combretaceae	Combretaceae	Combretaceae
Commelinaceae	Commelinaceae	Commelinaceae	Commelinaceae	Commelinaceae	Commelinaceae
Compositae	Compositae	as Asteraceae	as Asteraceae	Compositae	as Asteraceae
Connaraceae	Connaraceae	Connaraceae	Connaraceae	Connaraceae	Connaraceae
Conostylidaceae	Haemodoraceae	Conostylidaceae	Haemodoraceae	Haemodoraceae	Haemodoraceae
Convallariaceae	Liliaceae	Convallariaceae	Liliaceae	Convallariaceae	Convallariaceae
Convolvulaceae	Convolvulaceae	Convolvulaceae	Convolvulaceae	Convolvulaceae	Convolvulaceae
Cordiaceae	Boraginaceae	Cordiaceae	Boraginaceae	Boraginaceae	Boraginaceae
Coriariaceae	Coriariaceae	Coriariaceae	Coriariaceae	Coriariaceae	Coriariaceae
[Coridaceae]	Primulaceae	Primulaceae	Primulaceae	Primulaceae	Primulaceae
Cornaceae	Cornaceae	Cornaceae	Cornaceae	Cornaceae	Cornaceae
Corsiaceae	Corsiaceae	Corsiaceae	Corsiaceae	Corsiaceae	Corsiaceae
Corylaceae	Betulaceae	Betulaceae	Betulaceae	Corylaceae	Betulaceae
Corynocarpaceae	Corynocarpaceae	Corynocarpaceae	Corynocarpaceae	Corynocarpaceae	Corynocarpaceae
Costaceae	Zingiberaceae	Costaceae	Costaceae	Costaceae	Costaceae
Crassulaceae	Crassulaceae	Crassulaceae	Crassulaceae	Crassulaceae	Crassulaceae
[Croomiaceae]	Stemonaceae	Stemonaceae	Stemonaceae	Stemonaceae	Stemonaceae
Crossosomataceae	Crossosomataceae	Crossosomataceae	Crossosomataceae	Crossosomataceae	Crossosomataceae
Cruciferae	Cruciferae	as Brassicaceae	as Brassicaceae	Cruciferae	as Brassicaceae
Crypteroniaceae	Crypteroniaceae	Crypteroniaceae	Crypteroniaceae	Crypteroniaceae	Crypteroniaceae

Ctenolophonaceae	Linaceae	Ctenolophonaceae	? Linaceae	Ctenolophonaceae	Ctenolophonaceae
Cucurbitaceae	Cucurbitaceae	Cucurbitaceae	Cucurbitaceae	Cucurbitaceae	Cucurbitaceae
Cunoniaceae	Cunoniaceae	Cunoniaceae	Cunoniaceae	Cunoniaceae	Cunoniaceae
Curtisiaceae	Cornaceae	Curtisiaceae	Cornaceae	Cornaceae	Cornaceae
Cuscutaceae	Convolvulaceae	Cuscutaceae	Cuscutaceae	Convolvulaceae	Convolvulaceae
Cyanastraceae	Cyanastraceae	Cyanastraceae	Cyanastraceae	Cyanastraceae	Tecophilaeaceae
Cyclanthaceae	Cyclanthaceae	Cyclanthaceae	Cyclanthaceae	Cyclanthaceae	Cyclanthaceae
Cyclocheilaceae	? Verbenaceae	Acanthaceae	? Verbenaceae	Cyclocheilaceae	Cyclocheilaceae
Cymodoceaceae	Zanichelliaceae	Cymodoceaceae	Cymodoceaceae	Cymodoceaceae	Cymodoceaceae
Cynomoriaceae	Cynomoriaceae	Cynomoriaceae	Balanophoraceae	Cynomoriaceae	Cynomoriaceae
Cyperaceae	Cyperaceae	Cyperaceae	Cyperaceae	Cyperaceae	Cyperaceae
Cyphiaceae	Campanulaceae	Cyphiaceae	Campanulaceae	Campanulaceae	Campanulaceae
Cyphocarpaceae	Campanulaceae	Cyphocarpaceae	Campanulaceae	Campanulaceae	Campanulaceae
Cyrillaceae	Cyrillaceae	Cyrillaceae	Cyrillaceae	Cyrillaceae	Cyrillaceae
Cytinaceae	Rafflesiaceae	Cytinaceae	Rafflesiaceae	Rafflesiaceae	Cytinaceae
Dactylanthaceae	Balanophoraceae	Dactylanthaceae	Balanophoraceae	Balanophoraceae	Balanophoraceae
Daphniphyllaceae	Daphniphyllaceae	Daphniphyllaceae	Daphniphyllaceae	Daphniphyllaceae	Daphniphyllaceae
Dasypogonaceae	Xanthorrhoeaceae	Dasypogonaceae	Xanthorrhoeaceae	Dasypogonaceae	Dasypogonaceae
Datiscaceae	Datiscaceae	Datiscaceae	Datiscaceae	Datiscaceae	Datiscaceae
Davidiaceae	Davidiaceae	Davidiaceae	Cornaceae	Cornaceae	Cornaceae
Davidsoniaceae	Davidsoniaceae	Davidsoniaceae	Davidsoniaceae	Davidsoniaceae	Cunoniaceae
Degeneriaceae	Degeneriaceae	Degeneriaceae	Degeneriaceae	Degeneriaceae	Degeneriaceae
Desfontainiaceae	Desfontainiaceae	Desfontainiaceae	?	Loganiaceae	Columelliaceae
Dialypetalanthaceae	Dialypetalanthaceae	Dialypetalanthaceae	Dialypetalanthaceae	Dialypetalanthaceae	Rubiaceae
Diapensiaceae	Diapensiaceae	Diapensiaceae	Diapensiaceae	Diapensiaceae	Diapensiaceae
Dichapetalaceae	Dichapetalaceae	Dichapetalaceae	Dichapetalaceae	Dichapetalaceae	Dichapetalaceae
[Dicrastylidaceae]	Verbenaceae	Verbenaceae?	Verbenaceae?	Verbenaceae	Verbenaceae?
Didiereaceae	Didiereaceae	Didiereaceae	Didiereaceae	Didiereaceae	Didiereaceae
Didymelaceae	Didymelaceae	Didymelaceae	Didymelaceae	Didymelaceae	Didymelaceae
Diegodendraceae	?	Diegodendraceae	Ochnaceae	Diegodendraceae	in Bixaceae
Diervillaceae	Caprifoliaceae	Caprifoliaceae	Caprifoliaceae	Caprifoliaceae	Diervillaceae

(cont.)

Table 4.1. (*cont.*)

Names	Melchior (1964)	Takhtajan (1987)	Cronquist (1988)	Brummitt (1992)	APG (1998)
Dilleniaceae	Dilleniaceae	Dilleniaceae	Dilleniaceae	Dilleniaceae	Dilleniaceae
Dioncophyllaceae	Dioncophyllaceae	Dioncoophyllaceae	Dioncophyllaceae	Dioncophyllaceae	Dioncophyllaceae
Dioscoreaceae	Dioscoreaceae	Dioscoreaceae	Dioscoreaceae	Dioscoreaceae	Dioscoreaceae
Dipentodontaceae	Dipentodontaceae	Dipentodontaceae	Dipentodontaceae	Dipentodontaceae	Dipentodontaceae
Dipsacaceae	Dipsacaceae	Dipsacaceae	Dipsacaceae	Dipsacaceae	Dipsacaceae
Dipterocarpaceae	Dipterocarpaceae	Dipterocarpaceae	Dipterocarpaceae	Dipterocarpaceae	Dipterocarpaeae
Dirachmaceae	Geraniaceae	Dirachmaceae	Geraniaceae	Geraniaceae	Dirachmaceae
Donatiaceae	Stylidiaceae	Donatiaceae	Donatiaceae	Stylidiaceae	Donatiaceae
Doryanthaceae	Agavaceae	Doryanthaceae	Agavaceae	Doryanthaceae	Doryanthaceae
Dracaenaceae	Agavaceae	Dracaenaceae	Agavaceae	Dracaenaceae	Convallariaceae
Droseraceae	Droseraceae	Droseraceae	Droseraceae	Droseraceae	Droseraceae
Drosophyllaceae	Droseraceae	Droseraceae	Droseraceae	Droseraceae	Drosophyllaceae
Duabangaceae	Lythraceae	Duabangaceae	Lythraceae	Lythraceae	Lythraceae
Duckeodendraceae	Duckeodendraceae	Duckeodendraceae	Duckeodendraceae	Duckeodendraceae	Solanaceae
Dulongiaceae	Saxifragaceae	Dulongiaceae	Grossulariaceae	Escalloniaceae	Phyllonomaceae
Dysphaniaceae	Dysphaniaceae	Chenopodiaceae?	Chenopodiaceae?	Chenopodiaceae	Amaranthaceae
Ebenaceae	Ebenaceae	Ebenaceae	Ebenaceae	Ebenaceae	Ebenaceae
Ecdeiocoleaceae	? Restionaceae	Ecdeiocoleaceae	Restionaceae	Ecdeiocoleaceae	Ecdeicoleaceae
Ehretiaceae	Boraginaceae	Ehretiaceae	Boraginaceae	Boraginaceae	Boraginaceae
Elaeagnaceae	Elaeagnaceae	Elaeagnaceae	Elaeagnaceae	Elaeagnaceae	Elaeagnaceae
Elaeocarpaceae	Elaeocarpaceae	Elaeocarpaceae	Elaeocarpaceae	Elaeocarpaceae	Elaeocarpaceae
Elatinaceae	Elatinaceae	Elatinaceae	Elatinaceae	Elatinaceae	Elatinaceae
[Ellisiophyllaceae]	Scrophulariaceae	Scrophulariaceae?	Scrophulariaceae?	Scrophulariaceae	Scrophulariaceae
Emblingiaceae	Capparaceae	Emblingiaceae	Polygalaceae	Emblingiaceae	Emblingiaceae
Empetraceae	Empetraceae	Empetraceae	Empetraceae	Empetraceae	Ericaceae
Epacridaceae	Epacridaceae	Epacridaceae	Epacridaceae	Epacridaceae	Ericaceae
Eremolepidaceae	Loranthaceae	Eremolepidaceae	Eremolepidaceae	Eremolepidaceae	Santalaceae

Eremosynaceae	Saxifragaceae	Eremosynaceae	Saxifragaceae	Eremosynaceae	Eremosynaceae
Ericaceae	Ericaceae	Ericaceae	Ericaceae	Ericaceae	Ericaceae
Eriocaulaceae	Eriocaulaceae	Eriocaulaceae	Eriocaulaceae	Eriocaulaceae	Eriocaulaceae
Eriospermaceae	Liliaceae	Eriospermaceae	Liliaceae	Eriospermaceae	Convallariaceae
[Erythropalaceae]	Olacaceae	Olacaceae?	Olacaceae?	Olacaceae	Olacaceae?
Erythroxylaceae	Erythroxylaceae	Erythroxylaceae	Erythroxylaceae	Erythroxylaceae	Erythroxylaceae
Escalloniaceae	Saxifragaceae	Escalloniaceae	Grossulariaceae	Escalloniaceae	Escalloniaceae
Eucommiaceae	Eucommiaceae	Eucommiaceae	Eucommiaceae	Eucommiaceae	Eucommiaceae
Eucryphiaceae	Eucryphiaceae	Eucryphiaceae	Eucryphiaceae	Eucryphiaceae	Cunoniaceae
Euphorbiaceae	Euphorbiaceae	Euphorbiaceae	Euphorbiaceae	Euphorbiaceae	Euphorbiaceae
Euphroniaceae	Trigoniaceae	Trigoniaceae	Vochysiaceae	Euphroniaceae	Euphroniaceae
Eupomatiaceae	Eupomatiaceae	Eupomatiaceae	Eupomatiaceae	Eupomatiaceae	Eupomatiaceae
Eupteleaceae	Eupteleaceae	Eupteleaceae	Eupteleaceae	Eupteleaceae	Eupteleaceae
[Euryalaceae]	Nymphaeaceae?	Nymphaeaceae?	Nymphaeaceae?	Nymphaeaceae	Nymphaeaceae?
Fagaceae	Fagaceae	Fagaceae	Fagaceae	Fagaceae	Fagaceae
Flacourtiaceae	Flacourtiaceae	Flacourtiaceae	Flacourtiaceae	Flacourtiaceae	Flacourtiaceae
Flagellariaceae	Flagellariaceae	Flagellariaceae	Flagellariaceae	Flagellariaceae	Flagellariaceae
[Flindersiaceae]	Rutaceae?	Rutaceae?	Rutaceae?	Rutaceae	Rutaceae?
[Foetidiaceae]	Lecythidaceae?	Lecythidaceae?	Lecythidaceae?	Lecythidaceae	Lecythidaceae?
Fouquieriaceae	Fouquieriaceae	Fouquieriaceae	Fouquieriaceae	Fouquieriaceae	Fouquieriaceae
Francoaceae	Saxifragaceae	Frankoaceae	? Saxifragaceae	Saxifragaceae	Francoaceae
Frankeniaceae	Frankeniaceae	Frankeniaceae	Frankeniaceae	Frankeniaceae	Frankeniaceae
Fumariaceae	Papaveraceae	Fumariaceae	Fumariaceae	in Papaveraceae	in Papaveraceae
Garryaceae	Garryaceae	Garryaceae	Garryaceae	Garryaceae	Garryaceae
Geissolomataceae	Geissolomataceae	Geissolomataceae	Geissolomataceae	Geissolomataceae	Geissolomataceae
Gelsemiaceae	Loganiaceae	? Escalloniaceae	Loganiaceae	Loganiaceae	Gelsemiaceae
Gentianaceae	Gentianaceae	Gentianaceae	Gentianaceae	Gentianaceae	Gentianaceae

(cont.)

Table 4.1. (cont.)

Names	Melchior (1964)	Takhtajan (1987)	Cronquist (1988)	Brummitt (1992)	APG (1998)
Geosiridaceae	Geosiridaceae	Geosiridaceae	Geosiridaceae	Iridaceae	Iridaceae
Geraniaceae	Geraniaceae	Geraniaceae	Geraniaceae	Geraniaceae	Geraniaceae
Gesneriaceae	Gesneriaceae	Gesneriaceae	Gesneriaceae	Gesneriaceae	Gesneriaceae
Gisekiaceae	Molluginaceae	Phytolaccaceae	Phytolaccaceae	Gisekiaceae	Phytolaccaceae
Glaucidiaceae	Ranunculaceae	Glaucidiaceae	Ranunculaceae	Glaucidiaceae	Ranunculaceae
Globulariaceae	Globulariaceae	Globulariaceae	Globulariaceae	Globulariaceae	Plantaginaceae
Goetzeaceae	Solanaceae	Goetzeaceae	Solanaceae	Goetzeaceae	Solanaceae
Gomortegaceae	Gomortegaceae	Gomortegaceae	Gomortegaceae	Gomortegaceae	Gomortegaceae
[Gonystylaceae]	Thymelaeaceae	Thymelaeaceae?	Thymelaeaceae?	Thymelaeaceae	Thymelaeaceae?
Goodeniaceae	Goodeniaceae	Goodeniaceae	Goodeniaceae	Goodeniaceae	Goodeniaceae
Goupiaceae	Celastraceae	Goupiaceae	Celastraceae	Goupiaceae	Goupiaceae
Gramineae	Gramineae	as Poaceae	as Poaceae	Gramineae	Gramineae (Poaceae)
Greyiaceae	Melianthaceae	Greyiaceae	Greyiaceae	Greyiaceae	Greyiaceae
Griseliniaceae	Cornaceae	Griseliniaceae	Cornaceae	Griseliniaceae	Griseliniaceae
Grossulariaceae	Saxifragaceae	Grossulariaceae	Grossulariaceae	Grossulariaceae	Grossulariaceae
Grubbiaceae	Grubbiaceae	Grubbiaceae	Grubbiaceae	Grubbiaceae	Grubbiaceae
Gunneraceae	Haloragaceae	Gunneraceae	Gunneraceae	Gunneraceae	Gunneraceae
Guttiferae	Guttiferae	as Clusiaceae	as Clusiaceae	Guttiferae	as Clusiaceae
Gyrocarpaceae	Hernandiaceae	Gyrocarpaceae	Hernandiaceae	Hernandiaceae	Hernandiaceae
Gyrostemonaceae	Gyrostemonaceae	Gyrostemonaceae	Gyrostemonaceae	Gyrostemonaceae	Gyrostemonaceae
Haemodoraceae	Haemodoraceae	Haemodoraceae	Haemodoraceae	Haemodoraceae	Haemodoraceae
Halesiaceae	Styracaceae	Styraceae	Styraceae	Styraceae	Halesiaceae
Halophilaceae	Hydrocharitaceae	Halophilaceae	Hydrocharitaceae	Hydrocharitaceae	Hydrocharitaceae
Halophytaceae	? Chenopodiaceae	Halophytaceae	Halophytaceae	Halophytaceae	Amaranthaceae
Haloragaceae	Haloragaceae	Haloragaceae	Haloragaceae	Haloragaceae	Haloragaceae
Hamamelidaceae	Hamamelidaceae	Hamamelidaceae	Hamamelidaceae	Hamamelidaceae	Hamamelidaceae
Hanguanaceae	? Xanthorrhoeaceae	Hanguanaceae	Hanguanaceae	Hanguanaceae	Hanguanaceae

Hectorellaceae	Caryophyllaceae	Hectorellaceae	Portulacaceae	Hectorellaceae	Portulacaceae
Heliconiaceae	Musaceae	Heliconiaceae	Heliconiaceae	Heliconiaceae	Heliconiaceae
[Helleboraceae]	Ranunculaceae	Ranunculaceae	Ranunculaceae	Ranunculaceae	Ranunculaceae
Helwingiaceae	Helwingiaceae	Helwingiaceae	Cornaceae	Helwingiaceae	Helwingiaceae
Hemerocallidaceae	Hemerocallidaceae	Hemerocallidaceae	Liliaceae	Hemerocallidaceae	Hemerocallidaceae
Henriqueziaceae	Henriqueziaceae	Rubiaceae?	Rubiaceae?	Rubiaceae?	Rubiaceae?
Hernandiaceae	Hernandiaceae	Hernandiaceae	Hernandiaceae	Hernandiaceae	Hernandiaceae
Herreriaceae	Liliaceae	Herreriaceae	Liliaceae	Herreriaceae	Herreriaceae
Hesperocallidaceae	Liliaceae	Hesperocallidaceae	Liliaceae	in Hyacinthaceae	Hesperocallidaceae
Heteropyxidaceae	Myrtaceae	Heteropyxidaceae	? Myrtaceae	Myrtaceae	Heteropyxidaceae
Himantandraceae	Himantandraceae	Himantandraceae	Himantandraceae	Himantandraceae	Himantandraceae
Hippocastanaceae	Hippocastanaceae	Hippocastanaceae	Hippocastanaceae	Hippocastanaceae	Sapindaceae
Hippocrateaceae	Hippocrateaceae	Celastraceae	Hippocrateaceae	Celastraceae	Celastraceae
Hippuridaceae	Hippuridaceae	Hippuridaceae	Hippuridaceae	Hippuridaceae	Plantaginaceae
Hoplestigmataceae	Hoplestigmataceae	Hoplestigmataceae	Hoplestigmataceae	Hoplestigmataceae	Hoplestigmataceae
Hostaceae	Liliaceae	as Funkiaceae	Liliaceae	Hostaceae	Agavaceae
Huaceae	? Styracaceae	Huaceae	Huaceae	Huaceae	Huaceae
Hugoniaceae	Linaceae	Hugoniaceae	Hugoniaceae	Linaceae	Hugoniaceae
[Humbertiaceae]	Convolvulaceae	Convolvulaceae?	Convolvulaceae?	Convolvulaceae	Convolvulaceae?
Humiriaceae	Linaceae	Humiriaceae	Humiriaceae	Humiriaceae	Humiriaceae
Hyacinthaceae	Liliaceae	Hyacinthaceae	Liliaceae	Hyacinthaceae	Hyacinthaceae
Hydatellaceae	Centrolepidaceae	Hydatellaceae	Hydatellaceae	Hydatellaceae	Hydatellaceae
Hydnoraceae	Hydnoraceae	Hydnoraceae	Hydnoraceae	Hydnoraceae	Hydnoraceae
Hydrangeaceae	Saxifragaceae	Hydrangeaceae	Hydrangeaceae	Hydrangeaceae	Hydrangeaceae
Hydrastidaceae	Ranunculaceae	Hydrastidaceae	Ranunculaceae	Ranunculaceae	Ranunculaceae
Hydrocharitaceae	Hydrocharitaceae	Hydrocharitaceae	Hydrocharitaceae	Hydrocharitaceae	Hydrocharitaceae
[Hydrocotylaceae]	Umbelliferae	Umbelliferae	Umbelliferae	Umbelliferae	Umbelliferae
Hydroleaceae	Hydrophyllaceae	Hydrophyllaceae	Hydrophyllaceae	Hydrophyllaceae	[Hydroleaceae]
Hydrophyllaceae	Hydrophyllaceae	Hydrophyllaceae	Hydrophyllaceae	Hydrophyllaceae	Boraginaceae
Hydrostachyaceae	Hydrostachyaceae	Hydrostachyaceae	Hydrostachyaceae	Hydrostachyaceae	Hydrostachyaceae

(cont.)

Table 4.1. (*cont.*)

Names	Melchior (1964)	Takhtajan (1987)	Cronquist (1988)	Brummitt (1992)	APG (1998)
[Hymenocardiaceae]					
Hypecoaceae	Papaveraceae	Hypecoaceae	Papaveraceae	Papaveraceae	Papaveraceae
Hypoxidaceae	Hypoxidaceae	Hypoxidaceae	Liliaceae	Hypoxidaceae	Hypoxidaceae
Hypseocharitaceae	Oxalidaceae	Hypseocharitaceae	?	Oxalidaceae	in Geraniaceae
Icacinaceae	Icacinaceae	Icacinaceae	Icacinaceae	Icacinaceae	Icacinaceae
Idiospermaceae	? Calycanthaceae	Idiospermaceae	Idiospermaceae	Idiospermaceae	Calycanthaceae
Illecebraceae	Caryophyllaceae	Caryophyllaceae	Illecebraceae	Illecebraceae	Caryophyllaceae
Illiciaceae	Illiciaceae	Illiciaceae	Illiciaceae	Illiciaceae	Illiciaceae
Iridaceae	Iridaceae	Iridaceae	Iridaceae	Iridaceae	Iridaceae
Irvingiaceae	Simaroubaceae	Irvingiaceae	Simaroubaceae	Irvingiaceae	Irvingiaceae
Iteaceae	Saxifragaceae	Iteaceae	? Grossulariaceae	Escalloniaceae	Iteaceae
Ixerbaceae	Saxifragaceae	? Escalloniaceae	? Grossulariaceae	Escalloniaceae	Ixerbaceae
Ixioliriaceae	Amaryllidaceae	Ixioliriaceae	Liliaceae	Ixioliriaceae	Ixioliriaceae
Ixonanthaceae	Linaceae	Ixonanthaceae	Ixonanthaceae	Ixonanthaceae	Ixonanthaceae
Japonoliriaceae	Liliaceae	? Melanthiaceae	Liliaceae	in Melanthiaceae	Japonoliriaceae
Joinvilleaceae	Flagellariaceae	Joinvilleaceae	Joinvilleaceae	Joinvilleaceae	Joinvilleaceae
Juglandaceae	Juglandaceae	Juglandaceae	Juglandaceae	Juglandaceae	Juglandaceae
Julianiaceae	Julianiaceae	? Anacardiaceae	Julianiaceae	Anacardiaceae	Anacardiaceae
Juncaceae	Juncaceae	Juncaceae	Juncaceae	Juncaceae	Juncaceae
Juncaginaceae	Juncaginaceae	Juncaginaceae	Juncaginaceae	Juncaginaceae	Juncaginaceae
Kaliphoraceae	Cornaceae	Melanophyllaceae	? Cornaceae	Melanophyllaceae	Kaliphoraceae
Kiggelariaceae	Flacourtiaceae	Kiggelariaceae	Flacourtiaceae	Flacourtiaceae	Flacourtiaceae
Kingdoniaceae	Ranunculaceae	Ranunculaceae	? Ranunculaceae	in Ranunculaceae	in Circeastraceae
Kirkiaceae	Simaroubaceae	Kirkiaceae	? Simaroubaceae	Simaroubaceae	Kirkiaceae
Koeberliniaceae	Capparaceae	Capparaceae	? Capparaceae	Capparaceae	Koeberliniaceae
Krameriaceae	Krameriaceae	Krameriaceae	Krameriaceae	Krameriaceae	Krameriaceae
Labiatae	Labiatae	as Lamiaceae	as Lamiaceae	Labiatae	as Lamiaceae
Lacistemaceae	Flacourtiaceae	Lacistemataceae	Lacistemataceae	Lacistemaceae	Lacistemaceae

	Lactoridaceae	Lactoridaceae	Lactoridaceae	Lactoridaceae
Lactoridaceae	Lactoridaceae	Lactoridaceae	Lactoridaceae	Lactoridaceae
Lanariaceae	Tecophilaeaceae	? Liliaceae	Lanariaceae	Lanariaceae
Lardizabalaceae	Lardizabalaceae	Lardizabalaceae	Lardizabalaceae	Lardizabalaceae
Latraeophilaceae	Latraeophilaceae	Balanophoraceae	Balanophoraceae	Balanophoraceae
Lauraceae	Lauraceae	Lauraceae	Lauraceae	Lauraceae
Laxmanniaceae	Asphodelaceae	? Liliaceae	in Anthericaceae	Laxmanniaceae
Lecythidaceae	Lecythidaceae	Lecythidaceae	Lecythidaceae	Lecythidaceae
Ledocarpaceae	Ledocarpaceae	Geraniaceae	Geraniaceae	Ledocarpaceae
Leeaceae	Leeaceae	Leeaceae	Leeaceae	Vitaceae
Leguminosae	as Fabaceae	as Fabaceae	Leguminosae	as Fabaceae
Leitneriaceae	Leitneriaceae	Leitneriaceae	Leitneriaceae	Simaroubaceae
Lemnaceae	Lemnaceae	Lemnaceae	Lemnaceae	Araceae
Lennoaceae	Lennoaceae	Lennoaceae	Lennoaceae	Boraginaceae
Lentibulariaceae	Lentibulariaceae	Lentibulariaceae	Lentibulariaceae	Lentibulariaceae
Berberidaceae	Berberidaceae	Berberidaceae	Berberidaceae	Berberidaceae
[Leonticaceae]				
Lepidobotryaceae	Lepidobotryaceae	Lepidobotryaceae	Lepidobotryaceae	Lepidobotryaceae
Lepuropetalaceae	Lepuropetalaceae	Saxifragaceae	Parnassiaceae	in Parnassiaceae
Lilaeaceae	Lilaeaceae	Juncaginaceae	Lilaeaceae	Juncaginaceae
Liliaceae	Liliaceae	Liliaceae	Liliaceae	Liliaceae
Limnanthaceae	Limnanthaceae	Limnanthaceae	Limnanthaceae	Limnanthaceae
Limnocharitaceae	Limnocharitacaee	Limnocharitaceae	Limnocharitaceae	Limnocharitaceae
Linaceae	Linaceae	Linaceae	Linaceae	Linaceae
Linnaeaceae	? Caprifoliaceae	Caprifoliaceae	Caprifoliaceae	Linnaeaceae
Lissocarpaceae	Lissocarpaceae	Lissocarpaceae	Lissocarpaceae	Lissocarpaceae
Loasaceae	Loasaceae	Loasaceae	Loasaceae	Loasaceae
Lobeliaceae	Lobeliaceae	Campanulaceae	Campanulaceae	Campanulaceae
Loganiaceae	Loganiaceae	Loganiaceae	Loganiaceae	Loganiaceae
Lomandraceae	Dasypogonaceae	Xanthorrhoeaceae	Lomandraceae	Laxmanniaceae
Lophiraceae	Lophiraceae	Ochnaceae	Ochnaceae	Ochnaceae
Lophophytaceae	Lophophytaceae	Balanophoraceae	Balanophoraceae	Balanophoraceae
Lophopyxidaceae	Lophopyxidaceae	Celastraceae	Lophopyxidaceae	Lophopyxidaceae

(cont.)

Table 4.1. (*cont.*)

Names	Melchior (1964)	Takhtajan (1987)	Cronquist (1988)	Brummitt (1992)	APG (1998)
Loranthaceae	Loranthaceae	Loranthaceae	Loranthaceae	Loranthaceae	Loranthaceae
Lowiaceae	Lowiaceae	Lowiaceae	Lowiaceae	Lowiaceae	Lowiaceae
Luzuriagaceae	Liliaceae	Luzuriagaceae	? Smilacaceae	in Philesiaceae	Luzuriagaceae
Lythraceae	Lythraceae	Lythraceae	Lythraceae	Lythraceae	Lythraceae
Magnoliaceae	Magnoliaceae	Magnoliaceae	Magnoliaceae	Magnoliaceae	Magnoliaceae
Malesherbiaceae	Malesherbiaceae	Malesherbiaceae	Malesherbiaceae	Malesherbiaceae	Malesherbiaceae
Malpighiaceae	Malpighiaceae	Malpighiaceae	Malpighiaceae	Malpighiaceae	Malpighiaceae
Malvaceae	Malvaceae	Malvaceae	Malvaceae	Malvaceae	Malvaceae
Marantaceae	Marantaceae	Marantaceae	Marantaceae	Marantaceae	Marantaceae
Marcgraviaceae	Marcgraviaceae	Marcgraviaceae	Marcgraviaceae	Marcgraviaceae	Marcgraviaceae
Martyniaceae	Martyniaceae	Martyniaceae	Pedaliaceae	in Pedaliaceae	in Pedaliaceae
Mastixiaceae	Cornaceae	Mastixiaceae	Cornaceae	Cornaceae	Cornaceae
Maundiaceae	Juncaginaceae	Maundiaceae	Juncaginaceae	Juncaginaceae	Juncaginaceae
Mayacaceae	Mayacaceae	Mayacaceae	Mayacaceae	Mayacaceae	Mayacaceae
Medeolaceae	Liliaceae	Medeolaceae	Liliaceae	Convallariaceae	Liliaceae
Medusagynaceae	Medusagynaceae	Medusagynaceae	Medusagynaceae	Medusagynaceae	Medusagynaceae
Medusandraceae	Medusandraceae	Medusandraceae	Medusandraceae	Medusandraceae	Medusandraceae
Melanophyllaceae	Cornaceae	Melanophyllaceae	Cornaceae	Melanophyllaceae	Melanophyllaceae
Melanthiaceae	Liliaceae	Melanthiaceae	Liliaceae	Melanthiaceae	Melanthiaceae
Melastomataceae	Melastomataceae	Melastomataceae	Melastomataceae	Melastomataceae	Melastomataceae
Meliaceae	Meliaceae	Meliaceae	Meliaceae	Meliaceae	Meliaceae
Melianthaceae	Melianthaceae	Melianthaceae	Melianthaceae	Melianthaceae	Melianthaceae
Meliosmaceae	Sabiaceae	Sabiaceae	Sabiaceae	Meliosmaceae	Sabiaceae
Memecylaceae	Melastomataceae	Melastomataceae	Melastomataceae	Melastomataceae	Memecylaceae
Mendonciaceae	Acanthaceae	Mendonciaceae	Mendonciaceae	Acanthaceae	Acanthaceae
Menispermaceae	Menispermaceae	Menispermaceae	Menispermaceae	Menispermaceae	Menispermaceae
Menyanthaceae	Menyanthaceae	Menyanthaceae	Menyanthaceae	Menyanthaceae	Menyanthaceae
[Mesembryanthema-ceae]	Aizoaceae	Aizoaceae	Aizoaceae	Aizoaceae	Aizoaceae

Metteniusaceae	Icacinaceae	? Icacinaceae	? Icacinaceae	Icacinaceae	Metteniusaceae
Mimosaceae	Leguminosae	Leguminosae	Mimosaceae	Leguminosae	Leguminosae
Misodendraceae	Misodendraceae	Misodendraceae	Misodendraceae	Misodendraceae	Misodendraceae
Mitrastemonaceae	Rafflesiaceae	Mitrastemonaceae	Mitrastemonaceae	Mitrastemonaceae	Mitrastemonaceae
Molluginaceae	Molluginaceae	Molluginaceae	Molluginaceae	Molluginaceae	Molluginaceae
Monimiaceae	Monimiaceae	Monimiaceae	Monimiaceae	Monimiaceae	Monimiaceae
Monotaceae	Dipterocarpaceae	Monotaceae	Dipterocarpaceae	Dipterocarpaceae	Dipterocarpaceae
Monotropaceae	Ericaceae	Ericaceae	Monotropaceae	Ericaceae	Ericaceae
Montiniaceae	Saxifragaceae	Montiniaceae	Grossulariaceae	Montiniaceae	Montiniaceae
Moraceae	Moraceae	Moraceae	Moraceae	Moraceae	Moraceae
Morinaceae	Dipsacaceae	Morinaceae	Dipsacaceae	Morinaceae	Morinaceae
Moringaceae	Moringaceae	Moringaceae	Moringaceae	Moringaceae	Moringaceae
Muntingiaceae	Tiliaceae	? Tiliaceae	? Tiliaceae	Tiliaceae	Muntingiaceae
Musaceae	Musaceae	Musaceae	Musaceae	Musaceae	Musaceae
Myoporaceae	Myoporaceae	Myoporaceae	Myoporaceae	Myoporaceae	Myoporaceae
Myricaceae	Myricaceae	Myricaceae	Myricaceae	Myricaceae	Myricaceae
Myristicaceae	Myristicaceae	Myristicaceae	Myristicaceae	Myristicaceae	Myristicaceae
Myrothamnaceae	Myrothamnaceae	Myrothamnaceae	Myrothamnaceae	Myrothamnaceae	Myrothamnaceae
Myrsinaceae	Myrsinaceae	Myrsinaceae	Myrsinaceae	Myrsinaceae	Myrsinaceae
Myrtaceae	Myrtaceae	Myrtaceae	Myrtaceae	Myrtaceae	Myrtaceae
Najadaceae	Najadaceae	Najadaceae	Najadaceae	Hydrocharitaceae	Hydrocharitaceae
Nandinaceae	Berberidaceae	Nandinaceae	Berberidaceae	Berberidaceae	Berberidaceae
[Napoleonaeaceae]	Lecythidaceae	Lecythidaceae	Lecythidaceae	Lecythidaceae	Lecythidaceae
Nartheciaceae	Liliaceae	? Melanthiaceae	Liliaceae	in Melanthiaceae	Nartheciaceae
[Naucleaceae]	Rubiaceae	Rubiaceae	Rubiaceae	Rubiaceae	Rubiaceae
Nelumbonaceae	Nymphaeaceae	Nelumbonaceae	Nelumbonaceae	Nelumbonaceae	Nelumbonaceae
Nemacladaceae	Campanulaceae	Nemacladaceae	Campanulaceae	Campanulaceae	Campanulaceae
Nepenthaceae	Nepenthaceae	Nepenthaceae	Nepenthaceae	Nepenthaceae	Nepenthaceae
Nesogenaceae	Verbenaceae	Nesogenaceae	? Verbenaceae	Nesogenaceae	Cyclocheilaceae
[Neumanniaceae]	Flacourtiaceae	Flacourtiaceae?	Flacourtiaceae?	Flacourtiaceae	Flacourtiaceae?

(cont.)

Table 4.1. (*cont.*)

Names	Melchior (1964)	Takhtajan (1987)	Cronquist (1988)	Brummitt (1992)	APG (1998)
Neuradaceae	Neuradaceae	Neuradaceae	Neuradaceae	Neuradaceae	Neuradaceae
Nitrariaceae	Zygophyllaceae	Nitrariaceae	Zygophyllaceae	Zygophyllaceae	Nitrariaceae
Nolanaceae	Nolanaceae	Nolanaceae	Nolanaceae	Solanaceae	Solanaceae
Nolinaceae	Agavaceae	Nolinaceae	Agavaceae	Dracaenaceae	Convallariaceae
Nothofagaceae	Fagaceae	Fagaceae	Fagaceae	Fagaceae	Nothofagaceae
Nyctaginaceae	Nyctaginaceae	Nyctaginaceae	Nyctaginaceae	Nyctaginaceae	Nyctaginaceae
Nymphaeaceae	Nymphaeaceae	Nymphaeaceae	Nymphaeaceae	Nymphaeaceae	Nymphaeaceae
Nyssaceae	Cornaceae	Nyssaceae	Nyssaceae	Cornaceae	in Cornaceae
Ochnaceae	Ochnaceae	Ochnaceae	Ochnaceae	Ochnaceae	Ochnaceae
Octoknemataceae	Olacaceae	Octoknemataceae	Olacaceae	Olacaceae	Olacaceae
Olacaceae	Olacaceae	Olacaceae	Olacaceae	Olacaceae	Olacaceae
Oleaceae	Oleaceae	Oleaceae	Oleaceae	Oleaceae	Oleaceae
Oliniaceae	Oliniaceae	Oliniaceae	Oliniaceae	Oliniaceae	Oliniaceae
Onagraceae	Onagraceae	Onagraceae	Onagraceae	Onagraceae	Onagraceae
Oncothecaceae	? Theaceae	Oncothecaceae	Oncothecaceae	Oncothecaceae	Oncothecaceae
Opiliaceae	Opiliaceae	Opiliaceae	Opiliaceae	Opiliaceae	Opiliaceae
Orchidaceae	Orchidaceae	Orchidaceae	Orchidaceae	Orchidaceae	Orchidaceae
Orobanchaceae	Orobanchaceae	? Scrophulariaceae	Orobanchaceae	Scrophulariaceae	Orobanchaceae
Oxalidaceae	Oxalidaceae	Oxalidaceae	Oxalidaceae	Oxalidaceae	Oxalidaceae
[Oxystylidaceae]	Capparaceae	Capparaceae?	Capparaceae?	Capparaceae	Capparaceae?
Paeoniaceae	Paeoniaceae	Paeoniaceae	Paeoniaceae	Paeoniaceae	Paeoniaceae
Palmae	Palmae	as Arecaceae	as Arecaceae	Palmae	Palmae
					(Arecaceae)
Pandaceae	Pandaceae	Pandaceae	Pandaceae	Pandaceae	Pandaceae
Pandanaceae	Pandanaceae	Pandanaceae	Pandanaceae	Pandanaceae	Pandanaceae
Papaveraceae	Papaveraceae	Papaveraceae	Papaveraceae	Papaveraceae	Papaveraceae
Paracryphiaceae	?	Paracryphiaceae	Paracryphiaceae	Paracryphiaceae	Paracryphiaceae
Parnassiaceae	Saxifragaceae	Parnassiaceae	Saxifragaceae	Parnassiaceae	Parnassiaceae

Passifloraceae	Passifloraceae	Passifloraceae	Passifloraceae	Passifloraceae	Passifloraceae	Passifloraceae
Paulowniaceae	Bignoniaceae	? Scrophulariaceae	? Scrophulariaceae	? Scrophulariaceae	Scrophulariaceae	Paulowniaceae
Pedaliaceae	Pedaliaceae	Pedaliaceae	Pedaliaceae	Pedaliaceae	Pedaliaceae	Pedaliaceae
Peganaceae	Zygophyllaceae	Peganaceae	Zygophyllaceae	Zygophyllaceae	Zygophyllaceae	Nitrariaceae
Pellicieraceae	Theaceae	Pellicieraceae	Pellicieraceae	Pellicieraceae	Pellicieraceae	Pellicieraceae
Penaeaceae	Penaeaceae	Penaeaceae	Penaeaceae	Penaeaceae	Penaeaceae	Penaeaceae
Pentadiplandraceae	Capparaceae	? Capparaceae	? Capparaceae	Capparaceae	Capparaceae	Pentadiplandraceae
Pentaphragmataceae	Pentaphragmataceae	Pentaphragmataceae	Pentaphragmataceae	Pentaphragmataceae	Pentaphragmataceae	Pentaphragmataceae
Pentaphylacaceae	Pentaphylacaceae	Pentaphylacaceae	Pentaphylacaceae	Pentaphylacaceae	Pentaphylacaceae	Pentaphylacaceae
Penthoraceae	Saxifragaceae	Penthoraceae	Saxifragaceae	Penthoraceae	Penthoraceae	Penthoraceae
Peridiscaceae	Peridiscaceae	Peridiscaceae	Peridiscaceae	Peridiscaceae	Peridiscaceae	Peridiscaceae
Petermanniaceae	Liliaceae	Petermanniaceae	Smilacaceae	Petermanniaceae	Petermanniaceae	Colchicaceae
Petrosaviaceae	Liliaceae	? Melanthiaceae	Petrosaviaceae	in Melanthiaceae	in Melanthiaceae	Petrosaviaceae
Phellinaceae	Aquifoliaceae	Phellinaceae	Aquifoliaceae	Phellinaceae	Phellinaceae	Phellinaceae
[Philadelphaceae]	Saxifragaceae	Hydrangeaceae?	Hydrangeaceae?	Hydrangeaceae	Hydrangeaceae?	Hydrangeaceae?
Philesiaceae	Liliaceae	Philesiaceae	Smilacaceae	Philesiaceae	Philesiaceae	Philesiaceae
Philydraceae	Philydraceae	Philydraceae	Philydraceae	Philydraceae	Philydraceae	Philydraceae
Phormiaceae	Liliaceae	Phormiaceae	Agavaceae	Phormiaceae	Phormiaceae	Hemerocallidaceae
Phrymaceae	Phrymaceae	Verbenaceae	Verbenaceae	Verbenaceae	Phrymaceae	Phrymaceae
Phyllonomaceae	Saxifragaceae	? Escalloniaceae	? Grossulariaceae	Escalloniaceae	Escalloniaceae	Phyllonomaceae
Physenaceae	?	Physenaceae	?	Physenaceae	Physenaceae	Physenaceae
Phytolaccaceae	Phytolaccaceae	Phytolaccaceae	Phytolaccaceae	Phytolaccaceae	Phytolaccaceae	Phytolaccaceae
Picramniaceae	Simaroubaceae	Simaroubaceae	Simaroubaceae	Simaroubaceae	Simaroubaceae	Picramniaceae
Picrodendraceae	Picrodendraceae	Euphorbiaceae?	Euphorbiaceae?	Euphorbiaceae	Euphorbiaceae	Euphorbiaceae?
Piperaceae	Piperaceae	Piperaceae	Piperaceae	Piperaceae	Piperaceae	Piperaceae
[Pistaciaceae]	Anacardiaceae	Anacardiaceae	Anacardiaceae	Anacardiaceae	Anacardiaceae	Anacardiaceae
Pittosporaceae	Pittosporaceae	Pittosporaceae	Pittosporaceae	Pittosporaceae	Pittosporaceae	Pittosporaceae
Plagiopteraceae	? Flacourtiaceae	Flacourtiaceae	Flacourtiaceae	Plagiopteraceae	Plagiopteraceae	Plagiopteraceae

(cont.)

Table 4.1. (*cont.*)

Names	Melchior (1964)	Takhtajan (1987)	Cronquist (1988)	Brummitt (1992)	APG (1998)
Plantaginaceae	Plantaginaceae	Plantaginaceae	Plantaginaceae	Plantaginaceae	Plantaginaceae
Platanaceae	Platanaceae	Platanaceae	Platanaceae	Platanaceae	Platanaceae
Plocospermataceae		Plocospermataceae			Plocospermataceae
Plumbaginaceae	Loganiaceae	Plumbaginaceae	Loganiaceae	Loganiaceae	Plumbaginaceae
Podoaceae	Plumbaginaceae	Plumbaginaceae	Plumbaginaceae	Plumbaginaceae	Anacardiaceae
[Podophyllaceae]	Anacardiaceae	Anacardiaceae	Anacardiaceae	Podoaceae	
Podostemaceae	Berberidaceae	Berberidaceae	Berberidaceae	Berberidaceae	Berberidaceae
Polemoniaceae	Podostemaceae	Podostemaceae	Podostemaceae	Podostemaceae	Podostemaceae
Polygalaceae	Polemoniaceae	Polemoniaceae	Polemoniaceae	Polemoniaceae	Polemoniaceae
Polygonaceae	Polygalaceae	Polygalaceae	Polygalaceae	Polygalaceae	Polygalaceae
Polyosmaceae	Polygonaceae	Polygonaceae	Polygonaceae	Polygonaceae	Polygonaceae
Pontederiaceae	Saxifragaceae	Polyosmataceae	? Grossulariaceae	Escalloniaceae	Polyosmaceae
Portulacaceae	Pontederiaceae	Pontederiaceae	Pontederiaceae	Pontederiaceae	Pontederiaceae
Posidoniaceae	Portulacaceae	Portulacaceae	Portulacaceae	Portulacaceae	Portulacaceae
[Potaliaceae]	Potamogetonaceae	Posidoniaceae	Posidoniaceae	Posidoniaceae	Posidoniaceae
Potamogetonaceae	Loganiaceae	Loganiaceae	Loganiaceae	Loganiaceae	Loganiaceae
Pottingeriaceae	Potamogetonaceae	Potamogetonaceae	Potamogetonaceae	Potamogetonaceae	Potamogetonaceae
Primulaceae	Celastraceae	Pottingeriaceae	Celastraceae	Celastraceae	Pottingeriaceae
Prionaceae	Primulaceae	Primulaceae	Primulaceae	Primulaceae	Primulaceae
Proteaceae	Juncaceae	Juncaceae	Juncaceae	in Juncaceae	Prionaceae
Psiloxylaceae	Proteaceae	Proteaceae	Proteaceae	Proteaceae	Proteaceae
Ptaeroxylaceae	Myrtaceae	Psiloxylaceae	Myrtaceae	Myrtaceae	Psiloxylaceae
[Pteridophyllaceae]	Meliaceae	Ptaeroxylaceae	Sapindaceae	Ptaeroxylaceae	Rutaceae
Pterostemonaceae	Papaveraceae	Papaveraceae	Papaveraceae	in Papaveraceae	in Papaveraceae
Punicaceae	Saxifragaceae	Pterostemonaceae	Grossulariaceae	Pterostemonaceae	Pterostemonaceae
Putranjivaceae	Punicaceae	Punicaceae	Punicaceae	Lythraceae	Lythraceae
Pyrolaceae	Euphorbiaceae	Euphorbiaceae	Euphorbiaceae	Euphorbiaceae	Putranjivaceae
	Pyrolaceae	Ericaceae	Pyrolaceae	Ericaceae	Ericaceae

Quiinaceae	Quiinaceae	Quiinaceae	Quiinaceae	Quiinaceae
Quillajaceae	? Rosaceae	Rosaceae	Rosaceae	Quillajaceae
Rafflesiaceae	Rafflesiaceae	Rafflesiaceae	Rafflesiaceae	Rafflesiaceae
Ranunculaceae	Ranunculaceae	Ranunculaceae	Ranunculaceae	Ranunculaceae
Rapateaceae	Rapateaceae	Rapateaceae	Rapateaceae	Rapateaceae
Resedaceae	Resedaceae	Resedaceae	Resedaceae	Resedaceae
Restionaceae	Restionaceae	Restionaceae	Restionaceae	Restionaceae
Retziaceae	Retziaceae	Retziaceae	Retziaceae	Stilbaceae
Rhabdodendraceae	Rhabdodendraceae	Rhabdodendraceae	Rhabdodendraceae	Rhabdodendraceae
Rhamnaceae	Rhamnaceae	Rhamnaceae	Rhamnaceae	Rhamnaceae
Rhipogonaceae	Rhipogonaceae	Smilacaceae	Rhipogonaceae	Rhipogonaceae
Rhizophoraceae	Rhizophoraceae	Rhizophoraceae	Rhizophoraceae	Rhizophoraceae
Rhodoleiaceae	Rhodoleiaceae	Hamamelidaceae	Hamamelidaceae	Hamamelidaceae
Rhoipteleaceae	Rhoipteleaceae	Rhoipteleaceae	Rhoipteleaceae	Rhoipteleaceae
Rhopalocarpaceae	Sphaerosepalaceae	Sphaerosepalaceae	Sphaerosepalaceae	Sphaerosepalaceae
Rhynchocalycaceae	? Lythraceae	Rhynchocalycaceae	Rhynchocalycaceae	Rhynchocalycaceae
Rhynchothecaceae	Geraniaceae	Geraniaceae	Geraniaceae	Geraniaceae
Roridulaceae	Roridulaceae	Byblidaceae	Roridulaceae	Roridulaceae
Rosaceae	Rosaceae	Rosaceae	Rosaceae	Rosaceae
Rousseaceae	Saxifragaceae	? Grossulariaceae	Escalloniaceae	Rousseaceae
Rubiaceae	Rubiaceae	Rubiaceae	Rubiaceae	Rubiaceae
Ruppiaceae	Potamogetonaceae	Ruppiaceae	in Potamogetonaceae	Ruppiaceae
Ruscaceae	Liliaceae	Liliaceae	Ruscaceae	Convallariaceae
Rutaceae	Rutaceae	Rutaceae	Rutaceae	Rutaceae
Sabiaceae	Sabiaceae	Sabiaceae	Sabiaceae	Sabiaceae
Saccifoliaceae	? Gentianaceae	Saccifoliaceae	Saccifoliaceae	Gentianaceae

(cont.)

Table 4.1. (*cont.*)

Names	Melchior (1964)	Takhtajan (1987)	Cronquist (1988)	Brummitt (1992)	APG (1998)
Salicaceae	Salicaceae	Salicaceae	Salicaceae	Salicaceae	Salicaceae
[Salpiglossidaceae]	Solanaceae	Solanaceae	Solanaceae	Solanaceae	Solanaceae
Salvadoraceae	Salvadoraceae	Salvadoraceae	Salvadoraceae	Salvadoraceae	Salvadoraceae
Sambucaceae	Caprifoliaceae	Sambucaceae	Caprifoliaceae	Caprifoliaceae	Caprifoliaceae
[Samydaceae]	Flacourtiaceae	Flacourtiaceae	Flacourtiaceae	Flacourtiaceae	Flacourtiaceae
Santalaceae	Santalaceae	Santalaceae	Santalaceae	Santalaceae	Santalaceae
Sapindaceae	Sapindaceae	Sapindaceae	Sapindaceae	Sapindaceae	Sapindaceae
Sapotaceae	Sapotaceae	Sapotaceae	Sapotaceae	Sapotaceae	Sapotaceae
Sarcobataceae	Chenopodiaceae	Chenopodiaceae	Chenopodiaceae	Chenopodiaceae	Sarcobataceae
Sarcolaenaceae	Sarcolaenaceae	Sarcolaenaceae	Sarcolaenaceae	Sarcolaenaceae	Sarcolaenaceae
Sarcophytaceae	Balanophoraceae	Sarcophytaceae	Balanophoraceae	Balanophoraceae	Balanophoraceae
Sarcospermataceae	Sarcospermataceae	Sapotaceae?	Sapotaceae?	Sapotaceae	Sapotaceae
Sargentodoxaceae	Sargentodoxaceae	Sargentodoxaceae	Sargentodoxaceae	Sargentodoxaceae	Lardizabalaceae
Sarraceniaceae	Sarraceniaceae	Sarraceniaceae	Sarraceniaceae	Sarraceniaceae	Sarraceniaceae
[Saurauiaceae]	Actinidiaceae	Actinidiaceae	Actinidiaceae	Actionidiaceae	Actinidiaceae
Saururaceae	Saururaceae	Saururaceae	Saururaceae	Saururaceae	Saururaceae
Sauvagesiaceae	Ochnaceae	Sauvagesiaceae	Ochnaceae	Ochnaceae	Ochnaceae
Saxifragaceae	Saxifragaceae	Saxifragaceae	Saxifragaceae	Saxifragaceae	Saxifragaceae
Scheuchzeriaceae	Scheuchzeriaceae	Scheuchzeriaceae	Scheuchzeriaceae	Scheuchzeriaceae	Scheuchzeriaceae
Schisandraceae	Schisandraceae	Schisandraceae	Schisandraceae	Schisandraceae	Schisandraceae
Schlegeliaceae	Scrophulariaceae	Scrophulariaceae	Scrophulariaceae	Scrophulariaceae	Schlegeliaceae
Sclerophylacaceae	Solanaceae	Sclerophylacaceae	Solanaceae	Solanaceae	Solanaceae
Scrophulariaceae	Scrophulariaceae	Scrophulariaceae	Scrophulariaceae	Scrophulariaceae	Scrophulariaceae
Scyphostegiaceae	Scyphostegiaceae	Scyphostegiaceae	Scyphostegiaceae	Scyphostegiaceae	Scyphostegiaceae
Scytopetalaceae	Scytopetalaceae	Scytopetalaceae	Scytopetalaceae	Scytopetalaceae	Lecythidaceae
[Selaginaceae]	Scrophulariaceae	Scrophulariaceae	Scrophulariaceae	Scrophulariaceae	Scrophulariaceae
Setchellanthaceae	Capparaceae	Capparaceae	Capparaceae	Capparaceae	Setchellanthaceae
Simaroubaceae	Simaroubaceae	Simaroubaceae	Simaroubaceae	Simaroubaceae	Simaroubaceae

Simmondsiaceae	Buxaceae	Simmondsiaceae	Simmondsiaceae	Simmondsiaceae	Simmondsiaceae
Siparunaceae	Monimiaceae	Siparunaceae	Monimiaceae	in Monimiaceae	Siparunaceae
[Siphonodontaceae]	Celastraceae	Celastraceae	Celastraceae	Celastraceae	Celastraceae
Sladeniaceae	Theaceae	Sladeniaceae	Theaceae	Theaceae	Sladeniaceae
Smilacaceae	Liliaceae	Smilacaceae	Smilacaceae	Smilacaceae	Smilacaceae
Solanaceae	Solanaceae	Solanaceae	Solanaceae	Solanaceae	Solanaceae
Sonneratiaceae	Sonneratiaceae	Sonneratiaceae	Sonneratiaceae	Lythraceae	Lythraceae
Sparganiaceae	Sparganiaceae	Sparganiaceae	Sparganiaceae	in Typhaceae	Sparganiaceae
Sphaerosepalaceae	Rhopalocarpaceae	Sphaerosepalaceae	Sphaerosepalaceae	Sphaerosepalaceae	Sphaerosepalaceae
Sphenocleaceae	Sphenocleaceae	Sphenocleaceae	Sphenocleaceae	Campanulaceae	Sphenocleaceae
Sphenostemonaceae	Sphenostemonaceae	Aquifoliaceae	Aquifoliaceae	Sphenostemonaceae	Sphenostemonaceae
Spielmanniaceae	Spielmanniaceae	Scrophulariaceae	Scrophulariaceae	Scrophulariaceae	Scrophulariaceae
Spigeliaceae	Spigelieaceae	Loganiaceae	Loganiaceae	Loganiaceae	Loganiaceae
Stachyuraceae	Stachyuraceae	Stachyuraceae	Stachyuraceae	Stachyuraceae	Stachyuraceae
Stackhousiaceae	Stackhousiaceae	Stackhousiaceae	Stackhousiaceae	Stackhousiaceae	Stackhousiaceae
Staphyleaceae	Staphyleaceae	Staphyleaceae	Staphyleaceae	Staphyleaceae	Staphyleaceae
Stegnospermataceae	Phytolaccaceae	Stegnospermataceae	Phytolaccaceae	Stegnospermataceae	Stegnospermataceae
Stemonaceae	Stemonaceae	Stemonaceae	Stemonaceae	Stemonaceae	Stemonaceae
Stenomeridaceae	Dioscoreaceae	Stenomeridaceae	Dioscoreaceae	Dioscoreaceae	Dioscoreaceae
Sterculiaceae	Sterculiaceae	Sterculiaceae	Sterculiaceae	Sterculiaceae	Malvaceae
[Stilaginaceae]	Euphorbiaceae?	Euphorbiaceae?	Euphorbiaceae?	Euphorbiaceae?	Euphorbiaceae?
Stilbaceae	Verbenaceae	Stilbaceae	Verbenaceae	Stilbaceae	Stilbaceae
Strasburgeriaceae	Strasburgeriaceae	Strasburgeriaceae	Ochnaceae	Strasburgeriaceae	Strasburgeriaceae
Strelitziaceae	Musaceae	Strelitziaceae	Strelitziaceae	Strelitziaceae	Strelitziaceae
[Streptochaetaceae]	Gramineae	Gramineae	Gramineae	Gramineae	Gramineae
[Strychnaceae]	Loganiaceae	Loganiaceae	Loganiaceae	Loganiaceae	Loganiaceae
Stylidiaceae	Stylidiaceae	Stylidiaceae	Stylidiaceae	Stylidiaceae	Stylidiaceae
Stylobasiaceae	Chrysobalanaceae	Stylobasiaceae	Surianaceae	Stylobasiaceae	Surianaceae
Stylocerataceae	Buxaceae	Stylocerataceae	Buxaceae	Buxaceae	Buxaceae
Styracaceae	Styraceae	Styraceae	Styracaceae	Styracaceae	Styracaceae
Surianaceae	Simaroubaceae	Surianaceae	Surianaceae	Surianaceae	Surianaceae

(cont.)

Table 4.1. (*cont.*)

Names	Melchior (1964)	Takhtajan (1987)	Cronquist (1988)	Brummitt (1992)	APG (1998)
[Symphoremataceae]	Verbenaceae	Verbenaceae	Verbenaceae	Verbenaceae	Verbenaceae
Symplocaceae	Symplocaceae	Symplocaceae	Symplocaceae	Symplocaceae	Symplocaceae
Taccaceae	Taccaceae	Taccaceae	Taccaceae	Taccaceae	Taccaceae
Tamaricaceae	Tamaricaceae	Tamaricaceae	Tamaricaceae	Tamaricaceae	Tamaricaceae
Tapisciaceae	Staphyleaceae	Tapisciaceae	Staphyleaceae	Staphyleaceae	Tapisciaceae
Tecophilaeaceae	Haemodoraceae	Tecophilaeaceae	Liliaceae	Tecophilaeaceae	Tecophilaeaceae
Tepuianthaceae	?	Tepuianthaceae	?	Tepuianthaceae	Tepuianthaceae
Ternstroemiaceae	Theaceae	Theaceae	Theaceae	Theaceae	Ternstroemiaceae
Tetracarpaeaceae	Saxifragaceae	Tetracarpaeaceae	? Grossulariaceae	Escalloniaceae	Tetracarpaeaceae
Tetracentraceae	Tetracentraceae	Tetracentraceae	Tetracentraceae	Tetracentraceae	in Trochodendraceae
Tetrachondraceae	Labiatae	Labiatae	Labiatae	Tetrachondraceae	Tetrachondraceae
Tetradiclidaceae	Zygophyllaceae	Tetradiclidaceae	Zygophyllaceae	Zygophyllaceae	Zygophyllaceae
Tetragoniaceae	Aizoaceae	Tetragoniaceae	Aizoaceae	Aizoaceae	Aizoaceae
Tetramelaceae	Datiscaceae	Datiscaceae	? Datiscaceae	Datiscaceae	Tetramelaceae
Tetrameristaceae	Theaceae	Tetrameristaceae	Tetrameristaceae	Tetrameristaceae	Tetrameristaceae
Thalassiaceae	Hydrocharitaceae	Thalassiaceae	Hydrocharitaceae	Hydrocharitaceae	Hydrocharitaceae
Theaceae	Theaceae	Theaceae	Theaceae	Theaceae	Theaceae
Theligonaceae	Theligonaceae	Theligonaceae	Theligonaceae	Theligonaceae	Rubiaceae
Themidaceae	Liliaceae	Alliaceae	Liliaceae	Alliaceae	Themidaceae
Theophrastaceae	Theophrastaceae	Theophrastaceae	Theophrastaceae	Theophrastaceae	Theophrastaceae
Thismiaceae	Burmanniaceae	Burmanniaceae	? Burmanniaceae	in Burmanniaceae	Thismiaceae
Thunbergiaceae	Acanthaceae	Thunbergiaceae	Acanthaceae	Acanthaceae	Acanthaceae
Thurniaceae	Thurniaceae	Thurniaceae	Thurniaceae	Thurniaceae	Thurniaceae
Thymelaeaceae	Thymelaeaceae	Thymelaeaceae	Thymeleaeceae	Thymelaeaceae	Thymeleaceae
Ticodendraceae	?	?	?	Ticodendraceae	Ticodendraceae

Tiliaceae	Tiliaceae	Tiliaceae	Tiliaceae in Melanthiaceae	Malvaceae
Tofieldiaceae	?Melanthiaceae	Liliaceae		Tofieldiaceae
Torricelliaceae	Torricellliaceae	Cornaceae	Torricelliaceae	Torricelliaceae
Tovariaceae	Tovariaceae	Tovariaceae	Tovariaceae	Tovariaceae
Trapaceae	Trapaceae	Trapaceae	Trapaceae	Lythraceae
Trapellaceae	Trapellaceae	Pedaliaceae	Pedaliaceae	Pedaliaceae
Tremandraceae	Tremandraceae	Tremandraceae	Tremandraceae	Tremandraceae
Tribelaceae	Tribelaceae	? Grossulariaceae	Escalloniaceae	Tribelaceae
Trichopodaceae	Trichopodaceae	Dioscoreaceae	Trichopodaceae	Trichopodaceae
Trigoniaceae	Trigoniaceae	Trigoniaceae	Trigoniaceae	Trigoniaceae
Trilliaceae	Trilliaceae	Liliaceae	Trilliaceae	Melanthiaceae
Trimeniaceae	Trimeniaceae	Trimeniaceae	- Trimeniaceae	Trimeniaceae
Triplostegiaceae	Valerianaceae	Valerianaceae	Triplostegiaceae	Valerianaceae
Tristichaceae	Podostemaceae	Podostemaceae	Podostemaceae	Tristichaceae
Triuridaceae	Triuridaceae	Triuridaceae	Triuridaceae	Triuridaceae
Trochodendraceae	Trochodendraceae	Trochodendraceae	Trochodendraceae	Trochodendraceae
Tropaeolaceae	Tropaeolaceae	Tropaeolaceae	Tropaeolaceae	Tropaeolaceae
Turneraceae	Turneraceae	Turneraceae	Turneraceae	Turneraceae
Typhaceae	Typhaceae	Typhaceae	Typhaceae	Typhaceae
[Uapacaceae]	Ulmaceae	Ulmaceae	Ulmaceae	Ulmaceae
Ulmaceae	Ulmaceae	Ulmaceae	Ulmaceae	Ulmaceae
Umbelliferae	as Apiaceae	as Apiaceae	Umbelliferae	as Apiaceae
Urticaceae	Urticaceae	Urticaceae	Urticaceae	Urticaceae
[Vacciniaceae]	Ericaceae	Ericaceae	Ericaceae	Ericaceae
Vahliaceae	Vahliaceae	Saxifragaceae	Vahliaceae	Vahliaceae
Valerianaceae	Valerianaceae	Valerianaceae	Valerianaceae	Valerianaceae
Velloziaceae	Velloziaceae	Velloziaceae	Velloziaceae	Velloziaceae
Verbenaceae	Verbenaceae	Verbenaceae	Verbenaceae	Verbenaceae

(cont.)

Table 4.1. (*cont.*)

Names	Melchior (1964)	Takhtajan (1987)	Cronquist (1988)	Brummitt (1992)	APG (1998)
Viburnaceae	Caprifoliaceae	Viburnaceae	Caprifoliaceae	Caprifoliaceae	Caprifoliaceae
Violaceae	Violaceae	Violaceae	Violaceae	Violaceae	Violaceae
Viscaceae	Loranthaceae	Viscaceae	Viscaceae	Viscaceae	Santalaceae
Vitaceae	Vitaceae	Vitaceae	Vitaceae	Vitaceae	Vitaceae
Vivianaceae	Geraniaceae	Vivianaceae	Geraniaceae	Geraniaceae	Vivianaceae
Vochysiaceae	Vochysiaceae	Vochysiaceae	Vochysiaceae	Vochysiaceae	Vochysiaceae
Wellstediaceae	Boraginaceae	Wellstediaceae	Boraginaceae	Boraginaceae	Boraginaceae
Winteraceae	Winteraceae	Winteraceae	Winteraceae	Winteraceae	Winteraceae
Xanthophyllaceae	Polygalaceae	? Polygalaceae	Xanthophyllaceae	Polygalaceae	Polygalaceae
Xanthorrhoeaceae	Xanthorrhoeaceae	Xanthorrhoeaceae	Xanthorrhoeaceae	Xanthorrhoeaceae	Xanthorrhoeaceae
Xeronemataceae	Liliaceae	?	Liliaceae	in Phormiaceae	Xeronemataceae
Xyridaceae	Xyridaceae	Xyridaceae	Xyridaceae	Xyridaceae	Xyridaceae
Zannichelliaceae	Zannichelliaceae	Zannichelliaceae	Zannichelliaceae	Zannichelliaceae	Potamogetonaceae
Zingiberaceae	Zingiberaceae	Zingiberaceae	Zingiberaceae	Zingiberaceae	Zingiberaceae
Zosteraceae	Potamogetonaceae	Zosteraceae	Zosteraceae	Zosteraceae	Zosteraceae
Zygophyllaceae	Zygophyllaceae	Zygophyllaceae	Zygophyllaceae	Zygophyllaceae	Zygophyllaceae
Total family names: 662	Total: 344	Total: 533	Total: 383	Total: 454	Total: 474

Table 4.2. *This table takes its starting point from the comparison of families in (Table 4.1); it uses the families recognised in Melchior (1964) as a basis. Melchior's system contains 344 families, which, according to its own figures, contain 12 071 genera and 236 296 species; these figures, though rather low perhaps, can be used as a basis for comparisons (see text for further explanation). The results are summarised in Table 4.3 and Figure 4.2*

I: over 40 entries (1)

Liliaceae, 42: *Agapanthaceae* (also sometimes in *Alliaceae*), *Alliaceae, Alo[e]aceae, Alstroemeriaceae, Anemarrenaceae* (also sometimes in *Asphodelaceae, Anthericaceae*), *Anthericaceae* (also sometimes in *Asphodelaceae*), *Aphyllanthaceae, Asparagaceae, Asphodelaceae, Asteliaceae, Blandfordiaceae, Boryaceae* (also sometimes in *Anthericaceae* or *Asphodelaceae*), *Calochortaceae, Colchicaceae* (also sometimes in *Melanthiaceae*), *Colchicaceae* (also in *Melanthiaceae*), *Convallariaceae, Eriospermaceae* (also sometimes in *Convallariaceae*), *Hemerocallidaceae, Hesperocallidaceae* (also sometimes in *Hyacinthaceae*), *Hostaceae* (also sometimes in *Agavaceae*), *Hyacinthaceae, Japonoliriaceae* (also in *Melanthiaceae*), *Lanariaceae* (also in *Tecophilaeaceae*), *Laxmanniaceae* (also in *Anthericaceae* or *Asphodelaceae*), *Liliaceae sensu stricto, Lomandraceae* (also in *Dasypogonaceae, Xanthorrhoeaceae* or *Laxmanniaceae*), *Luzuriagaceae* (also in *Philesiaceae* or *Smilacaceae*), *Melanthiaceae, Nartheciaceae* (also in *Melanthiaceae*), *Petermanniaceae* (also in *Colchicaceae* or *Smilacaceae*), *Petrosaviaceae* (also in *Melanthiaceae*), *Philesiaceae* (also in *Smilacaceae*), *Phormiaceae* (also in *Agavaceae* or *Hemerocallidaceae*), *Rhipogonaceae* (also in *Smilacaceae*), *Ruscaceae* (also in *Convallariaceae*), *Smilacaceae, Themidaceae* (also in *Alliaceae*), *Tofieldiaceae* (also in *Melanthiaceae*), *Trichopodaceae* (also in *Dioscoreaceae*), *Trilliaceae* (also in *Melanthiaceae*), *Xeronemataceae* (also in *Phormiaceae*).

This group, the 'petaloid monocots' is by far the most split up of all, although Cronquist cuts the Gordian Knot by putting in almost everything that can be put in, in (including *Amaryllidaceae*, etc.). In Melchior's system this family contains 220 genera and 3500 species: thus it is 0.3% of the families, 1.8% of the total genera and 1.6% of the total species. Everyone (*pace* Cronquist) agrees that this group should be split up, but there is no agreement as to how this should be done. The monocot volumes of the Kubitzki account (1998), which has not been covered here, include several more families in this group.

II: over 20 entries (1)

Saxifragaceae, 24: *Argophyllaceae* (also in *Escalloniaceae* or *Grossulariaceae*), *Baueraceae* (also in *Cunoniaceae*), *Brexiaceae* (also in *Celastraceae, Escalloniaceae* or *Grossulariaceae*), *Carpodetaceae* (also in *Escalloniaceae* or *Grossulariaceae*), *Dulongiaceae* (also in *Escalloniaceae, Grossulariaceae* or *Phyllonomaceae*), *Eremosynaceae, Escalloniaceae* (also in *Grossulariaceae*), *Francoaceae, Grossulariaceae, Hydrangeaceae, Iteaceae* (also in *Escalloniaceae* or *Grossulariaceae*), *Ixerbaceae* (also in *Escalloniaceae* or *Grossulariaceae*), *Lepuropetalaceae* (also in *Parnassiaceae*), *Montiniaceae* (also in *Grossulariaceae*), *Parnassiaceae, Penthoraceae, Phyllonomaceae* (also *Escalloniaceae* or *Grossulariaceae*), *Polyosmataceae* (also in *Escalloniaceae* or *Grossulariaceae*), *Pterostemonaceae* (also in *Grossulariaceae*), *Rousseaceae* (also in *Escalloniaceae, Grossulariaceae*), *Saxifragaceae sensu stricto, Tetracarpaeaceae* (also in *Escalloniaceae* or *Grossulariaceae*), *Tribelaceae* (also in *Escalloniaceae* or *Grossulariaceae*), *Vahliaceae*.

(cont.)

85

Table 4.2. (*cont.*)

Another complicated family, which everyone agrees should be split but there is not much agreement as to how this should be done (although the situation is not as bad as with the Liliaceae). This family has, according to Melchior, 80 genera and 1200 species; i.e. it contains 0.3% of the families, 0.7% of the genera and 0.5% of the species.

III: 10 entries (1)
Cornaceae, 10: *Aralidiaceae, Aucubaceae, Cornaceae sensu stricto, Curtisiaceae,* [*Davidiaceae* – not separate in Melchior], *Griseliniaceae, Helwingiaceae, Kaliphoraceae* (also in *Melanophyllaceae*), *Mastixiaceae, Melanophyllaceae, Torricelliaceae.*

This family is less complicated than the two above, and there is more agreement as to how it should be divided. However, the family is quite small, with 12 genera and 95 species in the Melchior system: 0.3% of the families, 0.1% of the genera, <0.1% of the species. APG puts some of the splits back in the Cornaceae.

IV: 7 entries (2)
Caprifoliaceae (15/400): *Alseuosmiaceae, Caprifoliaceae sensu stricto, Carlemanniaceae, Diervillaceae, Linnaeaceae, Sambucaceae, Viburnaceae.*
Theaceae (35/600): *Asteropeiaceae, Oncothecaceae, Pellicieriaceae, Sladeniaceae, Ternstroemiaceae, Tetrameristaceae, Theaceae sensu stricto.*

These 2 families, with 50 genera and 1000 species, represent together 0.6% of the families, 0.4% of the genera and 0.5% of the species.

V: 6 entries (3)
Campanulaceae (70/2000): *Campanulaceae sensu stricto, Cyphiaceae, Cyphocarpaceae, Lobeliaceae, Nemacladaceae, Sphenocleaceae*
Flacourtiaceae (86/1300): *Aphloiaceae, Berberidopsidaceae, Flacourtiaceae sensu stricto, Kiggelariaceae, Lacistemaceae, Plagiopteraceae.*
Geraniaceae (11/780): *Biebersteiniaceae, Dirachmaceae, Geraniaceae sensu stricto, Ledocarpaceae, Rhynchothecaceae, Vivianaceae.*

These 3 families contain 167 genera and 4080 species; thus 0.9% of the families, 1.4% of the genera and 1.9% of the species.

VI: 5 entries (8)
Balanophoraceae (18/100): *Balanophoraceae sensu stricto, Dactylanthaceae, Latraeophilaceae, Lophophytaceae, Sarcophytaceae.*
Capparaceae (46/800): *Capparaceae sensu stricto, Emblingiaceae* (also in *Polygalaceae*), *Koeberliniaceae, Pentadiplandraceae, Setchellanthaceae.*
Celastraceae (60/850): *Canotiaceae, Celastraceae sensu stricto, Goupiaceae, Lophopyxidaceae, Pottingeriaceae.*
Linaceae (25/500): *Ctenolophonaceae, Hugoniaceae, Humiriaceae, Ixonanthaceae, Linaceae sensu stricto.*
Ranunculaceae (70/2000): *Circeasteraceae, Glaucidiaceae, Hydrastidaceae, Kingdoniaceae, Ranunculaceae sensu stricto.*
Simaroubaceae (24/100): *Irvingiaceae, Kirkiaceae, Picramniaceae, Simaroubaceae sensu stricto, Surianaceae.*
Verbenaceae (100/2600): *Avicenniaceae, Cyclocheilaceae* (also in *Acanthaceae*), *Nesogenaceae, Stilbaceae, Verbenaceae sensu stricto.*

Table 4.2. (*cont.*)

Zygophyllaceae (30/250): *Balanitaceae, Nitrariaceae, Peganaceae, Tetradiclidaceae, Zygophyllaceae sensu stricto.*

These 8 families contain 373 genera and 7200 species; hence 2.3% of the families, 3.1% of the genera and 3.3% of the species.

VII: 4 entries (10)

Agavaceae (18/500): *Agavaceae sensu stricto, Doryanthaceae, Dracaenaceae, Nolinaceae.*

Boraginaceae (100/2000): *Boraginaceae sensu stricto, Cordiaceae, Ehretiaceae, Wellstedtiaceae.*

Loganiaceae (18/500): *Gelsemiaceae* (also in *Escalloniaceae*), *Loganiaceae sensu stricto, Plocospermataceae, Spigeliaceae.*

Lythraceae (22/500): *Duabangaceae, Lythraceae sensu stricto, Rhynchocalycaceae, Sonneratiaceae.*

Nymphaeaceae (8/80): *Barclayaceae, Cabombaceae, Nelumbonaceae, Nymphaeaceae sensu stricto.*

Papaveraceae (47/700): *Fumariaceae, Hypecoaceae, Papaveraceae sensu stricto, Pteridophyllaceae.*

Phytolaccaceae (17/120): *Agdestidaceae, Barbeiuaceae, Phytolaccaceae sensu stricto, Stegnospermataceae.*

Potamogetonaceae (5/105): *Posidoniaceae, Potamogetonaceae sensu stricto, Ruppiaceae, Zosteraceae.*

Rafflesiaceae (9/55): *Apodanthaceae, Cytinaceae, Mitrastemonaceae, Rafflesiaceae sensu stricto.*

Xanthorrhoeaceae (8/50): *Calectasiaceae* (also *Dasypogonaceae*), *Dasypogonaceae, Hanguanaceae, Xanthorrhoeaceae sensu stricto.*

These 10 families contain 252 genera and 4610 species; thus 2.9% of the families, 2.1% of the genera and 2.1% of the species.

VIII: 3 entries (20)

Acanthaceae (250/2600): *Acanthaceae sensu stricto, Mendonciaceae, Thunbergiaceae.*
Aquifoliaceae (3/450): *Aquifoliaceae sensu stricto, Phellinaceae, Sphenostemonaceae.*
Buxaceae (6/60): *Buxaceae sensu stricto, Simmondsiaceae, Stylocerataceae.*
Caryophyllaceae(80/2000): *Caryophyllaceae sensu stricto, Hectorellaceae* (also *Portulacaceae*), *Illecebraceae.*
Chenopodiaceae (100/1500): *Chenopodiaceae sensu stricto, Halophytaceae, Sarcobataceae.*
Hamamelidaceae (26/115): *Altingiaceae, Hamamelidaceae sensu stricto, Rhodoleiaceae.*
Hydrocharitaceae (15/100): *Halophilaceae, Hydrocharitaceae sensu stricto, Thalassiaceae.*
Juncaginaceae (4/18): *Juncaginaceae sensu stricto, Lilaeaceae, Maundiaceae.*
Leguminosae (600/13000): *Caesalpiniaceae, Leguminosae sensu stricto, Mimosaceae.*
Loranthaceae (40/1400): *Eremolepidaceae, Loranthaceae sensu stricto, Viscaceae.*
Melianthaceae (3/38): *Campynemataceae, Greyiaceae, Melianthaceae sensu stricto.*
Monimiaceae (34/450): *Atherospermataceae, Monimiaceae sensu stricto, Siparunaceae.*
Myrtaceae (100/3000): *Heteropyxidaceae, Myrtaceae sensu stricto, Psiloxylaceae.*
Ochnaceae (28/400): *Lophiraceae, Ochnaceae sensu stricto, Sauvagesiaceae.*
Oxalidaceae (8/950): *Hypseocharitaceae, Lepidobotryaceae, Oxalidaceae sensu stricto.*

(*cont.*)

Table 4.2. *(cont.)*

Restionaceae (30/400): *Anarthriaceae, Ecdeiocoleaceae, Restionaceae sensu stricto.*
Scrophulariaceae (200/3000): *Schlegeliaceae, Scrophulariaceae sensu stricto,*
 Spielmanniaceae.
Solanaceae (85/2300): *Goetzeaceae, Sclerophylacaceae, Solanaceae sensu stricto.*
Styracaceae (11/150): *Halesiaceae, Huaceae, Styracaceae sensu stricto.*
Ulmaceae (15/150): *Barbeyaceae, Celtidaceae, Ulmaceae sensu stricto.*

These 20 families contain 1638 genera and 32 081 species; thus 5.8% of the families, 13.6% of the genera, 14.6% of the species.

These groups with between 3 and 7 entries, represent a middle ground: there are 43 families, 2480 genera and 48 971 species; thus 12.5% of the families, 20.6% of the genera and 22.3% of the species.

The last two categories are rather different from the others; in those with 2 entries, these are usually well-known splits which it is almost a matter of taste to adopt or reject; and in the final category there is no disagreement at all between the systems except, as noted in passing, where various families have been combined in systems other than Melchior.

IX: 2 entries (44)
Aizoaceae (120/2500): *Aizoaceae sensu stricto, Tetragoniaceae.*
Amaryllidaceae (65/860): *Amaryllidaceae sensu stricto, Ixioliriaceae.*
Anacardiaceae (79/600): *Anacardiaceae sensu stricto, Podoaceae.*
Araceae (110/1800): *Acoraceae, Araceae sensu stricto.*
Berberidaceae (14/650): *Berberidaceae sensu stricto, Nandinaceae.*
Betulaceae (6/100): *Betulaceae sensu stricto, Corylaceae.*
Bignoniaceae (120/800): Bignoniaceae *sensu stricto, Paulowniaceae*
 (also *Scrophulariaceae*).
Burmanniaceae (22/130): *Burmanniaceae sensu stricto, Thismiaceae.*
Butomaceae (4/13): *Butomaceae sensu stricto, Limnocharitaceae.*
Calycanthaceae (2/9): *Calycanthaceae sensu stricto, Idiospermaceae.*
Centrolepidaceae (7/40): *Centrolepidaceae sensu stricto, Hydatellaceae.*
Chrysobalanaceae (12/300): *Chrysobalanaceae sensu stricto, Stylobasiaceae*
 (also in *Surianaceae*).
Convolvulaceae (51/1600): *Convolvulaceae sensu stricto, Cuscutaceae.*
Datiscaceae (3/4): *Datiscaceae sensu stricto, Tetramelaceae.*
Dioscoreaceae (11/350): *Dioscoreaceae sensu stricto, Stenomeridaceae.*
Dipsacaceae (10/270): *Dipsacaceae sensu stricto, Morinaceae.*
Dipterocarpaceae (21/400): *Dipterocarpaceae sensu stricto, Monotaceae.*
Droseraceae (4/93): *Droseraceae sensu stricto, Drosophyllaceae.*
Ericaceae (82/2500): *Ericaceae sensu stricto, Monotropaceae.*
Euphorbiaceae (290/7500): *Euphorbiaceae sensu stricto, Putranjivaceae.*
Fagaceae (7/600): *Fagaceae sensu stricto, Nothofagaceae.*
Flagellariaceae (2/7): *Flagellariaceae sensu stricto, Joinvilleaceae.*
Guttiferae (49/900): *Bonnetiaceae, Guttiferae sensu stricto.*
Haemodoraceae (22/120): *Conostylidaceae, Haemodoraceae sensu stricto.*
Haloragaceae (8/160): *Gunneraceae, Haloragaceae sensu stricto.*
Hernandiaceae (4/65): *Gyrocarpaceae, Hernandiaceae sensu stricto.*
Hydrophyllaceae (20/270): *Hydroleaceae, Hydrophyllaceae sensu stricto.*
Icacinaceae (45/400): *Icacinaceae sensu stricto, Metteniusaceae.*
Juncaceae (8/300): *Juncaceae sensu stricto, Prionaceae.*

Table 4.2. (*cont.*)

Labiatae (200/3200): *Labiatae sensu stricto* (but sometimes including some
  Verbenaceae), *Tetrachondraceae*.
Meliaceae (50/1400): *Meliaceae sensu stricto, Ptaeroxylaceae* (also in *Sapindaceae*).
Molluginaceae (14/95): *Giseckiaceae, Molluginaceae sensu stricto*.
Moraceae (61/1500): *Cecropiaceae, Moraceae sensu stricto.*
Olacaceae (27/230): *Octoknemataceae, Olacaceae sensu stricto*.
Podostemaceae (43/200): *Podostemaceae sensu stricto, Tristichaceae*.
Polemoniaceae (18/320): *Cobaeaceae, Polemoniaceae sensu stricto*.
Polygalaceae (13/800): *Polygalaceae sensu stricto, Xanthophyllaceae*.
Rhizophoraceae (16/120): *Anisophyllaeaceae, Rhizoophoraceae sensu stricto*.
Rutaceae (150/1600): *Rhabdodendraceae, Rutaceae sensu stricto*.
Sabiaceae (4/90): *Meliosmaceae, Sabiaceae sensu stricto*.
Staphyleaceae (7/50): *Staphyleaceae sensu stricto, Tapisciaceae*.
Tiliaceae (45/400): *Muntingiaceae, Tiliaceae sensu stricto*.
Xyridaceae (4/270): *Abolbodaceae, Xyridaceae sensu stricto*.
Zingiberaceae (49/1500): *Costaceae, Zingiberaceae sensu stricto*.

These 44 families contain 1899 genera and 35 116 species: 12.8% of families, 15.7% of
genera, 15.9% of species.

X. Only 1 entry (254)

Aceraceae (2/150). Sunk in *Sapindaceae* by APG
Achariaceae (3/3)
Achatocarpaceae (2/8)
Actinidiaceae (3/320)
Adoxaceae (1/1)
Aextoxicaceae(1/1)
Akaniaceae (1/1)
Alangiaceae (1/18). Sunk in *Cornaceae* by APG
Alismataceae (10/70)
Amaranthaceae (60/900)
Amborellaceae (1/1)
Ancistrocladaceae (1/16)
Annonaceae (120/2100)
Apocynaceae (200/2000). Now some tendency to merge with *Asclepiadaceae*
Aponogetonaceae (1/40)
Araliaceae (70/700). Now some tendency to merge with *Umbelliferae*
Aristolochiaceae (7/600)
Asclepiadaceae (250/2000). Merged with *Apocynaceae* by APG
Austrobaileyaceae (1/2)
Balanopaceae (1/9)
Balsaminaceae (2/450)
Basellaceae (5/20)
Bataceae (1/2)
Begoniaceae (5/820)
Bixaceae (1/1)
Bombacaceae (28/200). Put in *Malvaceae* by APG
Bretschneideriaceae (1/1). Put in *Akaniaceae* by APG

(cont.)

Table 4.2. (*cont.*)

Bromeliaceae (46/1700)
Brunelliaceae (1/35). Put in *Cunoniaceae* by APG
Bruniaceae (12/75)
Brunoniaceae (1/1). Put in *Goodeniaceae* by Brummitt and APG
Buddlejaceae (19/160)
Burseraceae (20/600)
Byblidaceae (1/2)
Cactaceae (220/2000)
Callitrichaceae (1/44). Put in *Plantaginaceae* by APG
Calyceraceae (6/60)
Canellaceae (6/20)
Cannaceae (1/60)
Cardiopteridaceae (1/3)
Caricaceae (4/25)
Caryocaraceae (2/25)
Casuarinaceae (1/50)
Cephalotaceae (1/1)
Ceratophyllaceae (1/6)
Cercidiphyllaceae (1/2)
Chloranthaceae (5/70)
Cistaceae (8/175)
Clethraceae (1/30)
Cneoraceae (2/3)
Cochlospermaceae (2/20). Put in *Bixaceae* by Cronquist
Columelliaceae (1/4)
Combretaceae (18/500)
Commelinaceae (40/600)
Compositae (920/19000)
Connaraceae (24/400)
Coriariaceae (1/10)
Corsiaceae (2/9)
Corynocarpaceae (1/5)
Crassulaceae (30/1400)
Crossosomataceae (1/4)
Cruciferae (350/3000)
Crypteroniaceae (1/4)
Cucurbitaceae (100/850)
Cunoniaceae (25/350)
Cyanastraceae (1/6). Put in *Tecophilaeaceae* by APG
Cyclanthaceae (11/180)
Cynomoriaceae (1/1). Put in *Balanophoraceae* by Cronquist
Cyperaceae (70/3700)
Cyrillaceae (3/14)
Daphniphyllaceae (1/35)
Davidiaceae (1/1). Put in *Cornaceae* by Cronquist, Brummitt and Takhtajan
Davidsoniaceae (1/1). Put in *Cunoniaceae* by APG
Degeneriaceae (1/1)
Desfontainiaceae (1/5). Put in *Loganiaceae* by Brummitt and in *Columelliaceae* by APG

Table 4.2. (*cont.*)

Dialypetalanthaceae (1/1). Put in *Rubiaceae* by APG
Diapensiaceae (6/18)
Dichapetalaceae (4/250)
Didiereaceae (4/11)
Didymelaceae (1/2)
Dilleniaceae (10/350)
Dioncophyllaceae (3/3)
Dipentodontaceae (1/1)
Duckeodendraceae (1/1). Put in *Solanaceae* by APG
Dysphaniaceae (1/6)
Ebenaceae (4/450)
Elaeagnaceae (3/65)
Elaeocarpaceae (10/400)
Elatinaceae (2/45)
Empetraceae (3/9). Put in *Ericaceae* by APG
Epacridaceae (30/400). Put in *Ericaceae* by APG
Eriocaulaceae (13/1200)
Erythroxylaceae (4/200)
Eucommiaceae (1/1)
Eucryphiaceae (1/5). Put in *Cunoniaceae* by APG
Eupomatiaceae (1/2)
Eupteleaceae (1/2)
Fouquieriaceae (1/7)
Frankeniaceae (4/50)
Garryaceae (1/15)
Geissolomataceae (1/1)
Gentianaceae (70/1100)
Geosiridaceae (1/1)
Gesneriaceae (140/1800)
Globulariaceae (2/27). Put in *Plantaginaceae* by APG
Gomortegaceae (1/1)
Goodeniaceae (14/320)
Gramineae (700/8000)
Grubbiaceae (1/5)
Gyrostemonaceae (5/16)
Henriqueziaceae (2/7)
Himantandraceae (1/3)
Hippocastanaceae (2/15). Put in *Sapindaceae* by APG
Hippocrateaceae (18/300). Put in *Celastraceae* by Takhtajan, Brummitt and APG
Hippuridaceae (1/1). Put in *Plantaginaceae* by APG
Hoplestigmataceae (1/2)
Hydnoraceae (2/18)
Hydrostachyaceae (1/30)
Hypoxidaceae (5/140). Put in *Liliaceae* by Cronquist
Illiciaceae (1/42).
Iridaceae (70/1500)
Juglandaceae (8/58)

(*cont.*)

Table 4.2. (*cont.*)

Julianaceae (2/5). Put in *Anacardiaceae* by Takhtajan, Brummitt and APG
Krameriaceae (1/20)
Lactoridaceae (1/1)
Lardizabalaceae (8/30)
Lauraceae (31/2250)
Lecythidaceae (24/450)
Leeaceae (1/70). Put in *Vitaceae* by APG
Leitneriaceae (1/1). Put in *Simaroubaceae* by APG
Lemnaceae (4/25). Put in *Araceae* by APG
Lennoaceae (3/4). Put in *Boraginaceae* by APG
Lentibulariaceae (5/300)
Limnanthaceae (2/9)
Lissocarpaceae (1/1)
Loasaceae (15/250)
Lowiaceae (1/5)
Magnoliaceae (10/215)
Malesherbiaceae (1/25)
Malpighiaceae (63/800)
Malvaceae (85/1500). Much enlarged by APG
Marantaceae (32/350)
Marcgraviaceae (5/120)
Martyniaceae (5/16). Put in *Pedaliaceae* by Cronquist, Brummitt and APG
Mayacaceae (1/10)
Medusagynaceae (1/1)
Medusandraceae (1/1)
Melastomataceae (200/4000)
Menispermaceae (67/425)
Menyanthaceae (5/40)
Misodendraceae (1/11)
Moringaceae (1/10)
Musaceae (6/220)
Myoporaceae (5/180)
Myricaceae (3/56)
Myristicaceae (15/250)
Myrothamnaceae (1/2)
Myrsinaceae (33/1000)
Najadaceae (1/35). Put in *Hydrocharitaceae* by Brummitt and APG
Nepenthaceae (1/79)
Neuradaceae (3/10).
Nolanaceae (2/77). Put in *Solanaceae* by Brummitt and APG
Nyctaginaceae (30/300)
Nyssaceae (2/9)
Oleaceae (27/600)
Oliniaceae (1/8)
Onagraceae (20/650)
Opiliaceae (7/60)
Orchidaceae (700/20000)
Orobanchaceae (13/150). Put in *Scrophulariaceae* by Takhtajan and Brummitt

Table 4.2. (*cont.*)

Paeoniaceae (1/33)
Palmae (236/3400)
Pandaceae (1/1)
Pandanaceae (3/880)
Passifloraceae (12/600)
Pedaliaceae (16/55)
Penaeaceae (5/21)
Pentaphragmataceae (1/25)
Pentaphylacaceae (1/4)
Peridiscaceae (2/2)
Philydraceae (4/5)
Phrymaceae (1/4). Put in *Verbenaceae* by Takhtajan and Cronquist
Picrodendraceae (1/3)
Piperaceae (12/1400)
Pittosporaceae (9/240)
Plantaginaceae (3/265). Enlarged by APG
Platanaceae (1/7)
Plumbaginaceae (10/500)
Polygonaceae (40/800)
Pontederiaceae (17/30)
Portulacaceae (19/500)
Primulaceae (28/800)
Proteaceae (62/1400)
Punicaceae (1/2). Put in *Lythraceae* in Brummitt
Pyrolaceae (16/75)
Quiinaceae (3/37)
Rapateaceae (16/80)
Resedaceae (6/70)
Rhamnaceae (58/900)
Rhoipteleaceae (1/1)
Roridulaceae (1/2). Put in *Byblidaceae* by Cronquist
Rosaceae (100/3000). Includes *Quillajaceae* of APG
Rubiaceae (500/7000)
Salicaceae (2/350). Recently proposed enlargement to include *Flacourtiaceae*
Salvadoraceae (3/12)
Santalaceae (35/400)
Sapindaceae (140/1500)
Sapotaceae (50/800)
Sarcolaenaceae (8/33)
Sarcospermataceae (1/8)
Sargentodoxaceae (1/1). Put in *Lardizabalaceae* by APG
Sarraceniaceae (3/16)
Saururaceae (4/5)
Scheuchzeriaceae (1/1)
Schisandraceae (2/47)
Scyphostegiaceae (1/1)
Scytopetalaceae (5/32). Put in *Lecythidaceae* by APG

(*cont.*)

Table 4.2. (*cont.*)

Sonneratiaceae (2/7)
Sparganiaceae (1/20). Put in *Typhaceae* in Brummitt
Sphaerosepalaceae (2/14)
Sphenocleaceae (1/2)
Stachyuraceae (1/6)
Stackhousiaceae (3/22)
Stemonaceae (3/30)
Sterculiacaceae (70/1000). Put in *Malvaceae* by APG
Strasburgeriaceae (1/1). Put in *Ochnaceae* by Cronquist
Stylidiaceae (6/140)
Symplocaceae (1/300)
Taccaceae (2/30)
Tamaricaceae (4/100)
Tetracentraceae (1/1). Put in *Trochodendraceae* by APG
Theligonaceae (1/3). Put in *Rubiaceae* by APG
Theophrastaceae (4/110)
Thurniaceae (1/3)
Thymelaeaceae (48/650)
Tovariaceae (1/2)
Trapaceae (1/11). Put in *Lythraceae* by APG
Tremandraceae (3/28)
Trigoniaceae (4/35)
Trimeniaceae (2/7)
Triuridaceae (7/80)
Trochodendraceae (1/1). Includes *Tetracentraceae* by APG
Tropaeolaceae (1/80)
Turneraceae (8/120)
Typhaceae (1/15). Includes *Sparganiaceae* by Brummitt
Umbelliferae (300/3000)
Urticaceae (42/700)
Valerianaceae (13/300)
Velloziaceae (3/190)
Violaceae (16/850)
Vitaceae (12/700). Includes *Leeaceae* by APG
Vochysiaceae (6/200)
Winteraceae (6/95)
Zannichelliaceae (5/20). Put in *Potamogetonaceae* by APG

This group, by far the largest, contains 254 families, 7380 genera and 131 414 species; that is, 73.8% of the (Melchior) families, 61.1% of the genera and 59.7% of the species. All of these families, except for the various mergers noted as appropriate, are essentially the same in all 5 systems.

Table 4.3. *Summary of results of Table 4.2 (and shown in Figure 4.2)*

Group	No. of entries	No. of families	No. of genera	No. of species	% families	% genera	% species
I	42	1	220	3500	0.3	1.8	1.6
II	24	1	80	1200	0.3	0.7	0.5
III	10	1	12	95	0.3	0.1	0.0
IV	7	2	50	1000	0.6	0.4	0.5
V	6	3	167	4080	0.9	1.4	1.9
VI	5	8	373	7200	2.3	3.1	3.3
VII	4	10	252	4610	2.9	2.1	2.1
VIII	3	20	1638	32081	5.8	13.6	14.6
IX	2	44	1899	35116	12.8	15.7	15.9
X	1	254	7380	131414	73.8	61.1	59.7
Totals		344	12071	220296			

5

Taxonomy, Floras and conservation

Santiago Castroviejo

It stands to reason that the task of describing and recognising species comes before that of working for their preservation, and unquestionably it forms the basis for the work of conservationists. Without a thorough knowledge of the distinct species, it is impossible to decide which should be preserved, since this knowledge – like nearly every principle of biology – is full of complexities that require some clarification.

But what depth of knowledge is required before serious conservation efforts can be justified? Should we be content with an old, simple list of species? Or should we not rather demand wider information about the geographical distribution, the density and even some genetic or biological data of the populations? The simplest answer is the obvious one: we must always act on whatever information is available, whether it is a great deal or not much at all. Here again, this answer is limited by considerations that are not easily determined.

This chapter presents data that may help to clarify these questions; data based on the experience of 22 years of the *Flora iberica* project.

Flora iberica (Castroviejo *et al.*, 1986–2003) encompasses the whole peninsula of Spain and Portugal and also the Balearic Islands. No other Flora has this territorial demarcation, although some earlier Floras of the region made separate studies of the three areas: as in the well-known *Flora Europaea* (Tutin *et al.*, 1964–80) which includes the Spanish Peninsula (Hs), the Balearic Islands (BI) and Peninsular Portugal (Lu) as separate territories. The same plan was adopted in the *Med-Checklist* (Greuter *et al.*, 1984–9), whereas Willkomm and Lange (1861–80) include in their well-known *Prodomus* the plants of the peninsula of Spain and the Balearic Islands but not those of Portugal. Since there are very few species that are endemic to Portugal, we can use this *Prodomus* for our comparative analyses. Two other treatments can also be compared: *Elenco de la Flora Vascular Española* (Guinea & Ceballos, 1974), which covers the Peninsular and the Balearics; and the *Checklist of Vascular Plants* by Smythies (1984–6), which covers Spain and the Balearics.

Table 5.1. *Iberian floristic diversity as shown by Floras and checklists*

	Prodr.	Fl. Eur.	Elenco	Checklist	Fl. iber.
Species	5089	5250	5926	5597	—
Subspecies + varieties	500	698	611	1541	—
Total	5589	5948	6537	7138	7344

Sources: Prodr., *Prodromus florae hispanicae* (Willkomm & Lange, 1880); Fl. Eur., *Flora Europaea* (Tutin *et al.*, 1964–80); Elenco, *Elenco de la Flora Española* (Guinea & Ceballos, 1974); Checklist, *Flora of Spain and the Balearic Islands. Checklist of Vascular Plants* (Smythies, 1986); Fl. iber., *Flora iberica* (Castroviejo *et al.*, 1986–2003).

Table 5.2. *Number of accepted taxa in* Flora Europaea *(Tutin et al., 1964–80) and* Flora iberica *(Castroviejo et al., 1986–2003). (Varieties are not included. Families are taken from* Flora Europaea *to correspond with the* Flora iberica *volumes.)*

	Volumes										
	I	II	III	IV	V	VI	VII	VIII	X	XIV	Total
Flora Europaea	327	535	240	342	168	172	475	175	210	125	2768
Flora iberica	398	758	275	406	186	237	579	207	254	118	3418
% increase	21.7	41.7	14.6	18.7	10.7	37.8	21.9	18.3	20.9	−5.6	23.5

As we see, the territory covered by *Flora iberica*, as compared with many diversity-rich regions of the world, was acceptably well documented. Table 5.1 summarises the numerical data of the species recognised by the various studies, and shows that each new Flora presents a larger number of species than were hitherto recognised in the same territory.

The figures given for *Flora iberica* are calculated from an extrapolation that is explained in Table 5.2. It is precisely on this point that I wish to comment. In a territory that was relatively well studied, covered by a modern Flora (*Flora Europaea*, Tutin *et al.* (1964–80)), we proposed to start almost immediately on another work that involved an original taxonomic study of all the groups, or at least most of them. We started in 1980, and so far we have completed and published almost 50% of the whole study. We, therefore, have enough data for a family-by-family comparison of the findings of the various studies. The figures of this comparison are given in Table 5.2 and they show that with the exception of volume XIV (Myoporaceae–Campanulaceae), a larger number of taxa – specific and infraspecific – are accepted or recognised in *Flora iberica* than in the ample *Flora Europaea*. The increase in the number of recognised species ranges from the 41.7%

Table 5.3. *New species and subspecies described for* Flora iberica

| | Volumes | | | | | | | | | |
	I	II	III	IV	V	VI	VII	VIII	X	XIV	Total
New species	8	5	33	1	2	53	9	2	1	0	114
New subspecies	3	9	2	3	3	0	0	2	8	0	30
Total	11	14	35	4	5	53	9	4	9	0	144

in volume II (Platanaceae–Plumbaginaceae (partim)) to the 14% in volume III (Plumbaginaceae (*partim*)–Capparaceae). Altogether, even with the negative data of the above-mentioned volume XIV, the average increase in the number of species or subspecies recognized is 23.5%. In other words, for every five species mentioned in *Flora iberica*, one will be an addition to the territory when compared with the data given in *Flora Europaea*.

This increase is simply a consequence of the taxonomic methodology adopted in writing each Flora. While *Flora Europaea* was compiled as a synthesis of what was known at the time about vascular plants across Europe, *Flora iberica* – dealing with a much smaller territory – allowed a more detailed approach based on original taxonomic investigation, a close study of a great deal of herbarium material, expeditions to solve taxonomic problems in the field and so on.

This taxonomic investigation produced not only an increase in the number of species recognized, but also the discovery and description of taxa new to science. Most of these taxa were very localised species, critical and often requiring special attention for their conservation. The numbers of new taxa are given in Table 5.3.

A good example of the consequences of working from catalogues or lists of species, copied indiscriminately and without the backing of taxonomic studies, was given by Kirschner and Kaplan (2002) with reference to the families Juncaceae and Potamogetonaceae. These authors compared the names of the species of the two families that were given in The *1997 IUCN Red List of Threatened Plants* (Walter & Gillett, 1998) with data from recent monographic studies of the same families. The comparison showed that a considerable number of the names in the *IUCN Red List* are synonyms that are normally applied to widely distributed plants, while other names are ambiguous. In fact, only 10 to 25% of the names would be maintained in a reasonable Red List. The authors conclude by saying that 'the comparison clearly shows the importance of detailed taxonomic monographs . . . for the management of plant resources of the world'.

Taxonomy also has to provide clues to solve many other problems in the populations of some plants that are designated for conservation. A good example is *Carex*

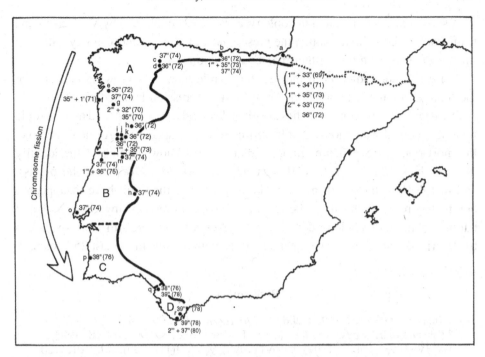

Figure 5.1 Geographic distribution of the different cytotypes of *C. laevigata*. Note that chromosome number increases in the north–south direction. A, unstable $n = 36$ area; B, stable $n = 37$ area; C, stable $n = 38$ area; D, stable $n = 39$ area. Thick line shows the limit of the distribution of the *C. laevigata* in the Iberian Peninsula; dashed lines show the limits of the different cytogeographic areas.

laevigata L., relatively common in Western Europe and considered as a species with little variation, whether morphological or taxonomic. However, a cytotaxonomic study (Luceño & Castroviejo, 1991) showed this to be far from true. The species, supposedly stable, was found to be a mosaic of populations with different chromosome numbers. The number of chromosomes found in the Iberian Peninsula varied between $2n = 69, 72, 80$ (and recently a value of 82 was found). The distribution of these cytotypes is not haphazard; due to a mechanism of chromosome fission, it follows a north–south orientation, the lower numbers being found in the more northerly populations and the higher numbers in those further south (Figure 5.1). When it comes to preserving this plant, which populations should be selected for conservation? From the ecological and evolutionary points of view, a plant with $2n = 69, 71$ or 72 chromosomes is not the same as one with $2n = 80$ or 84. The more southerly populations of *Carex* laevigata are threatened with extinction, while the more northerly ones are not. Is this a species that must be preserved? Taxonomy and common sense tell us that it is the southernmost populations that require attention on account of their singularity and biological difference.

A similar problem is found in zoology, as is clearly explained by Valdecasas and Camacho (2003), who insist that the present lack of interest in taxonomy will have dire consequences in fields other than conservation.

There is a contradiction in the fact that while there is a growing concern for the 'conservation of biological diversity' – an expression bandied about, even in political jargon – taxonomic studies are given less and less support, and many people no longer consider taxonomy as a front-line scientific speciality. At the same time, the most optimistic estimates indicate that we still know nothing of the identity of more than a million and a half organisms (50 000 to 100 000 vascular plants), and that 40% of them will never be known because they will have disappeared before they reach the hands of the scientists capable of recognising, describing and naming them. In this field of describing biodiversity, there are still more species to be described than all that have been recognised up to now, in the last 250 years.

References

Castroviejo, S. *et al.*, (eds) (1986–2003). *Flora iberica. Plantas vasculares de la Península Ibérica e Islas Baleares*, Vol. I (1986); vol. II (1990); vol. III (1993); vol. IV (1993); vol. V (1997); vol. VI (1998); vol. VII(1) (1999); vol. VII(2) (2000); vol. VIII (1997); vol. X (2003); vol. XIV (2001). Madrid, Spain: CSIC.

Greuter, W. R., Burdet, H. M. & Long, G. (1984–9). *Med-Checklist*, vol. 1 (Pteridophyta (edn. 2). Gymnospermae. Dicotyledones (Acanthaceae–Cneoraceae)) (1984); vol. 3 (Convolvulaceae–Labiatae), (1986); vol. 4 (Lauraceae–Rhamnaceae) (1989). Genève Conservatoire et Jardin Botaniques de la Ville de Genève and Berlin: Botanic Garden and Botanical Museum Berlin-Dahlem.

Guinea, E. & Ceballos, A. (1974). *Elenco de la Flora vascular Española (Península y Baleares)*. Madrid, Spain: ICONA.

Kirschner, J. & Kaplan, Z. (2002). Taxonomic monographs in relation to global Red Lists. *Taxon*, **51**, 155–8.

Luceño, M. & Castroviejo, S. (1991). Agmatoploidy in *Carex laevigata* (Cyperaceae). Fusion and fission of chromosomes as the mechanism of cytogenetic evolution in Iberian populations. *Plant Systematics and Evolution*, **177**, 149–59.

Smythies, B. E. (1984–6). *Flora of Spain and the Balearic Islands. Checklist of Vascular Plants*. 1: *Englera*, **3**(1), 1–212 (1984); 2: *Englera*, **3**(2), 213–486 (1984); *Englera*, **3**(3), 487–880 (1986).

Tutin, T. G., Heywood, V. H., Burges, N. A. *et al.* eds. (1964–80). *Flora Europaea*. 5 vols. Cambridge, UK: Cambridge University Press.

Valdecasas, A. G. & Camacho, A. I. (2003). Conservation to the rescue of taxonomy. *Biodiversity and Conservation*, **12**, 1113–17.

Walter, K. S. & Gillett, H. J. eds. (1998). *1997 IUCN Red List of Threatened Plants*. Gland, Switzerland and Cambridge, UK: IUCN.

Willkomm, H. M. & Lange, J. M. Ch. (1861–80). *Prodromus florae hispanicae: seu synopsis methodica omnium plantarum in Hispania*. Stuttgart, Germany: Schweizerbart.

6

The democratic processes of botanical nomenclature

R. K. Brummitt

Introduction

The stated aim of botanical nomenclature is to produce a stable set of names for all plant taxa. In any one classification, every family, genus, species, subspecies and so on should have only one correct name. The taxonomic system that has developed over the last two and a half centuries, and the nomenclatural rules that have been steadily refined over the last century, have interacted sufficiently well for us to be able to say that this aim of stability has been achieved with considerable success. There will always be a few who complain that the name of their favourite plant has been changed by meddling nomenclaturists yet again, but on investigation these complaints usually prove to be ill-founded. Serious inconvenience can be avoided by the conservation and rejection mechanisms described below. Changes of names in the twenty-first century are usually due to changing taxonomy rather than to the rules of nomenclature.

To those not familiar with the rules governing plant names, the **International Code of Botanical Nomenclature** (ICBN, or here referred to simply as 'the Code') may seem a formidable document. The mechanisms by which the botanical taxonomic community operates in this field may seem complex, obscure and even arcane. The rules and processes are, however, governed by very democratic internationally agreed procedures, and represent an outstanding example of how democracy can work in a scientific context. It may be noted also that virtually all the work described here is performed on a voluntary basis without any salaried staff positions, but, as noted below, often with remarkable efficiency.

The Code is a steadily evolving document with new editions being published regularly (since 1951). The process by which the Code is changed is laid down in Division III of each Code (pp. 103–4 of the current 2000 St Louis Code, see Appendix), but this has also evolved somewhat since the early part of the last

Submitted for publication October 2003.

century. A fairly standard procedure has been operated for most of the last half a century, both for changing the Code and for conserving or rejecting names. The following notes are based on the present writer's experience of practices operating over the last four decades. It is hoped that this summary will be helpful to those who have an interest in this subject but have not previously been involved, and perhaps may show how they themselves may play a part in the process.

Authority of the ICBN

Our modern Code has developed through a series of editions of what was originally called the 'Rules', published at intervals through the twentieth century. These Rules have their roots in de Candolle's *Lois de la nomenclature botanique* (1867), which in turn may be seen as a development from Linnaeus' *Critica botanica* (1737) (see Nicolson (1991) for historical discussion and bibliographic details). However, in the twentieth century the rules were formalised by adoption at the series of the **International Botanical Congresses** (IBC), which since 1969 have met at intervals of six years. These Congresses meet under the aegis of the International Union of Biological Sciences (IUBS), their continuing existence being ensured by the International Association of Botanical and Mycological Societies (IABMS), an umbrella organisation comprising all the international organisations in the fields of botany and mycology that are scientific members of the IUBS. The IABMS has no staff of its own, its officers having full-time research, teaching or administrative duties in other institutions.

An earlier review of the formal structures of nomenclature has been given by McNeill and Greuter (1986). Each Congress is a separate organisational entity, and there are no continuing staff to run the extensive nomenclatural work involved both at and between the Congresses. Between Congresses, botanical nomenclature is governed by the General Committee for Botanical Nomenclature (commonly known merely as the General Committee), a voluntary body without direct funding, composed entirely of professional botanists with nomenclatural interests (J. McNeill, personal communication, 2003). The **International Association for Plant Taxonomy** (IAPT; current secretary, Tod Stuessy at the University of Vienna) therefore acts as the agent, and provides limited financial support, for the nomenclatural activities of botany and mycology including those associated with the Congresses; the IAPT also publishes each new edition of the Code. All decisions concerning the Code, however, must be ratified at the final plenary session of each Congress, which is the effective authority for the Rules (see Division III.1 of the present Code). This is normally done merely by the Congress approving recommendations made to it by the Nomenclature Section, but such recommendations could, in principle, be overturned by the plenary session of the Congress. So although it is closely associated with the process, and holds its General Assembly during the week of the Nomenclature Section at the Congress (see for example Nicolson (1999a)), the IAPT does not have authority for decisions made or for officers or committees appointed there.

The ICBN governs the nomenclature of all plant groups, including algae and all fossil plants, and also of the fungi – which are not regarded as plants by some. It does not cover the bacteria, which have their own separate International Code of Nomenclature of Bacteria (ICNB) (Lapage *et al.*, 1992); nor does it cover the zoological kingdom, which falls under the International Code of Zoological Nomenclature (ICZN) (Ride *et al.*, 1999). There are, however, some ongoing problems in unicellular flagellates in deciding whether these should be treated as plants under the ICBN or as animals under the ICZN. Names peculiar to cultivated plants are not controlled by the ICBN but by the International Code of Nomenclature for Cultivated Plants (ICNCP) – see final section in this chapter.

One of the very satisfactory aspects of the modern nomenclatural scene is that the Code's authority is accepted without question by virtually all plant taxonomists in all countries of the world. The democratic way it has evolved, together with the associated committee system, is no doubt a major contributory factor to this universal acceptance. The community should be aware of the advantage that such global unanimity brings to nomenclature, and should jealously guard it in future. Any attempts to disrupt the democratic processes, to overthrow the authority of the Code or to replace it by something else, should be viewed with great concern.

Changing the Rules in the ICBN

Preliminary thoughts

More or less any change to the Code can be proposed by anybody. Proposals ranging from a minor adjustment of the punctuation to re-writing the whole document have been made in recent decades. However, the voting is usually fairly conservative, and it is probably fair to say that since 1950 the great majority of proposals that have been accepted have not had a revolutionary effect on plant nomenclature but have been more of the nature of clarifications or extensions to what has been accepted before. Some might see the acceptance of conservation of specific names at the 1981 and 1993 Congresses (see 'Conservation of specific names' in section on 'Matters considered by Permanent Committees for plant groups') as being revolutionary, but it was merely an extension to species level of what had been accepted at generic level since 1905. In fact, each Congress rejects more proposals than it accepts. Those who have just seen the defeat of their favourite proposal, which they considered the best improvement to the Code since de Candolle, are sometimes moved to say that it is impossible to get a Congress to vote for any change at all, however sensible. But a little conservatism may be a good thing.

Those making a proposal to change the Code may often find it relevant to trace the history of the present wording and to present a historical perspective of the matter. This may be done by checking back to previous editions to see when the present wording was introduced, and then it is possible to consult the arguments made at that time. As noted below, detailed records are kept of what is said in the Nomenclature Sessions at the Congresses and this is published as a permanent record. The literature of this record is quite diverse; therefore, as an aid to locating it, an Appendix to this chapter includes all the bibliographic references.

What happens before a Congress

Proposals must be submitted for publication in *Taxon*, the journal of the IAPT. The proposals will be accepted only during a window of about two to three years half-way between two Congresses. The closing date will usually be about 18 months before the next Congress. Guidelines on when proposals will be accepted, and on their length and format, are published in *Taxon* about two years after each Congress (i.e. four years before the next Congress) – for example, those published by McNeill and Stuessy (2001) for the 2005 Vienna Congress. Proposals are subject to normal editing procedures regarding format and accuracy, and advice may be given on how better to achieve the intent of the proposal, but no opinion on the merits of the proposal is offered at this stage. All proposals submitted for each Congress are numbered by the editors in the sequence in which they are published in *Taxon*.

Each Congress elects one person to act as **Rapporteur-Général** for the following Congress; this person will nominate a **Vice-Rapporteur** to assist them. (Technically, according to Division III.3 of the Code, the Vice-Rapporteur is elected by the organising committee of the forthcoming Congress on the proposal of the Rapporteur-Général; in practice, it comes down to the latter's personal choice). The task of these two Rapporteurs is to bring together all the proposals for changes to the existing Code and to give their best opinions on the merits of each. This is done in one extensive report, again published in *Taxon*, usually in the February issue in the year of the relevant Congress. All proposals relevant to each Article of the Code are grouped together – see for example Greuter and Hawksworth (1999) in the report of the Rapporteurs for the 1999 St Louis Congress. The precise wording of each proposal is repeated here, but the arguments originally given by the proposers are not repeated, only a reference being given to where they were first published. The Rapporteurs give their views on how they think voters should react, indicating the likely practical results of acceptance or rejection. They also assign each proposal under an Article, a reference letter, from A to as far as is necessary, so that at this stage each proposal has two reference points, a serial number relating to the sequence of publication in *Taxon* and the Rapporteurs' letter.

Soon after publication of the Rapporteurs' report, the IAPT office sends a voting form for a preliminary '**mail vote**' to every member of the Association, and to members of permanent committees and authors of proposals who are not members of the IAPT. On the form a reference is given to the Rapporteurs' comments, and four options for voting on each proposal are given: 'Yes', 'No', 'Sp. Comm.' and 'Ed. Comm.' A vote for 'Sp. Comm.' would be a vote for the meeting to set up a special committee (see section on 'Appointment and functions of committees') to investigate the matter and report back to the following Congress. A vote for 'Ed. Comm.' would be in favour of merely referring the proposal to the Editorial Committee, and might suggest approval of the proposal in principle but allowing the Editorial Committee to decide the precise wording. Recipients of the form are invited to indicate how they would like to see each vote resolved in the final ballot and return the signed form by post or by fax. This preliminary postal vote acts merely as a guide to the opinions of a wide range of people and does not count at all in the final vote at the Congress, although a heavy defeat in the postal vote may have significance (see 'What happens at a congress' in this section). A summary of all the postal votes is distributed at the Congress to all those who actually register at the Nomenclature Sessions, and will be published in *Taxon* later. Further interesting comments on the most recent postal vote were given by Barrie and Greuter (1999), who reported approximately 20% of ballot forms being returned. Their analysis of regional tendencies on some issues was quite revealing.

The organising committee of the whole Congress also has a small part to play. It elects a **President of the Nomenclature Section**, appoints a **Recorder** for this Section – whose job it is to record the recommendations made and see that they are referred to the final plenary session – and 'elects' a Vice-Rapporteur on the personal recommendation of the Rapporteur-Général (see comment above). These persons, together with the Rapporteur-Général elected by the previous Congress, are the officers of a body known as the **Bureau of Nomenclature**. For the last Congress (St Louis 1999), the President was H. M. Burdet, the Rapporteur-Général was W. Greuter, the Vice-Rapporteur was D. L. Hawksworth and the Recorder was F. R. Barrie. The officers usually co-opt further members to form the full Bureau (see Barrie and Greuter (1999), p. 774). This body has responsibility for the smooth running of the Nomenclature Sessions at the Congress, and has one or two other functions, but is probably outside the awareness of most persons attending the meeting.

Before the Congress, the Bureau of Nomenclature draws up a list of botanical centres with a stake in taxonomic matters and assigns **institutional votes**. A sliding scale from one to seven is implemented, the small institutions being allotted one vote at the next Nomenclature Sessions while the biggest or most active ones are given seven votes, with others in between. The list of institutions, and the number of votes they used at the previous Congress, is circulated by the Bureau of Nomenclature to members of the General Committee and suggestions of changes are invited. The list is later published, and appeals can be made. Cries of 'class distinction' are occasionally heard, and some opinions on the voting rights of institutions with fewer than 100 000 specimens in their herbarium have been expressed recently by Filgueiras *et al.* (1999). The most recent list at the time of writing was published in *Englera* (Greuter *et al.*, 2000, pp. 247–53) which will be revised before the 2005 Vienna Congress. See below, 'What happens at a Congress', for notes on how institutional votes operate.

What happens at a Congress

The Nomenclature Sessions are held in the week immediately preceding the main Congress. Attendance is open to anybody who is registered as a member of the Congress. At recent Congresses the number attending has been between 150 and 300. The sessions are chaired by the President of the Nomenclature Section, but in close cooperation with the Rapporteur-Général and Vice-Rapporteur, while the Recorder appointed by the Bureau (see above) sits with them and records the recommendations made. Tellers to count votes are also appointed for each session.

Two procedural matters are put to a vote before the proposals on each Article are considered. First, the meeting is asked to approve formally the Code that the Editorial Committee has produced following the previous Congress, and technically this Code is not accepted until that vote is taken. It is exceptional for the printed Code to be challenged at this point, but it did happen (unsuccessfully) at the 1964 Edinburgh Congress. Secondly, the majority required to pass a proposal to amend the Code in the following voting has to be decided. In all recent Congresses it has been proposed and accepted that a 60% majority of the votes cast is required. On rather rare occasions a proposal to take an action other than amend the Code is made; in this case, a simple majority is required – see, for example, comment by McNeill, where a Standing Committee on Names in Current Use was set up (Greuter *et al.*, 1994, p. 31). Occasionally also, after the principle of changing a particular point in the Code has been agreed by a 60% majority, the selection between two competing methods of making the change is determined by a simple majority (see, for example, Greuter *et al.* (2000), pp. 197–9).

Proposals are normally considered in the numerical sequence of the Articles affected and then in the alphabetical sequence assigned by the Rapporteurs. The preliminary postal vote comes into play to some extent here. If a proposal has received a 75% or greater 'No' vote, it is considered rejected and is not discussed unless somebody on the floor of the meeting requests that it be considered and this is seconded. For all other proposals, the President invites comments from the floor of the meeting. Discussion is conducted along the lines of formal debating procedure, although it may be fair to say that not everybody is completely *au fait* with this. However, it does not matter if one does not know the difference between 'tabling the motion' and 'asking the question', or between a 'point of order' and a 'point of information', as long as the Chair can direct proceedings successfully.

The voting procedure is at times complicated by the setting aside of the principle of 'one person one vote' and the introduction of **institutional votes** (see above, 'What happens before a Congress' in this section). Each institution may appoint someone – either from its own staff or from outside, according to its wishes – to use these votes. These appointed people are asked to carry a written statement authorising them to claim these institutional votes when they register for the Nomenclature Section, and they are then given cards of different colours indicating how many votes they have. It has been the practice also to grant representatives of institutions not on the official list one institutional vote if they attend the nomenclature meetings and request an institutional vote. A number which may be assigned to any proposal is printed on a detachable portion which may be torn off for voting purposes. Any one person may carry up to 15 votes (including his or her own personal vote), but not more.

When discussion of a proposal has been terminated, at first a vote by a show of hands is asked for, and if this immediately appears decisive it is accepted. If the vote appears close, however, anybody present may ask for a **card vote**, when everybody displays their different-coloured cards. The first time this happens, great interest is shown in how many votes other people are carrying! If this vote is close again, the tear-off section for the relevant proposal is collected from each person and a count of these is made by the tellers, the number of votes being indicated by the colour of the card. This can often swing what might have seemed a comfortable majority, on the first show of hands, in the opposite direction. At the St Louis Congress there were 297 individual members with one vote each, and in addition a total of 494 institutional votes claimed. Great is the anguish on the faces of those who thought they had won a hand vote only for it to be defeated on a card vote! The writer speaks from agonised personal experience.

At recent Congresses a **recording** has been made of the entire proceedings. To assist those trying to interpret this later (see below in 'What happens after a Congress' in this section), each speaker is also asked to write down on a numbered piece of paper their name and a summary of what he or she thinks (!) he or she has said. In the heat of the moment, and in the at times rapid exchanges, not everyone manages to achieve this every time, but one hopes that at least one's voice may be distinctive. In the post-war Congresses before recording voices became an option, a stenographer – Wil Keuken of the botanical institute in Utrecht – sat at the front and took down everything in shorthand using a stenotype machine, a formidable task which she accomplished with what appeared to be a remarkable lack of problems (see also J. Lanjouw's comments (1996, p. 14) on the 1966 Edinburgh Code).

The Recorder records the voting on every proposal, and, often very importantly, also records the exact wording of any modification made to the original wording of a proposal during the meeting. The Rapporteur and Vice-Rapporteur may also keep their own records of votes taken. Consideration of the recommendations of the Nomenclature Section by the **plenary meeting** at the end of the Congress is usually a brief 'rubber stamp'; proposals are usually accepted en bloc, although there have been occasions when a challenge has been made (unsuccessfully) concerning an individual proposal at this stage.

What happens after a Congress

A summary of the decisions on each proposal is prepared and rapidly published in *Taxon*. For example, the nomenclatural decisions of the St Louis Congress held in July to August 1999 were published in the November issue of *Taxon* the same year (Barrie & Greuter, 1999). These days little or no time is lost in preparing a new edition of the Code, a task that falls to the Editorial Committee elected at the Congress. Thus, just three months after the St Louis Congress the Chairman of the Editorial Committee – in this case Werner Greuter – had circulated a first draft of the revised Code to the rest of the committee, and the committee met in Berlin for a week during the last week of January 2000 to finalise the text (Greuter & Hawksworth, 2000, p. xiv). Over the last 50 years, each new edition of the Code has been published by the IAPT as a part of its occasional series *Regnum Vegetabile*. The St Louis Code appeared as volume 138 of this series later in 2000, and this edition will then be ratified by the next Congress in Vienna in 2005. The production of the new edition of the Code on recent occasions has been an example of great efficiency by a rather small group of people, and the botanical community owes them much for their dedication and hard work.

It was noted above that a recording is made of the entire proceedings of the Nomenclature Sessions at each Congress. A working group is appointed to transcribe the tape into a text that records in the third person the gist of what every speaker said, and this is then published in hard copy as the **Report on Botanical Nomenclature**. For the last four Congresses this most interesting document has appeared in the occasional publication of the Berlin Botanic Garden, *Englera*. For details of these four Congresses and transcriptions of earlier Congress discussions, see the Appendix to this chapter. For the 1999 St Louis Congress, five days of discussion – often vigorous and perhaps occasionally heated – was condensed into 231 pages of text by W. Greuter, J. McNeill, D. L. Hawksworth and F. R. Barrie; this was published by 2000, another remarkable achievement. Those whose utterances are recorded there may sometimes be astonished and swear that they have no recollection of ever saying such a thing, but the evidence of the tape recording is always there. Again, the present writer speaks from personal experience.

Appointment and functions of Committees

Apart from voting on amendments to the Code, the Nomenclature Section also appoints the next Rapporteur-Général and establishes a series of committees to function until the next Congress. These are of three kinds: **Permanent Committees, Special Committees** and (more rarely) **Standing Committees**. The Permanent Committees include the Editorial Committee, the General Committee and committees for each of the five major groups of extant plants (Fungi, Algae, Bryophyta, Pteridophyta and Spermatophyta) and one for fossils of all groups. They continue through a succession of Congresses and are reappointed by each successive Congress, their membership usually being included in the formal report of the recommendations made by the Nomenclature Section and ratified by the plenary session of the whole Congress. The Special Committees are set up by one Congress to consider particular topics and to report back to the following Congress, after which they cease to exist unless reconstituted again. The topics designated for each of these Special Committees are determined by votes of the Nomenclature Section and are included in its recommendations to the Congress plenary meeting. The membership of the Special Committees, however, is by appointment by the General Committee (not by a Congress) and is announced separately, usually a year or so after the Congress. Occasionally a Standing Committee may also be recommended by the Nomenclature Section. Its function and operation are similar to that of a Special Committee, but it is charged with continuing its work from one Congress to another. Further notes on these different sorts of committee are given below.

Permanent Committees

As noted above, a body known as the Bureau of Nomenclature is appointed for each Congress following the initiative of the organising committee of the whole Congress. The President of the Bureau, in consultation with its other members, then appoints a **Nominating Committee** which will be active behind the scenes at the Nomenclature Sessions. This committee seeks nominations for the next Rapporteur-Général (normally only one name is put forward) and for membership of all the Permanent Committees. In practice, the secretaries and chairs of these committees will usually let the Nominating Committee know who is willing to continue from the previous committee and will add names of any likely new recruits. It is probably good tactics by anyone wishing to be elected to one of the Permanent Committees to let the secretary or chairman of that committee, or a member of the Nominating Committee, know of his or her willingness to be elected.

The rules for replacing members of Permanent Committees between Congresses are laid down by the General Committee. Members who do not reply to correspondence or do not vote three times in sequence over a period of nine months are treated as having resigned. Members may of course also retire voluntarily. Replacements are appointed through consultation between the chairman and the secretary of the committee concerned and the secretary of the General Committee.

The **General Committee** has a wide range of activities, at least in theory. It has a Chairman, Vice-Chairman and Secretary, and a number of ex-officio members from other committees. The ex-officio members at present are the Rapporteur-Général, the President and Secretary of the IAPT, and the secretaries of all the seven other Permanent Committees. It then also has a number of members at large. At the time of writing, the General Committee has 25 members (see Greuter and Hawksworth (1999), p. 775). It serves to vet the recommendations of the other committees and bring them forward (or not) to be ratified at the plenary session of the next congress.

The function of the **Editorial Committee** is self-evident: to produce the new edition of the Code in the light of decisions taken in the voting at each Congress. Some comments on this are made above. It operates at first by correspondence and then at a meeting lasting for most of a week. By tradition, members of the Editorial Committee must have attended the Nomenclature Sessions of the Congress.

The work of the **Permanent Committees for each of the five major groups** of extant plants and for fossils is focused on processing the proposals for conservation or rejection of names as published in *Taxon* (for further details see section on 'Matters considered by Permanent Committees for plant groups'). They make a recommendation to the General Committee on each of these proposals, and the General Committee in turn makes its recommendation through the Nomenclature Section to the plenary meeting of the next Congress. In recent years they have also made recommendations on whether two names are so similar that they should be treated as homonyms. Requests for such recommendations are currently not published in *Taxon* but are simply submitted to the secretary of the appropriate committee (for further discussion, see 'Questions of homonymy' in section on

'Matters considered by the Permanent Committees for plant groups'). Whether this is desirable, or whether space should be taken up in *Taxon* for such requests, may be a matter of opinion.

These six Permanent Committees work very largely by correspondence, although some may conduct some formal business when their members meet together during the Nomenclature Sessions of a Congress. Their eventual recommendations on each case for conservation or rejection, or homonymy (see note above), are published in *Taxon* in a series of reports by their secretary, as for example that by Zijlstra (1999) for the Committee for Bryophyta. The committees receiving relatively few cases to consider may publish a report only once or twice in the six-year period between Congresses, but those receiving a heavier workload may publish such a report on average once a year or more. Some more detailed comments on the functioning of the Committee for Spermatophyta are given in the section on 'How the Committee for Spermatophyta operates'.

Special Committees

A Special Committee may be set up to examine any topic relevant to possible proposals to the next Congress. On usually the last day of the Nomenclature Sessions at each Congress, notices are posted advertising each proposed Special Committee that has been established by the Nomenclature Section, and those present are invited to sign up their own names or those of others who they think would be appropriate as potential members. The General Committee then considers these lists and other potential candidates – and appoints a convener, a secretary and members of each committee. Usually each will have between about five and ten members (for examples, see Barrie and Nicolson (2001)). These committees operate by correspondence only. In the past some have been very successful while others have failed to produce anything, usually a reflection of the vigour with which the appointed secretary pursues the task, although some secretaries of the past would argue that the other members never replied to correspondence sent to them!

The Special Committees set up by the last Congress (St Louis, 1999) were listed in *Taxon* (Greuter & Hawksworth, p. 783, note 1 1999), and are as follows: (1) Electronic Publishing; (2) Early (pre-Cambridge Congress) Generic Typifications; (3) Suprageneric Names; (4) Effective Publication, Especially Theses; (5) Intercode ICBN/ICNCP to Coordinate and Harmonise Provisions on Nomenclature of Hybrids; (6) Liaison with other Nomenclatural codes; and (7) Division III of ICBN, Particularly Voting Procedures. The membership of these committees is given by Barrie and Nicolson (2001). The Rapporteur-Général and the Secretary of the General Committee are ex-officio members of all Special Committees. The secretary of each Special Committee is expected to submit a report to *Taxon*, including any formal proposals required, in the same way as others making proposals.

Standing Committees

The only current Standing Committee is the Standing Committee on Lists of Names in Current Use (NCU) established by the 1993 Tokyo Congress and appointed by the General Committee 'to initiate, assist, coordinate and vet production of further lists and of updating of the existing lists of NCU and to report to each subsequent International Botanical Congress' (Nicolson, 1994, p. 283). A previous committee was the Standing Committee on Stabilisation, referred to in *Taxon* (Moore *et al.*, 1970, p. 51).

Matters considered by Permanent Committees for plant groups

Conservation and rejection of names

As with proposals to change the Code, any person may make a proposal to conserve or reject a name. This is done by submitting a proposal for publication in *Taxon*, which is automatically treated as a submission to the General Committee (Articles 14.12 and 56.2) and a reference for consideration by the appropriate committees; if successful, the proposal will eventually be ratified by an International Botanical Congress. Conservation and rejection of names are sanctioned by the ICBN, and basically allow other regulations of the Code to be put aside in the interests of nomenclatural stability. Under Article 14 of the present Code it is possible to conserve names of families, genera and species, but not of the intermediate ranks; i.e. not subdivisions of families (tribes etc.), subdivisions of genera (subgenera, sections, etc.), or subdivisions of species (subspecies, varieties, etc.). Under Article 56, however, it is possible to reject any name at all, and if conservation of one name is not permitted the same result may sometimes be achieved by proposing rejection of another. Lists of all conserved or rejected names are given in the ICBN in a series of appendices (which now take up very much more space in the Code than do the Articles themselves).

The **conservation of family names** was introduced in the 1961 Montreal Code, when a long list of such names of Spermatophyta (gymnosperms and flowering plants) was included in Appendix II. The principle here was different from that adopted for conservation of generic names, for there were no names listed as rejected against them and all names on the list were said to be conserved over all names not listed. The list of spermatophyte families was intended to be comprehensive for all family names then in use. (It may be noted in passing that this principle of blanket conservation over non-included names is the same as that proposed at the

1999 St Louis Congress, but not accepted, for names at lower ranks in the Names in Current Use initiative.) A few family names of algae and bryophytes were added on the same principle in subsequent Codes, but in the 1988 Berlin Code a change was made. Here, the few names in algae and pteridophytes were put into Appendix IIA, with the names conserved only over those names listed as rejected against them, while in Appendix IIB names of a few bryophyte families and the long list of spermatophyte families were given blanket conservation over names not listed. Since 1988 a relatively few family names in Fungi and one for a Fossil have been added to Appendix IIA.

Conservation of generic names has been permitted ever since the 1905 Vienna Congress. The impetus for this came from the publication by Otto Kuntze of his *Revisio Generum Plantarum* (1891, 1893), in which he proposed the displacement of over 1000 generic names then in current use, with 30 000 new specific combinations. In response, a first list of generic names proposed for conservation was published in Ascherson (1892, pp. 332–4). The mechanism for generic names operates as described above for family names listed in Appendix IIA, names being conserved only against those names listed against them (except in cases of homonyms). Generic names may be conserved for the following reasons.

1. Over an earlier taxonomic synonym; e.g. *Capsella* Medik. 1792 is conserved over *Bursa-pastoris* Ség. 1754. In such cases the name is conserved only against those names which are listed as rejected against it. A rejected name may be restored if it is regarded as taxonomically different from the genus bearing the conserved name.
2. Over an earlier homonym; e.g. *Welwitschia* Hook. f. 1862 (Gymnospermae) is conserved over *Welwitschia* Reichb. 1837 (Polemoniaceae). In such cases a name is also automatically conserved over any other uses of the same name which may have been validly published, whether listed or not.
3. To maintain a particular spelling; e.g. *Helichrysum* is conserved over the original spelling *Elichrysum*.
4. To conserve a new type; e.g. *Chrysanthemum* L. is conserved with *C. indicum* L. (the garden Chrysanthemum) as its type, the other original species (including the first type species chosen) having been transferred to other genera (see Brummitt (1997) for discussion).
5. To conserve a particular gender if this is in doubt; e.g. *Sapindus* is conserved as masculine, past usage having often treated it as feminine.
6. To confer legitimacy on an originally superfluous illegitimate name; e.g. *Prismatocarpus* L'Hér. 1789, which was illegitimate when published.

The list of conserved generic names, with their corresponding names rejected against them, constitutes Appendix IIIA of the present Code. As noted below, a few generic names have also been rejected *'utique'* in Appendix IV (*'utique'* meaning 'completely' or 'in every circumstance').

Conservation of specific names under Article 14 is a relatively recent introduction, although it was a matter of discussion for much of the last century. It was eventually allowed for names of species of major economic importance at the 1981 Sydney Congress, where *Triticum aestivum* L. was the classic example used. It was then extended to names that indicated types of generic names at the 1987 Berlin Congress, and at the 1993 Tokyo Congress it was opened to any name. The principles are the same as listed above for generic names. Examples so far adopted include *Erica carnea* L. conserved over *E. herbacea* L., *Galanthus elwesii* Hook. f. conserved with a new type when the original type was found to belong to a species different from that to which the name had generally been applied, and *Anemone narcissiflora* L. conserved in preference to the original (accidental) spelling *A. uarcissiflora*. Specific names conserved under Article 14 are currently listed in Appendix IIIB.

Informal **rejection of names** that had become used in different senses, and so a long-persistent source of error, has been permitted since as far back as the 1906 Vienna Rules (Article 54.4), but formal lists of these were not agreed until the 1975 Leningrad Congress. (D. Nicolson, personal communication, 2003, informs the present author that there was a move around 1930 to produce a formal list of rejected specific names but it was never implemented. Nonetheless he has included these proposals in his website (http://persoon.si.edu/codes/props)). The list of such formally rejected names was opened as Appendix IV ('*Nomina utique rejicienda*') in the 1983 Sydney Code, where it consisted of a single name, *Bromus purgans* L., and has grown steadily since then with many specific names now listed. These rejected names listed in Appendix IV of the Code cannot be revived under any circumstances. However, the option of rejecting a name '*utique*' under Article 56 is not restricted to specific names, and a few generic names of uncertain application now also appear in Appendix IV rather than Appendix IIIA where they would have had to be rejected against a certain conserved name. One name of an infrageneric subdivision in Fungi has also been listed.

The current **guidelines for making proposals** for conserving or rejecting names are given by McNeill *et al.* (2003), where a lot of technical and other advice is given that need not be repeated in full here. Those who wish to propose conservation or rejection of a name should follow the format of any such proposal published in a recent issue of *Taxon*. The title should be a simple statement of what is being proposed, and the running number preceding this will be added by the nomenclature editor of the journal. A statement of the bibliographic details, including types, of all names to be conserved or rejected should appear beneath this title. Arguments should include a concise indication of why conservation or rejection is needed. The geographical range of the plants to which the name or names apply should be stated. An indication should be given of how widely the names concerned have been used in the literature, and citation of up to perhaps 12 of these publications will normally be desirable. A conservation proposal will stand an improved chance of success if usage of the name outside the strictly taxonomic community can be demonstrated. Emphasis of the importance of the group concerned from an economic or ecological point of view, or for any other reason, will thus be helpful.

The question often arises as to whether it is better to try **to conserve one name or to reject another**. If a name has been long misidentified or has no obvious type, and so is not likely to be wanted in the future, it is probably better to propose its complete rejection under Article 56 rather than conservation of another name over it under Article 14. If a name has any chance of becoming accepted later as a result of a taxonomic reassessment, it would be preferable to propose conservation of another name that it appears to threaten. Committees will reserve the right to switch a proposal from conservation to rejection if they consider a proposer has chosen the wrong option.

A **recommendation** on each proposal is made by the committee for the major group concerned. Some detailed notes of how the Committee for Spermatophyta functions are offered below. Results of the voting are published in a report from the committee in *Taxon*. The requirements for further action are controlled by the General Committee and are reviewed by it from time to time. At present there is a requirement that a 60% majority of the potential votes of the committee for the major group is attained in order for the General Committee to act further, so if a committee has 15 members it needs nine in favour of a proposal in order to recommend acceptance. (Until a few years ago the required majority was two-thirds, or 66%, rather than 60%.) Normally all those proposals with 60% in favour will also be accepted by the General Committee and then forwarded to a Congress for ratification.

Until a few years ago it was agreed that if a proposal did not get a 60% majority in favour in the first committee, it was considered to have failed. However, a new vote in the General Committee required that if a 60% vote *against* a proposal was not attained, the proposal was not rejected but was held over until a 60% vote one way or the other was gained. A number of recent proposals that did not gain 60% either positive or negative in the first committee have been held in limbo pending further action. Since it is the General Committee that reports to the Nomenclature Section – and hence to the Congress itself – this is the body that makes the ultimate recommendation, although it generally accepts the recommendation of the Permanent Committees for the different groups. Recently, suggestions that the General Committee should dispute some recommendations of the other committees have been made. The way the General Committee operates in relation to the other committees is thus currently under discussion. The General Committee takes a vote on all recommendations from the other committees and publishes its own report on what should be presented to the next Congress (see, for example, Nicolson (1999b, 1999c)). Once a proposal has been approved by the

General Committee, retention or rejection of the names involved is authorised by Article 14.14.

Names ratified by a Congress as conserved or rejected are added to the appendices of the following Code by the Editorial Committee in consultation with the group committee concerned. A very useful comprehensive website of all conservation and rejection proposals ever made has been made available recently by Dan Nicolson, giving the place of publication and the committee recommendations and actions (http://persoon.si.edu/codes/props).

Questions of homonymy

Where heterotypic names that are identical in spelling exist, the later one is ruled illegitimate by Article 53.1 of the present Code. However, sometimes two names are very similar but not identical in spelling, so-called **paranyms** or **parahomonyms**. Article 53.3 then allows that if they are likely to be confused they are to be treated as homonyms, i.e. the later one is illegitimate. However, names that could be confused by one person would not necessarily be confused by another. Article 53.5 therefore says that when it is doubtful whether they are likely to be confused, a request for a decision may be submitted to the General Committee which will refer it to an appropriate other committee for examination. A recommendation may then be put to a Congress for ratification as a binding decision. This provision has only been in the Code since the 1987 Berlin Congress. Since then a mechanism for achieving this has been established by accepted practice. A written request is sent to the Secretary of the Permanent Committee for the group concerned, and after taking a vote this committee then passes its recommendation on to the General Committee which will convey it to the next Congress. It has not been found necessary to publish such requests in *Taxon*, but the recommendation will be published in *Taxon* in the reports of one of the committees.

Recent cases have shown that there is a rather wide range of attitudes to such requests among members. The wording of Article 53.5 indicates that the criterion on which committees should focus is whether the two names are likely to be confused 'because they are applied to related taxa or for any other reason'. But as a well-known nomenclaturist – the late Ray Fosberg – said in this connection, 'you can confuse some of the people all of the time or all of the people some of the time'. Some committee members may argue that if one name is not in current use, the two names are not likely to be confused; others will argue that even if both names are in use, if they apply to widely different groups (for example one an alga and the other a flowering plant) or if the plants occur on different continents, again the names are not likely to be confused. Still others may argue simply from the similarity of the names or their etymological derivation and ignore other factors.

It is important that all factors are presented when a committee is asked to make a recommendation. Requests should therefore include bibliographic details of the names concerned, indication of their taxonomic group (preferably to family), a statement of how widely the names are or have been in general use, some idea of the geographical distribution of the taxa involved, and preferably some notes on the derivation of the names. Recent examples of recommendations made by the Committee for Spermatophyta include (earlier name first): *Agathaea* Cass. (currently a synonym in Compositae) and *Agatea* A.Gray (currently accepted in Violaceae, named after Agate) – not homonymous; *Cornera* Furtado (Palmae, usually a synonym) and *Corneria* A. V. Bobrov & Melikyan (Gymnospermae, recently proposed), both named after E. J. H. Corner – homonymous; *Faurea* Harv. (accepted name in Proteaceae, named after Faure) and *Fauria* Franch. (Menyanthaceae, good genus, named after Faurie, also known as *Nephrophyllidium*) – homonymous; *Tuomeya* Harv. (red alga, named after Tuomey) and *Toumeya* Britton & Rose (Cactaceae, named after Toumey) – homonymous; and *Centaurea epirota* and *C. epirotica* (species in same genus) – homonymous. It appears that the committee has a tendency to be confused.

At the time of writing there is no cumulated list of recommendations on homonymy passed by Congresses. The only way to find them is to go through individual committee reports in *Taxon*. However, plans are in hand to produce a list and to make it available for quick reference (D. Nicolson, personal communication, 2003).

'Completely suppressed publications'

At the 1993 Tokyo Congress it was agreed to rule formally that certain names in a specified limited range of publications should be considered not validly published. These works are currently listed as '*Opera utique oppressa*' in Appendix V of the Code; they are, with only two exceptions, works from the eighteenth century in which names have traditionally been disregarded since their publication. (One should note that only those names at ranks specified in square brackets for each work are suppressed, so the heading title given to the list may not be entirely accurate (see note on meaning of '*utique*' above)). This list is currently sanctioned by Articles 32.7 and 32.8 of the Code. Proposals to add more works to this list should be treated like proposals to conserve or reject names; i.e. addressed to the General Committee by publication in *Taxon*, after which they will be referred to other relevant committees and eventually a recommendation will be made to the next Congress for ratification.

How the Committee for Spermatophyta operates

The correspondence among members of the Permanent Committees is private and more or less confidential. However, since the present writer has been a member of the Committee for Spermatophyta since 1969 and its secretary since 1975 (and so ex-officio member of the General Committee from that date), it may be possible

to give some idea of how that committee operates without giving away too many
close secrets. At the expressed wish of the members, the secretary still sends his
correspondence to all members in hard copy. This may include xerox copies of
more obscure literature relevant to a case. On average, a letter of anything up to
30 pages is sent to members about every three months. The pages are numbered
in a running sequence so that reference back to previous discussions of a case
can easily be made. The members only meet in person during the Nomenclature
Sessions of a Congress, when it is to some extent an informal social occasion
although some serious discussion may also take place about how the committee
functions. Members then have a particular opportunity of letting the secretary know
if they want things done differently.

Since the 1999 St Louis Congress this committee has had 15 members, whereas
previously it had only 12. The present geographical distribution of its members
includes: 6 from Europe (Denmark 1, United Kingdom 2, Russia 1, Sweden 1 and
Germany 1); 4 from North America (Missouri 1, Washington DC 1, New York
State 1, Massachusetts 1); and 1 each from Brazil, South Africa, the Philippines,
Japan and Australia. Informally, qualifications for membership might be seen as an
interest in the subject, a reasonable understanding of the provisions of the Code,
hopefully access to a reasonable botanical library (although the secretary can often
provide copies of essential references if necessary), a willingness to devote a few
hours every three months to examining cases, and a willingness to reply to correspondence and vote when asked to do so. The proceedings of the committee are also
sent to the Secretary and Chairman of the General Committee, to the Rapporteur-
Général and to the IAPT for its archives; complete sets are probably deposited in
several places. For example, a complete set from the committee's inception in 1952
onwards through the successive secretaryships of R. E. G. Pichi Sermolli (1952–4,
including Pteridophyta), H. W. Rickett (1954–65), R. McVaugh (1965–75) and the
present author (1975–) – bound in indexed volumes – is available for *bona fide*
consultation in the Kew library.

After publication of each issue of *Taxon*, the secretary takes up each case for
conservation or rejection of names in Spermatophyta published there – as indexed
under 'Proposals to conserve or reject' on the back cover – and writes a short report
on each. This is likely to be only one paragraph, except in a complicated situation.
The secretary will hopefully make particularly sure that the nomenclatural facts
are correct, indicate why action is needed and perhaps offer some mild opinion
on whether the arguments seem compelling or not. Such reports are included in a
letter to members soon after the publication of the issue of *Taxon*. Members are
invited to submit their reactions to this during the next six weeks (if the secretary
is busy with other things they may often have longer in practice); which they
may do as briefly as a simple 'OK' or as lengthily as several pages, as the mood

takes them. Comments from members are sent to the secretary by email. Quite frequently, comments are also submitted from outside the committee by people either supporting or opposing the proposal; such submissions are always welcomed and are circulated together with the responses from the members. One of the reasons for publication of proposals in *Taxon* is to publicise the case and allow any interested party to try to influence the committee's decision.

The secretary then gathers together all the comments he has received on each proposal – usually between six and ten – and put them all together verbatim and give a summary of members' reactions. This may be just 'All in favour', or may involve extensive reconsideration on controversial issues. These comments are sent out in the next letter, with an invitation to comment again if there seems to be a need to do so. Where the matter is not unanimous, a second round of correspondence may be circulated, and sometimes even more. The more difficult proposals in recent years have generated up to 20 pages of comments. Sometimes the committee feels it cannot win and it will receive criticism whatever recommendation it makes. An account of such a case, written for the layman and concerning the name of the garden Chrysanthemum, has been given by Brummitt (1997). However, the majority of cases currently being received do not seem too controversial to committee members, and narrow votes are the exception rather than the rule.

When all members have had their say, the secretary will include the case on a hard-copy ballot form, usually with 10 or 20 other cases, requesting a vote on each and a signature as soon as possible. As soon as a firm recommendation is reached, the secretary sends a quick note of the result to the proposer in order to provide as rapid a service as possible. From publication of a proposal in *Taxon* to a vote being available will probably take around 6 months, but if the matter is controversial (or if the secretary is slow in the job) it may take longer. All returned ballot forms are archived; the secretary can then substantiate any vote if it should be queried. After completion of a ballot, the secretary will then write a report on all cases voted on and submit it for publication in *Taxon*; see, for example, Brummitt (2002). The responsibility for the process then passes to the General Committee.

The 'Cultivated Plant Code'

The **International Code of Nomenclature for Cultivated Plants** (ICNCP, or 'Cultivated Plant Code' is not under the control of the International Botanical Congresses and is independent of the ICBN, being produced by another International Union of Biological Sciences Commission for cultivated plants. Its first edition was in 1953, and later editions have been published in 1958, 1961, 1969, 1980, 1995 and 2004. The publication details of the 2004 edition are given in the reference list (Bricknell *et al.*, 2004).

The history and organisation of the 'Cultivated Code' are well explained in the foreword to the 2004 edition by C. D. Brickell and in the preface by C. D. Brickell, A. C. Leslie and P. Trehane. Those needing more information are referred there for details.

Acknowledgements

The author is much indebted to the following: John McNeill (current Rapporteur-Général) for much information and corrections on various technical points in the original draft, and for general encouragement and improvement of the text; Dan Nicolson (former Secretary of the General Committee) for further valuable comments; and Piers Trehane (editor of the Cultivated Plant Code) for information on that publication. Their kind interest and assistance are much appreciated.

References

Ascherson, P. (1892). Vorläufige Bericht über die von Berliner Botanikern unternommenen Schritte zur Ergänzung der 'Lois de la nomenclature botanique'. *Berichte der Deutschen Botanischen Gesellschaft*, **10**, 327–59.

Barrie, F. R. & Greuter, W. (1999). XIVI International Botanical Congress: preliminary mail vote and report of Congress action on nomenclatural proposals. *Taxon*, **48**, 771–84.

Barrie, F. R. & Nicolson, D. H. (2001). Announcement: Special Nomenclature Committees. *Taxon*, **50**, 893–6.

Bricknell, C. D., Baum, B. R., Hetterscheid, W. L. A. *et al.* (eds.) (2004). *International Code of Nomenclature for Cultivated Plants*, 7th edn. Regnum Vegetabile **144**, Vienna, Austria: IAPT, pp. I–xxii, 1–124. *Acta Horticulturae* 647, Leuven, Belgium: ISHS, pp. I–xxii, 1–124.

Brummitt, R. K. (1997). *Chrysanthemum* once again. *The Garden*, **122**, 662–3.
    (2002). Report of the Committee for Spermatophyta: 53. *Taxon*, **51**, 795–9.

Camp, W. H., Rickett, H. W. & Weatherby, C. A. (1947). International rules of botanical nomenclature. Formulated by the International Botanical Congresses of Vienna, 1905, Brussels, 1910, Cambridge, 1930; adopted and rev. by the International Botanical Congress of Amsterdam, 1935. *Britonnia*, **6**, 1–120.

Candolle, A. L. P. P. de (1867) *Lois de la nomenclature botanique*. Paris: V. Masson et fils. English translation, Weddell, H. A. (1868), London: L. Reeve & Co.

Filgueiras, T. S., Davidse, G., Kirkbride, J. H. *et al.* (1999). Should small herbaria have voting rights? *Taxon*, **48**, 767–70.

Greuter, W. & Hawksworth, D. L. (1999). Synopsis of proposals on botanical nomenclature: St Louis 1999. A review of the proposals concerning the International Code of Botanical Nomenclature submitted to the XVII International Botanical Congress. *Taxon*, **48**, 69–128 and 775.
    (2000). *International Code of Botanical Nomenclature (St Louis Code)*. Regnum Vegetabile **138**. Vienna, Austria: IAPT, pp. vii–xviii.

Greuter, W., McNeill, J. & Barrie, F. R. (1994). *Report on Botanical Nomenclature, Yokohama 1993. XV International Botanical Congress, Tokyo: Nomenclature Section. Englera* **14**. Berlin: Berlin Botanic Garden, p. 31.

Greuter, W., McNeill, J., Hawksworth, D. L. *et al.* (2000). *Report on Botanical Nomenclature – Saint Louis 1999. XVI International Botanical Congress, Saint Louis: Nomenclature Section, 1999. Englera* **20**. Berlin: Berlin Botanic Garden.

Kuntze, O. (1891, 1893). *Revisio Generum Plantarum*. Leipzig, Germany: A. Felix.

Lanjouw, J. (1950a). *Chronica Botanica*, 12, 9–53.

(1950b). *Synopsis of Proposals concerning the International Rules of Botanical Nomenclature Submitted to the Seventh International Botanical Congress – Stockholm 1950*. Utrecht, the Netherlands: A. Oosthoek's Uitg.

(1951). The Stockholm Rules of Botanical Nomenclature [with] Nomenclature Committees appointed at Stockholm. *Taxon*, **1**, 7–11.

(1953). *Proceedings of the Seventh International Botanical Congress, Stockholm 1950*. Stockholm, Sweden: Almqvist & Wiksell and Waltham, MA: The Chronica Botanica Co.

Lanjouw, J., Baehni, C., Merrill, E. D. *et al.* (1952) *International Code of Botanical Nomenclature Adopted by the Seventh International Botanical Congress, Stockholm, July 1950. Regnum Vegetabile* **3**. Vienna: IAPT and Waltham, MA: The Chronica Botanica Co., pp. 1–228.

Lanjouw, J., Mamay, S. H., McVaugh, R. *et al.* (1996) *International Code of Botanical Nomenclature Adopted by the Tenth International Botanical Congress, Edinburgh, August 1964. Regnum Vegetabile* **46**. Utrecht, the Netherlands: IAPT, pp. 1–402.

Lapage, S. P., Sneath, P. H. A., Lessel, E. F. *et al.* (eds.) (1992). *International Code of Nomenclature of Bacteria (Bacteriological Code 1990 Revision)*. Washington, DC: American Society for Microbiology.

Linnaeus, C. (1737). *Critica botanica*. Leyden: Conradum Wishoff.

McNeill, J. & Greuter, W. (1986). Botanical nomenclature. In *Biological Nomenclature Today*, eds. W. D. L. Ride & T. Younès. IUBS Monograph Series 2: Oxford, UK: IRL Press and Miami, FL: ICSU Press, pp. 3–26.

McNeill, J. & Stuessy, T. F. (2001). Procedures and timetable for proposals to amend the International Code of Botanical Nomenclature. *Taxon*, **50**, 557–8.

McNeill, J., Redhead, S. A. & Wiersema, J. (2003). Guidelines for proposals to conserve or reject names. *Taxon*, **52**, 182–4.

Moore, R. J. Stafleu, F. A. & Voss, E. G. (1970). Final mail vote and Congress action on nomenclatural proposals. *Taxon*, **19**, 43–51.

Nicolson, D. H. (1991). A history of botanical nomenclature. *Annals of the Missouri Botanical Garden*, **78**, 33–56.

(1994). Announcement: Standing and Special Nomenclature Committees. *Taxon*, **43**, 283–5.

(1999a). Report on the IAPT General Assembly. *Taxon*, **48**, 785–92.

(1999b). Report of the General Committee: 8. *Taxon*, **48**, 373–8.

(1999c). Report of the General Committee: 9. *Taxon*, **48**, 821–2.

Ride, W. D. L., Cogger, H. G., Dupuis, C. *et al.* eds. (1999). *International Code of Zoological Nomenclature*, 4th edn. London: International Trust for Zoological Nomenclature.

Sprague, T. A. (1936). *Bulletin of Miscellaneous Information, Royal Gardens, Kew*, 185–8.

Zijlstra, G. (1999). Report of the Committee for Bryophyta: 5. *Taxon*, **48**, 563–5.

Appendix: Rules and Codes of the last 100 years

Those wishing to change the present Code these days may often find it necessary to refer back to the wording of previous editions and to the arguments that have been put forward in the past. A comprehensive history of botanical nomenclature from the times of Linnaeus up to 1989 has been published by Nicolson (1991), and the serious reader is referred to this for an enormous amount of invaluable bibliographic detail and informative discussion. At the same time, however, it seems that a checklist of the main editions of the rules of nomenclature, together with their essential related references, may be of interest and assistance to those who require merely quick access to the main events; a detailed bibliography is presented in this Appendix.

In this chronological list, the date of publication is given first together with the place of the Congress, and the date of the latter follows. It is convenient to refer to each edition by its date of publication and the seat of the Congress where it was adopted, as for example 'the 1906 Vienna Rules' or the '1978 Leningrad Code'. For each edition since 1905, four references are given in the listings below: (1) the bibliographic details of the Rules or Code itself; (2) the places where all the proposals made can be found, which are conveniently collated within the Rapporteurs' comments from 1950 onwards; (3) the transcription of the discussions which took place at the Congress; and (4) the formal reports published to record recommendations made. An attempt is made to give the more or less full bibliographic citation that modern journals may require when citing the various publications, details of which are sometimes rather obscure. For the early Congresses the proposals were widely scattered in numerous different publications, and the enquirer here is simply referred to the excellent bibliography already given by Nicolson (1991).

The change of title to 'Code' instead of 'Rules' was apparently a late decision after the 1950 Stockholm Congress by the Rapporteur, J. Lanjouw (1951), who referred to the 'Stockholm Rules' in his report but published the new edition as the 'Stockholm Code' in 1952 (Lanjouw *et al.*, 1951).

The early post-war Codes were mostly published with a dark navy-blue cover, and because they all look very similar they are often difficult to pick out on a shelf. Later the practice of giving each edition its own distinctive colour was introduced, and this is noted after the bibliographic details to assist those looking for different editions.

1906 Vienna Rules, from Congress of 1905

Briquet, J. (1906). Règles internationales de la Nomenclature botanique. Adoptées par le
    Congrès international de Botanique de Vienne 1905 et publiées au nom de la
    Commission de rédaction du Congrès/International Rules of Botanical
    Nomenclature. Adopted by the International Botanical Congress of Vienna
    1905/Internationale Regeln de Botanischen Nomenclatur. Angennomen von
    Internationalen Botanischen Kongress zu Wien 1905. In *Verhandlungen des
    Internationalen Botanischen Kongresses in Wien 1905*, eds. R. von Wettstein, J.
    Wiesner & A. Zahlbruckner. Jena, Germany: Gustav Fischer. Reprinted with
    pagination 1–99 in 1906. (Text in French, English and German).
For an extensive bibliography and commentary on proposals made to the Vienna Congress,
    see Nicolson (1991, pp. 38–9, 47–8).

1912 Brussels Rules, from the Congress of 1910

Briquet, J. (1912). *Règles internationales de la Nomenclature botanique adoptées par le Congrès international de Botanique de Vienne 1905 deuxième édition mise au point d'après les décisions du Congrès international de Botanique de Bruxelles 1910/International Rules of Botanical Nomenclature adopted by the International Botanical Congresses of Vienna 1905 and Brussels 1910/Internationale Regeln de Botanischen Nomenclatur. Angennomen von Internationalen Botanischen Kongress zu Wien 1905 und Brüssel 1910.* Jena, Germany: Gustav Fischer. (Text given in French, English and German).

Again, for comments and details of proposals made to the Brussels Congress and report following the Congress, both documents by J. Briquet, see Nicolson (1991, pp. 39, 49).

1935 Cambridge Rules, from Congress of 1930

Briquet, J. (1935). *International Rules of Botanical Nomenclature adopted by the International Botanical Congresses of Vienna 1905 and Brussels 1910 revised by the International Botanical Congress of Cambridge 1930/Règles internationales de la Nomenclature botanique adoptées par le Congrès international de Botanique de Vienne 1905, Bruxelles 1910 et Cambridge 1930/Internationale Regeln de Botanischen Nomenclatur angennomen von den Internationalen Botanischen Kongress zu Wien 1905, Brüssel 1910 und Cambridge 1930.* Jena, Germany: Gustav Fischer. (Text in French, English and German, the first edition in which the official version is the English rather than French.)

The First World War caused a major disruption to the development of the Rules. Differences of opinion between Europe and America came to a head at the Ithaca Congress of 1926, but no new edition was agreed (see Nicolson (1991), pp. 39–40 for commentary). The next formal discussions were held at Cambridge, UK, 20 years after the Brussels Congress. Again, it does not seem necessary to repeat Nicolson's enumeration of the many different sets of proposals that were considered at Cambridge. The unexpected death of John Briquet in 1931 at the age of only 62 caused a delay of five years before publication of the Cambridge Rules. This edition was particularly important for the formal introduction of the type method and the requirement of Latin diagnoses from 1935 onwards, among other things. Perhaps one of the most significant aspects of the Cambridge Rules was the move towards agreement between the European and American camps on major issues.

Further debate was held at the Amsterdam Congress of 1935, but the outbreak of the Second World War intervened and no new edition of the Code was produced. The Stockholm Congress originally planned for 1940 was inevitably cancelled, to be revived again ten years later in 1950. After the Second World War, American botanists produced an unofficial new version of the Rules based on the Amsterdam discussions which was published in the journal *Brittonia* (Camp *et al.*, 1947) and became known as the 'Brittonia Rules' (or sometimes incorrectly known as the 'Brittonia Code'). A reprint was published by the Chronica Botanica Co., Waltham, Massachusetts in 1948.

The changes that had been agreed at Amsterdam in 1935 had been recorded by the Rapporteur, T. A. Sprague (1936), who published a brief and unofficial synopsis of them. After the war, on the initiative of J. Lanjouw, a group of 13 botanists attended, by personal invitation, a conference in Utrecht on 14–19 June 1948 to discuss the results from Amsterdam and proposals that should be made to the following Congress in Stockholm in 1950. The detailed discussions of these proposals were published as 'Minutes of the Utrecht Conference' by M. L. Sprague (wife of T. A. Sprague), revised by J. Lanjouw (1950a) from minutes taken. In the same volume, a more detailed account of the changes agreed at Amsterdam in 1935 was published by T. A. Sprague as 'International Rules of Botanical Nomenclature: Supplement' in *Chronica Botanica*, **12**; 63–77 (1950). This included only the new wordings of Articles that had been changed, and was not a complete new edition of the Rules. According to an annotation on the cover of this publication at Kew, it was received there in June 1950 (Nicolson (1991) suggested August 1950 for the possible publication date), scarcely a month before the next Congress was held in Stockholm in July 1950. The present writer is indebted to John McNeill for pointing out that Sprague's 'Supplement' was given formal 'general approval' by the Stockholm Congress as recorded in the Stockholm proceedings published in 1953 (Lanjouw, 1953, p. 461). The recommendations of the Utrecht Conference were incorporated into the Synopsis of Proposals (Lanjouw, 1950b, p. viii) at the Stockholm Congress.

1952 Stockholm Code, from the Congress of 1950

Lanjouw, J., Baehni, C., Merrill, E. D., Rickett, H. W., Robyns, W., Sprague, T. A. & Stafleu, F. A. (1952). *International Code of Botanical Nomenclature Adopted by the Seventh International Botanical Congress, Stockholm, July 1950*. Regnum Vegetabile 3. Vienna: IAPT and Waltham, MA: The Chronica Botanica Co., pp. 1–228. (Text in French, German and English, bound in navy blue cover.)

Proposals with Rapporteur's comments

Lanjouw, J. (1950). *Synopsis of Proposals Concerning the International Rules of Botanical Nomenclature Submitted to the Seventh International Botanical Congress: Stockholm 1950*. Utrecht, the Netherlands: A. Oosthoek's Uitg.

Congress discussions

Lanjouw, J. (1953). *Proceedings of the Seventh International Botanical Congress, Stockholm 1950*. Stockholm, Sweden: Almqvist & Wiksell and Waltham, MA: The Chronica Botanica Co. pp. 457–550. Reprinted as *Regnum Vegetabile* 1 with same pagination, 1954.

Post-Congress report: Lanjouw, J. (1951). The Stockholm Rules of Botanical
Nomenclature [with] Nomenclature Committees appointed at Stockholm. *Taxon*, **1**,
7–11.

1956 Paris Code, from Congress of 1954

Lanjouw, J., Baehni, C., Robyns, W., Rollins, R. C., Ross, R., Rousseau, J., Schulze,
G. M., Smith, A. C., de Vilmorin, R. & Stafleu, F. A. (1956). *Code International
de la Nomenclature Botanique adopté par le Huitième Congrès International de
Botanique, Paris, Juillet 1954/International Code of Botanical Nomenclature
Adopted by the Eighth International Botanical Congress, Paris, July 1954.* (Title
page also in German and Spanish). Regnum Vegetabile 8. Utrecht: IAPT. Text in
English, French, German and Spanish, bound in red cover.

Proposals with Rapporteur's comments

Lanjouw, J. (1954). *Recueil Synoptique des Propositions concernant le Code
International de la Nomenclature Botanique soumises a la Section de Nomenclature
du Huitième Congrès International de Botanique, Paris: 1954.* Regnum Vegetabile 4.
Utrecht: IAPT. Also Additions and Corrections in *Taxon*, **3**, 156–7 (1954).
Preliminary mail vote: Anon. (1954). Preliminary mail vote: *Taxon*, **3**, 157–62.

Congress discussions

Pichon, M. & Stafleu, F. A. (1955). Discussions. *Taxon*, **4**, 123–77.

Post-Congress reports

Stafleu, F. A. (1954). VIIIth International Botanical Congress, Paris 1954, Nomenclature
Section. *Taxon*, **3**, 184–91 (decisions, committees). Nomenclature at the Paris
congress. *Taxon*, **3**, 217–225 (discussion).

1961 Montreal Code, from Congress of 1959

Lanjouw, J., Baehni, C., Robyns, W., Ross, R., Rousseau, J., Schopf, J. M., Schulze,
G. M., Smith, A. C., de Vilmorin, R. & Stafleu, F. A. (1961). *Code International
de la Nomenclature Botanique adopté par le Neuvième Congrès International de
Botanique, Montreal, Août 1959/International Code of Botanical Nomenclature
Adopted by the Ninth International Botanical Congress, Montreal, August 1954.*
(Title page also in German.) Regnum Vegetabile 23. Utrecht: IAPT. (Navy-blue
cover.)

Proposals and Rapporteur's comments

Lanjouw, J. (1959). *Synopsis of Proposals concerning the International Code of Botanical
Nomenclature submitted to the Ninth International Botanical Congress, Montreal –
1959/Recueil Synoptique des Propositions concernant le Code International de la*

*Nomenclature Botanique soumises au Neuvième Congrès International de
Botanique, Montreal – 1959. Regnum Vegetabile* **14**. Utrecht, the Netherlands: IAPT.

Congress discussions

Anon. (1960). Nomenclature Section. In *Proceedings of the IX International Botanical
Congress, Montreal, 19–29 August 1959,* vol. 3, *Plenary Sessions.* Montreal, Canada:
The Congress. Reprinted as *Regnum Vegetabile* **20**, with same pagination. Utrecht,
the Netherlands: IAPT.

Post-Congress report

Anon. (1959). IXth International Botanical Congress, Montreal 1959. Resolutions
accepted by the plenary session. *Taxon*, **8**, 245–6.

1966 Edinburgh Code, from Congress of 1964

Lanjouw, J., Mamay, S. H., McVaugh, R., Robyns, W., Ross, R., Rousseau, J., Schulze,
G. M., de Vilmorin, R. & Stafleu, F. A. (1966). *International Code of Botanical
Nomenclature Adopted by the Tenth International Botanical Congress, Edinburgh,
August 1964.* (Title page also in French and German.) Regnum Vegetabile 46.
Utrecht, the Netherlands: IAPT. (Text in English, French and German, bound in
navy-blue cover.)

Proposals and Rapporteurs' comments

Lanjouw, J. & Stafleu, F. A. (1964). *Synopsis of Proposals concerning the International
Code of Botanical Nomenclature submitted to the Tenth International Botanical
Congress, Edinburgh – 1964/Recueil Synoptique des Propositions concernant le
Code International de la Nomenclature Botanique soumises au Dixième Congrès
International de Botanique, Edimbourg – 1964.* Regnum Vegetabile 30. Utrecht, the
Netherlands: IAPT.
Preliminary mail vote: Anon. (1964). Nomenclature proposals Xth Congress, preliminary
vote. *Taxon*, **13**, 183–7.

Congress discussions

Stafleu, F. A. (1966). *Tenth International Botanical Congress, Edinburgh 1964,
Nomenclature Section.* Regnum Vegetabile 44. Utrecht, the Netherlands: IAPT.

Post-Congress reports

Stafleu, F. A. (1964). Nomenclature at Edinburgh. *Taxon*, **13**, 273–82. Anon. (1964).
Tenth International Botanical Congress, Edinburgh – August 1964. Resolutions.
Taxon, **13**, 282–92.

1972 Seattle Code, from Congress of 1969

Stafleu, F. A., Bonner, C. E. B., McVaugh, R., Meikle, R. D., Rollins, R. C., Ross, R.,
Schopf, J. M., Schulze, G. M., de Vilmorin, R. & Voss, E. G. (1972). *International
Code of Botanical Nomenclature Adopted by the Eleventh International Botanical*

Congress, Seattle, August 1969. (Title page also in French and German.) Regnum Vegetabile 82. Utrecht, the Netherlands: A. Oosthoek's Uitg. (Text in English, French and German, bound in navy-blue cover.)

Proposals and Rapporteurs' comments

Stafleu, F. A. & Voss, E. G. (1969). *Synopsis of Proposals on Botanical Nomenclature, Seattle 1969/Recueil Synoptique des Propositions concernant la Nomenclature Botanique, Seattle 1969.* Regnum Vegetabile 60. Utrecht, the Netherlands: IAPT.

Congress discussions

Stafleu, F. A. & Voss, E. G. (1972). *Report on Botanical Nomenclature, Seattle 1969.* Regnum Vegetabile 81. Utrecht, the Netherlands: A. Oosthoek's Uitg.

Post-Congress reports

Stafleu, F. A. (1970). Nomenclature at Seattle. *Taxon*, **19**, 36–42.
Moore, H. E., Stafleu, F. A. & Voss, E. G. (1970). Final mail vote and Congress action on nomenclatural proposals. *Taxon*, **19**, 43–51.

1978 Leningrad Code, from Congress of 1975

Stafleu, F. A., Demoulin, V., Hiepko, P., Linczevski, I. A., McVaugh, R., Meikle, R. D., Rollins, R. C., Ross, R., Schopf, J. M. & Voss, E. G. (1978). *International Code of Botanical Nomenclature Adopted by the Twelfth International Botanical Congress, Leningrad, July 1975.* Regnum Vegetabile 97. Utrecht, the Netherlands: Bohn, Scheltema & Holkema. (Text in English, French and German, bound in red cover).

Proposals and Rapporteurs' comments

Stafleu, F. A. & Voss, E. G. (1975). Synopsis of proposals on botanical nomenclature, Leningrad 1975. *Taxon*, **24**, 201–54.

Congress discussions

Voss, E. G. (1979). Section 1, Nomenclature. In *Proceedings XII International Botanical Congress, Leningrad, 3–10 July 1975*, eds. D. V. Lebedev. Leningrad: Nauka, pp. 129–86.

Post-Congress report

E. G. Voss. (1976). XII International Botanical Congress: mail vote and final Congress action on nomenclatural proposals. *Taxon*, **25**, 169–74.

1983 Sydney Code, from Congress of 1981

Voss, E. G., Burdet, H. M., Chaloner, W. G., Demoulin, V., Hiepko, P., McNeill, J., Meikle, R. D., Nicolson, D. H., Rollins, R. C., Ross, R., Silva, P. C. & Greuter, W. (1983). *International Code of Botanical Nomenclature Adopted by the Thirteenth International Botanical Congress, Sydney, August 1981.* Regnum Vegetabile 111.

Utrecht, the Netherlands: Bohn, Scheltema & Holkema and the Hague, W. Junk (Text in English, French and German, bound in green cover).

Proposals and Rapporteurs' comments

Voss, E. G. & Greuter, W. (1981). Synopsis of proposals on botanical nomenclature, Sydney 1981. *Taxon*, **30**, 95–293.

Congress discussions

Greuter, W. & Voss, E. G. (1982). Report on botanical nomenclature – Sydney 1981. *Englera*, **2**, 1–124.

Post-Congress reports

Greuter, W. (1981). XIII International Botanical Congress: mail vote and final Congress action on nomenclatural proposals. *Taxon*, **30**, 904–11. Voss, E. G. (1982). Nomenclature at Sydney. *Taxon*, **31**, 151–4.

1988 Berlin Code, from Congress of 1987

Greuter, W., Burdet, H. M., Chaloner, W. G., Demoulin, V., Grolle, R., Hawksworth, D. L., Nicolson, D. H., Silva, P. C., Stafleu, F. A., Voss, E. G. & McNeill, J. *International Code of Botanical Nomenclature Adopted by the Fourteenth International Botanical Congress, Berlin, July–August 1987.* Regnum Vegetabile 118. Königstein, Germany: Koelz Scientific Books (Text in English only for first time, bound in yellow cover).

Proposals and Rapporteurs' comments

Greuter, W. & McNeill, J. (1987). Synopsis of proposals on botanical nomenclature, Berlin (1987). *Taxon*, **36**, 174–281.

Congress discussions

Greuter, W., McNeill, J. & Nicolson, D. H. (1989). Report on botanical nomenclature – Berlin 1987. *Englera*, **9**, 1–228.

Post-Congress report

McNeill, J. (1987). XIV International Botanical Congress: mail vote and final Congress action on nomenclatural proposals. *Taxon*, **36**, 858–68.

1994 Tokyo Code, from Congress of 1993

Greuter, W., Barrie, F. R., Burdet, H. M., Chaloner, W. G., Demoulin, V., Hawksworth, D. L., Jrgensen, P. M., Nicolson, D. H., Silva, P. C., Trehane, P. & McNeill, J. *International Code of Botanical Nomenclature (Tokyo Code) Adopted by the Fifteenth International Botanical Congress, Yokohama, August–September 1993.*

Regnum Vegetabile 131. Königstein, Germany: Koelz Scientific Books. (Text in English only, bound in purple cover).

Proposals and Rapporteurs' comments

Greuter, W. & McNeill, J. (1993). Synopsis of proposals on botanical nomenclature, Tokyo 1993. A review of the proposals concerning the International Code of Botanical Nomenclature submitted to the XV International Botanical congress. *Taxon*, **42**, 193–271.

Congress discussions

Greuter, W., McNeill, J. & Barrie, F. R. (1994). Report on botanical nomenclature – Yokohama 1993. *Englera*, **14**, 1–265.

Post-Congress reports

McNeill, J. (1993). XV International Botanical Congress. Preliminary mail vote and report of Congress action on nomenclatural proposals. *Taxon*, **42**, 907–22. Nicolson, D. H. (1993). Report of the Nominating Committee for Rapporteur-Général and permanent nomenclature committees. *Taxon*, **42**, 923–4.

2000 St Louis Code, from Congress of 1999

Greuter, W., McNeill, J., Barrie, F. R., Burdet, H. M., Demoulin, V., Filgueiras, T. S., Nicolson, D. H., Silva, P. C., Skog, J. E., Trehane, P., Turland, N. J. & Hawksworth, D. L. *International Code of Botanical Nomenclature (Saint Louis Code) Adopted by the Sixteenth International Botanical Congress, St Louis, Missouri, July–August 1999*. Regnum Vegetabile 138. Königstein, Germany: Koelz Scientific Books. (Text in English only, bound in black cover).

Proposals and Rapporteurs' comments

Greuter, W. & Hawksworth, D. L. (1999). Synopsis of proposals on botanical nomenclature, St Louis 1999. A review of the proposals concerning the International Code of Botanical Nomenclature submitted to the XVI International Botanical congress. *Taxon*, **48**, 69–128.

Congress discussions

Greuter, W., McNeill, J., Hawksworth, D. L. & Barrie, F. R. (2000). Report on botanical nomenclature – Saint Louis 1999. *Englera*, **20**, 1–253.

Post-Congress report

Barrie, F. R. & Greuter, W. (1999). XVI International Botanical Congress: preliminary mail vote and report of Congress action on nomenclatural proposals. *Taxon*, **48**, 771–84.

7

Bringing taxonomy to the users

Ghillean T. Prance

Introduction

We recommend that the systematic biology community, especially via the Systematics Association and the Linnean Society, should increase efforts to demonstrate the relevance and importance of systematic biology.

Agriculture, forestry and horticulture all depend in some way on products of systematic biology.

House of Lords, 2002

These two quotes from the recent publication from the UK House of Lords Select Committee on Science and Technology, both illustrate the problem and make an appeal to the systematic biology community to make their science more relevant to the user. Systematists are good at producing scholarly monographs and Floras and Faunas, and these are indeed the most important part of our work. There is a shortage of monographs of tropical plant and animal groups, especially in areas such as Amazonia, where I have gained most of my experience as a systematist. This is in no way a plea to divert efforts from the scholarly work that is the basis of taxonomy, but rather to go further and make secondary publications and to place material on the internet that takes into account the need of the user. For birds, admittedly a comparatively small and well-studied group, there are new clearly illustrated field guides for most parts of the world. As a bird watcher, wherever I travel, I am usually able to find a book with which to identify the local birds. The same is certainly not true for plants, yet it is plants that agriculturists, foresters, horticulturists, researchers on medicinal uses of plants and many other people need to be able to identify to carry out their work. Plants, like birds, have a large amateur constituency such as travellers and ecotourists and we do not cater enough to their needs. Because of the large number of interested people, Floras and field guides are an economic possibility and need to be increased. There are more than half a dozen commercial CD-ROM field guides for birds (see www.cs.umb.edu/efg/dbi). We need the same for plants because these guides are good learning tools and can include databases for monitoring biodiversity. This chapter is a gentle encouragement to all of us to make a greater effort to make our science more readily available because it is basic to so many other areas of science and of practical applied work.

Professor Vernon Heywood's contribution

This book is published to honour Professor Vernon Heywood, a person who has devoted his life to taxonomy and is co-author of one of the most-used basic texts on plant taxonomy (Davis & Heywood, 1973). However, he has set an example of taking his science far beyond textbooks and *Flora Europaea* (Tutin *et al.*, 1964–80). His book on the families of flowering plants (Heywood, 1978), which has had several reprints and is presently being completely revised, is a fine example of making taxonomy accessible. I have seen this book on the shelves of many users of taxonomy as I have travelled around the developing world. With this book it is easy to find out the basic information and distribution of plant families. Professor Heywood has also used his data extremely well for the promotion of plant conservation, as can be seen in other contributions to this volume. The topic of this chapter, 'Bringing taxonomy to the users', is most appropriate in a volume dedicated to Heywood because he has set an example in this very area. *Flora Europaea* (Tutin *et al.*, 1964–80) was in fact a commercial success with over 5000 sets sold. It remains the basic text for European law on plant conservation largely through the good efforts of Professor Heywood.

Academic recognition of popularising science

One of the impediments to the greater production of more popularly accessible publications, is that they are often not given the same recognition as more scholarly works, either by universities to their students or museums to their employees. One of my Ph.D. students in economic botany did a fine academic study on the use of four species of wild fruit trees in eastern Amazonia. She also produced a popular, well-illustrated guide to the local fruits with details of their identification, their uses, the way to prepare them and their sustainable conservation (Shanley *et al.*, 1997). I, as a co-supervisor of this work with considerable knowledge of the region where the work was done, had a hard time persuading the university to allow this work while she was working on a Ph.D. dissertation. The end result was both a thesis that satisfied the rigours of the examiners and a book in Portuguese that brought the work to the villagers who had taken so much time to help her study.

One area that needs to be worked on is getting academic recognition for this kind of work, rather than it being frowned upon. Field guides and popular publications aimed at local peoples are one of the most important end products of taxonomy, since they put the data to good practical use. They also justify the work to a public that has frequently paid for the research and who find the 'ivory-tower' approach difficult to comprehend.

Florulas and field guides

It is taking a long time to complete the major tropical Floras such as *Flora Malesiana* (see Roos, 1993, 1997) and *Flora Neotropica* (see Prance, 1984; Mori 1992). In Africa, the *Flora of Tropical East Africa* is nearing completion, but it has taken 50 years to write. The users of tropical plants cannot wait for these important foundation works to be completed, and so it is encouraging that the number of florulas of small areas is increasing. Some good examples include: Barro Colorado, Panama (Croat, 1979); Río Palenque, Equador (Dodson & Gentry, 1978); the Flora of the biological reserves around Iquitos, Peru (Martínez, 1997); and the Flora of Pico das Almas in Brazil (Stannard, 1995).

These florulas are usually well illustrated and because they cover a relatively small area, the keys are usually much easier to use than those of a monograph. For example there are 218 species of *Licania* in our recent monograph of Chrysobalanaceae (Prance & Sothers, 2003) which involves a lengthy series of keys, but only 30 species in the field guide to the Ducke Reserve (Ribeiro *et al.*, 1999) which makes identifications around Manaus easier. An exceptionally good local Flora is the *Guide to the Plants of Central French Guiana* (Mori *et al.*, 1997, 2002). This abundantly illustrated guide is easy to use and has much introductory material to aid the user.

Guide to the Ducke Forest Reserve, Brazil

A good example of making systematics more available is the field guide to the Reserva Florestal Adolpho Ducke, near to Manaus, Brazil (Ribeiro *et al.*, 1999). Initially, a conventional florula of the reserve was planned; however, as it developed and as imaging technology also improved, the project leader, Michael Hopkins, had the vision to change it from a florula to a profusely illustrated field guide (Figure 7.1). This guide to the 2200 species in the 100-km^2 forest reserve is useful for identifying plants throughout the central Amazon region. Each species is illustrated by photographs of the leaves, bark and other vegetative features that aid identification. The technology used was standard 35-mm slides that were later scanned into the computer system. It is possible to identify most species with the guide, even without the flowers and fruit. This was made possible by a large grant from the UK Government Department for International Development (DFID) that enabled a large

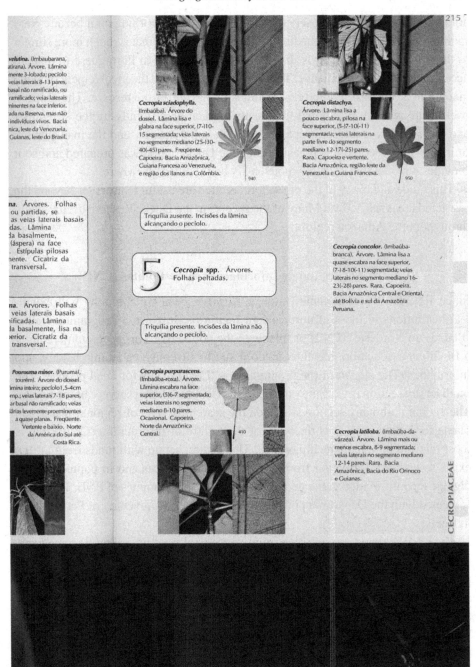

Figure 7.1 Illustration of the *Field Guide to the Reserva Florestal Adolpho Ducke* (Ribeiro *et al.*, 1999).

amount of field work and observations so that the team working on it became very familiar with the species in the field. The result is a guide that is much more similar to those in existence for birds. It was published initially in Portuguese to make it accessible to the local audience. The project grant allowed the involvement of many taxonomic specialists in the plant families that occur in the reserve. The guide is backed up by more traditional treatments of each plant family by the specialists and these will be published separately. Digital photography and the management of images has become much easier since the Ducke guide was prepared and so it provides a model that can now more easily be followed elsewhere.

This guide and many of the other florulas cited are useful to users far beyond the restricted areas they cover. Many of the species in the Ducke Reserve occur widely throughout the Amazon region.

Centro Nordestino de Informações sobre Plantas (CNIP)

Since 1992 the Royal Botanic Gardens, Kew has been collaborating with botanical institutions in the arid northeastern region of Brazil in a programme called 'Plantas do Nordeste' (PNE) or 'plants of the northeast'. This programme, run by a Brazilian association, has three main parts: biodiversity, economic botany, and informatics. The slogan of the programme is 'Local plants for local people'. The biodiversity section is working on systematics and vegetation surveys of the region. The economic-botany section works on the search for uses of regional plants and plants for honey bees. It is the work of the Centre of Information (CNIP) that is most relevant here (cnip@cnip.org.br). This centre, set up with funds from DFID, exists to gather information from a wide range of sources and to popularise and divulge the information generated by the other two programmes of PNE. The CNIP is located with the Department of Botany of the Federal University of Pernambuco in Recife.

The CNIP's stated mission is to improve the quality of life of low-income communities of the rural northeast through better use and conservation of the regional plant resources. The Centre has gathered together much bibliographic information and its database is a wonderful resource on the use, management and conservation of regional plants. In addition to providing much information to users from its library and database, the Centre is producing many other products to make its information widely available. This includes publication of leaflets, games for children, videos, radio programmes, street theatre and, of course, access to much information through the Internet. The Centre has produced 52 leaflets about medicinal plants authored by Professor Franciso J. de Abreu Matos (a regional expert on this topic), an instruction manual on making a medicinal plant garden, a memory game with attractive cards depicting 48 of the commonest trees of the region and a card game.

All this material has originated from the work of systematists, ecologists and eco-nomic botanists. Information is now reaching many people throughout northeast Brazil about how to manage their plant resources better. CNIP has formed a valuable bridge between the academics and the local community.

The Brazilian information centre is just one of many projects that are bring-ing taxonomy closer to the user through the use of electronic data and the Internet. The availability of *Index Kewensis* combined with two other plant-name databases (The Harvard University Herbaria and The Australian National Herbarium) through the International Plant Names Index (IPNI, 2005), has been greatly welcomed by researchers in the developing world, who did not have the resources to accumulate large libraries and who do not have access to the type specimens of taxa. The imaging of type specimens, such as is being done by the Netherlands National Herbarium (www.nationalherbarium.nl) and the New York Botanical Garden (www.nybg.org/bsci/hcol), are invaluable tools for researchers. The study developing the *Species Plantarum* project (www.anbg.gov.au/abrs/flora/splant/spplant.htm) intends to place all the monographs on the Internet. An excellent example of the use of the Internet is the elegant website on Neotropical Ericaceae of Dr Jim Luteyn (www.nybg.org/bsci/res/lut2). There are also a huge amount of data available on the Missouri Botanical Garden, VAST (VAScular Tropicos) (2004) website, the CONABIO (Comisión nacional para conocimiento y uso de la bio-diversidad or National Commission for the Knowledge and Use of Biodiversity) (2004) database in Mexico and the INBIO (Instituto Nacional de Biodiversidad or National Biodiversity Institute) (2004) website in Costa Rica. An internet key to the tree and shrub genera of Borneo (www.phylodiversity.net/borneo/delta) is another good example of the use of electronic data. Making our data available through the Internet is one of the best ways we can benefit the users and other researchers of taxonomy.

Plant blindness

The newsletter of the Botanical Society of America, *Plant Science Bulletin*, has recently published two articles about 'plant blindness' (Wandersee & Schussler, 2001; Hershey, 2002). These articles were aimed primarily at those who teach botany, but they have good advice for all systematists about making their subject of interest and of greater relevance to the user community. These articles are all about stimulating interest in plants, which is surely one of the roles of the taxonomist, and it is often systematists that teach botany at many different levels. It is not just through more popular publications, but also in the way we teach that we bring tax-onomy to the user community. Plants are the vital basis of all life, yet conservation organisations have tended to emphasise large furry animals. We need teachers and

Figure 7.2 Visitors to the Eden Project – a 'show-case of botany' (photo: Ghillean Prance, The Eden Project, UK).

publications that encourage interest in plants in order to promote interest in plant conservation as well.

Plant show-cases

Taxonomists have been involved in two plant exhibits that have brought many aspects of plants to the public. In the United States in 2001 the Bonfante Gardens in California (www.bonfantegardens.com) opened a plant-themed amusement park. This garden includes: a large greenhouse; educational exhibits; and plant-themed rides such as Garlic Twirl, Banana Split, Strawberry Sundae and Artichoke Dip! Also in 2001, in the UK, the Eden Project (www.edenproject.com) opened and attracted 2,000,000 visitors in its first year of operation, to see an exhibition that is all about the importance of plants to people and the sustainable use of plants (see Figure 7.2). It has been a huge success and has done more to stimulate interest in plants, plant taxonomy and conservation than almost any other institution or initiative. The Eden Project has not promoted itself as a botanic garden, but rather as a 'showcase of botany' and has succeeded. Its innovative education programmes for schools are certainly stimulating interest in plants. Behind the scenes, I and other taxonomists are constantly checking the presentations for accuracy and for excitement of the way in which the message is presented. These types of projects are surely going to stimulate future systematists to study this fascinating discipline and to project it to the public and to the user in a better way than past generations of systematists have done.

Conclusion

Part of the evaluation of taxonomic research should be based on the ways in which it is made available to the user. This would do a lot to increase the relevance of the discipline, to increase public perception and understanding of systematics and to open up new sources of funding. This chapter suggests several different ways in which taxonomy can be brought to a much wider audience.

References

CONABIO (Comisión nacional para conocimiento y uso de la biodiversidad) (2004). www.conabio.gob.mx.

Croat, T. B. (1979). *Flora of Barro Colorado Island*. Stanford, CA: University of California Press.

Davis, P. H. & Heywood, V. H. (1973). *Principles of Angiosperm Taxonomy*, 2nd edn. Huntington, NY: Robert E. Krieger.

Dodson, C. H. & Gentry, A. H. (1978). Flora of the Río Palenque Science Center, Los Ríos Province, Ecuador. *Selbyana*, **4**, 1–628.

Hershey, D. R. (2002). Plant blindness: 'we have met the enemy and he is us'. *Plant Science Bulletin*, **48**(3), 78–85.

Heywood, V. H. (ed.). (1978). *Flowering Plants of the World*. Oxford, UK: Oxford University Press.

House of Lords (2002). *What on Earth? The Threat to the Science Underpinning Conservation*. House of Lords, Select Committee on Science and Technology. London: HMSO.

INBIO (Instituto Nacional de Biodiversidad) (2004). www.inbio.ac.cr.

IPNI (International Plant Names Index) (2005). http://www.ipni.org/ik_blurb.html (accessed 2005).

Martínez, R. V. (1997). *Flórula de las Reservas Biológicas de Iquitos, Peru. Allpahuago–Mishana Explornapo Camp, Explorana Lodge. Monographs in Systematic Botany*, **63**, 1–1046. Missouri: Missouri Botanical Garden Press, pp. 1–1046.

Missouri Botanical Garden (2004). VAST (VAScular Tropicos). www.mobot.org/W3T/Search/vast.html.

Mori, S. A. (1992). Neotropical floristics and inventory: who will do the work? *Brittonia*, **44**, 372–5.

Mori, S. A., Cremers, G., Gracie, C. *et al.* (1997). *Guide to the Vascular Plants of Central French Guiana. Part 1. Pteridophytes, Gymnosperms and Monocotyledons. Memoirs of the New York Botanical Garden*, **76**, 1–422.

(2002). *Guide to the Vascular Plants of Central French Guiana. Part 2. Dicotyledons. Memoirs of the New York Botanical Garden,* **76** (2), 1–776.

Prance, G. T. (1984). Completing the inventory. In *Current Concepts in Plant Taxonomy,* eds. V. H. Heywood & D. M. Moore. London and Orlando, FL: Academic Press, pp. 365–96.

Prance, G. T. & Sothers, C. A. (2003). *Chrysobalanaceae 1 & 2. Species Plantarum: Flora of the World,* vols. 9 and 10. Canberra, Australia: Australian Biological Resources Study for the International Organization for Plant Information.

Ribeiro, J. E. L. da S., Hopkins, M. G., Vincentini, A. *et al.* (1999). *Flora da Reserva Ducke: Guia de identificação das plantas Vasculares de uma floresta de terra-firme na Amazônia Central. Manaus, Brazil: INPA.*

Roos, M. C. (1993). State of affairs regarding Flora Malesiana: progress in revision work and publication schedule. *Flora Malesiana Bulletin,* **11**, 133–42.

(1997). Flora Malesiana: progress, needs and prospects. In *Plant Diversity in Malesia III. Proceedings of the Third Flora Malesiana Symposium 1995,* eds. J. Dransfield, M. J. E. Coode & D. A. Simpson. Kew, UK: The Royal Botanic Gardens, pp. 231–46.

Shanley, P., Cymerys, M. & Gacuão, J. (1997). *Frutíferas da Mata na Vida Amazônica.* Belém, Brazil: Editora Supercores.

Stannard, B. L. (ed.) (1995). *Flora of the Pico das Almas, Chapada Diamantina: Bahia, Brazil.* Kew, UK: The Royal Botanic Gardens.

Tutin, T. G., Heywood, V. H., Burges, N. A. *et al.* (1964–80). *Flora Europaea,* 5 vols. Cambridge, UK: Cambridge University Press.

Wandersee, J. H. & Schussler, E. E. (2001). Toward a theory of plant blindness. *Plant Science Bulletin,* **47**, 2–9.

Part III

Establishing priorities: the role of taxonomy

8

Measuring diversity

Chris J. Humphries

Introduction

In a world where the human population is now so vast, it is no exaggeration to say that every aspect of biodiversity has to be managed – from the most intensive farmland to the wildest wildernesses and the deepest of oceans. It has become clear since the biodiversity crisis became manifest during the mid 1980s that measures of biodiversity are needed to determine: (i) what it is; (ii) where we might find it and, (iii) the combinations of available areas that can represent and help sustain the greatest biodiversity value for particular future goals. This raises many questions, including the many varied human concerns such as what are the best estimates of biodiversity value (genes), what popular estimates of biodiversity value (species) exist, which practical estimates of biodiversity value (higher taxa) actually work; in addition, a scheme is required to show the relationship among estimates (surrogacy scale) so that there is some reality in the prescriptions for conserving biodiversity. It is clear that for any scheme to work, it has to encompass the many and varied notions of taxonomy and ecology, dovetail them into one system and make an assessment of the usual surrogates of biodiversity.[1]

Biodiversity value means identifying fundamental currencies so that people can see the total (and irreducible) complexity of all life, including not only the great variety of organisms but also their varying behaviour and interactions. There can be no single objective measure of biodiversity, only measures relating to particular applications. For conservationists, therefore, a measure of biodiversity should quantify both its value to the people whom they serve and that is considered in

---

[1] Diversity is a property of sets of organisms, usually defined by the volume of space in which they occur, although in many studies the third dimension can be ignored, so that sets can be defined by an area of land or sea such as grid cells (or polygons). For purposes other than choosing among land-management units for conservation, equal-area (Williams *et al.*, 1997) or nearly equal-area grid cells are often used because they reduce the species-area effects on diversity and rarity measures. Above all, the studies described here try to report unbiased occurrences of organisms, or at the very least, to report the best-available knowledge of their distributions (Williams *et al.*, 1996a).

need of protection. For example, it might be crop plants in one instance, native trees in another. One of the more broadly shared and economically defensible reasons for conserving wholesale biodiversity (rather than just the few components or 'biospecifics' with obvious high use value at present) may be seen to lie in ensuring continued possibilities both for changes in landscape, and for future use by people in a changing and uncertain world. Consequently, biologists have argued that this value in biodiversity is likely to be associated with the variety of different genes that can be expressed by organisms as potentially useful phenotypic characters (different chemicals, morphological features, functional behaviour). Because we will never really know precisely which genes or characters will be of value in the future, in the first instance they must all be treated as having equal value; secondly, the greatest value for conservation will come from ensuring the persistence of as many different genes or characters as possible, as a form of insurance (Williams *et al.*, 1994a; Humphries *et al.*, 1995).

A dandelion and a giant redwood can be seen to represent a richer collection of characters in total – and so greater diversity value – than just a pair of more similar species, such as two species of dandelion (Vane-Wright *et al.*, 1991). This shows how the phenotypic characters (or the genes that code for them) could provide a 'currency' of value for biodiversity. Pursuing this idea, we will then need to maximise richness in the character currency within managed or protected areas. This arguably provides a unified view of the traditional three levels (genetic, specific and ecosystem) at which biodiversity has been described. In practice, it uses genetic diversity as a basis for valuing both species diversity (for their relative richness in different genes) and ecosystem diversity (for the relative richness in the different processes to which the genes ultimately contribute). This should provide justification for multi-level approaches and surrogate methods.

A particular strength of the single-currency approach is that it avoids the problems that arise when using compound measures or indices trying to trade off measures for different properties that really cannot be compared or inter-converted (species richness and relative abundance), which can lead to confusion and a loss of accountability. The advantage of accountability when using one currency becomes particularly important when faced with the problem of choosing areas for biodiversity conservation, when many other factors may be involved in decision-tree analyses. The difficulty with a single-currency approach is that numbers of genes or characters cannot be counted directly except for the most trivial questions, so the problem is how best to estimate them? The research at the Natural History Museum, London has considered that the best estimates come from using genealogy to predict genetic or character richness. Popular estimates use species richness and perhaps the most practical estimates use higher taxa or environmental surrogates. These will be

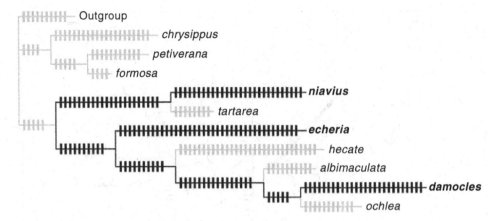

Figure 8.1 The phylogenetic classification of Milkweed butterflies. Vertical bars are the characters. Given a sample of any three taxa, then those shown in bold (*niavius*, *escheria* and *damocles*) are the most different taxa.

covered under three headings: (1) Principles, (2) Phylogenetic and taxonomic measures and (3) Lack of phylogenetic information.

Principles

Biologists needing to estimate richness in terms of different genes or characters are usually unable to measure this directly, or at best are only able to study small samples of genes or characters. However, because genes and characters are inherited, biologists have been able to respond by proposing phylogenetic or taxonomic measures of diversity (Williams & Humphries, 1996). These measures predict the biodiversity value of different biotas, using knowledge of the genealogical (or hierarchical) relationships among organisms in combination with models of gene or character evolution (Williams *et al.*, 1994a; Humphries *et al.*, 1995). For example, in Milkweed butterflies (Figure 8.1), the branching pattern of the classification is derived by analysis of 217 morphological and chemical characters. The branch lengths are scaled by the number of supposed character changes found within the sample. Considering all combinations of three species, the most diverse set of three species is *niavius*, *echeria* and *damocles*, because these three have the longest total branch lengths with the largest numbers of character differences between them. This approach can affect the relative values of different faunas or floras, for example, species richness (as seen in bumble bees (Figure 8.2)).

In bumble bees, character-difference data are not available except for a genealogical classification. However, a simple evolutionary model can be used to estimate the

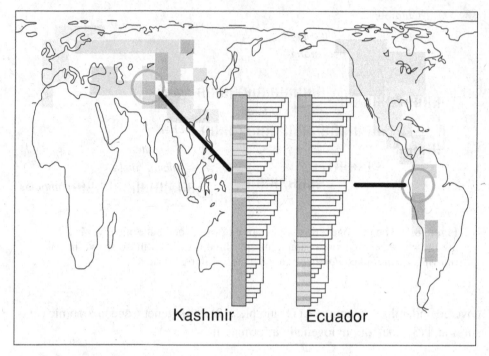

Figure 8.2   Bumble bee diversity. Ecuador is the most species rich, with ten species, but the fauna shown in grey on the cladogram are all closely related terminal species. For character richness, Kashmir has only nine species, but at least four of the major clades from the modelled genealogy are represented.

accumulated characters for these bees, given a specific cladogram. When combined with a measure of tree length (to count the relative expected number of character differences within each fauna), the example shows an improved prediction of relative diversity value for the same bee faunas (to take into account the expected numbers of gene or character differences) but completely changes the most rich area from Ecuador to Kashmir in the Himalayas (Figure 8.2).

Phylogenetic and taxonomic measures

In the absence of complete knowledge of the valued genetic or character differences among organisms, the most direct approach to estimating relative value is to use a phylogenetic measure. This is used to scale the branch lengths on the tree for use by the measure in the third step. It can use sample data for gene or character variation, or it can be based on just a taxonomic classification, if that is all that is available and if it is believed to represent genealogical relationships. These measures require three elements: cladograms, models of character change and some kind of clock (Williams *et al.*, 1994a).

The phylogenetic approach is based on the general evolutionary model of descent with modification: that genes and characters are inherited, with few alterations or changes. Some of these changes are reversals (homoplasy), so that data often conflict with any one tree (although such conflicts are minimised in the construction of cladograms). If the resulting trees are good estimates of genealogical relation-ships, then they should be more reliably predictive of the unsampled genetic or character variation, which always remains the greatest majority. An explicit choice has to be made as to the particular type of evolutionary model used for linking the branching pattern of the tree with the way genes or characters change along the tree (Williams *et al.*, 1994a). Three extreme options for this model cover the range of possibilities: the anagenetic clock, the sample (empirical) model and the saltatory (cladogenetic) model. The anagenetic clock assumes that changes occur at random and are subject to little constraint by selection. Consequently, in effect, changes accumulate more or less in proportion to the time elapsed along the branches. If sample data are unavailable or are expected to be biased and unrepresentative, all lineages are scaled to a common length. The sample (empirical) model assumes that the distribution of changes in a small sample of variation is representative of the great majority of unsampled variation. The pattern of changes is usually expected to be intermediate between those from the other two models. If sample data are available and are expected to be representative, all branches are scaled in proportion to sample changes. Finally, the saltatory (cladogenetic) model assumes that most changes are associated with speciation or divergence events, for example if strong selection constraints are relaxed at these times. Although the numbers of changes associated with each branching event may differ, in effect changes accumulate more or less in proportion to the number of branching events (including those to extinct branches). If sample data are unavailable or are expected to be biased and unrepresentative, all branches are scaled to unit length.

A measure of relative richness in different genes or characters is needed, which sums the relative degree of change along the tree using the branch lengths scaled by the chosen special evolutionary model. The form of this measure, is the length of the subtree that spans any given set of species of interest on the tree (Figure 8.3). The distance, added to species A + B by C can be calculated simply as $\Delta = (\delta AC - \delta CB - \delta AB)/2$.

For the Milkweed butterfly example (see Figure 8.1), branch lengths can be scaled on the horizontal axis using the three evolutionary models described above (Figure 8.4). For each model, the set of three species that is estimated to have the greatest genetic or character richness is sought. For this tree, the unique result

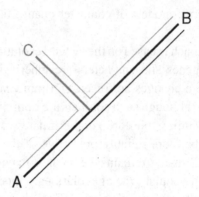

Figure 8.3 Measuring branch lengths of cladograms. Diversity increase when adding species $C = (\delta AC + \delta CB - \delta AB)/2$.

is with the sample-model solution. With the other models, there are at least two equivalent choices for each species. The choice of special evolutionary model gives most disagreement (Williams *et al.*, 1994a). The sample model is the easiest to use, and its consequences (as seen in this example) show that it may give less equivocal answers concerning which species represent the greatest relative diversity. However, the apparent advantages of the sample model occur only when the sample truly represents the overall pattern of variation. If not, a severe bias is introduced into the measure and the results. Consequently, any apparent increase in resolution arising from using information from the sample in this way could actually be misleading.

Lack of phylogenetic information

Detailed and reliable phylogenetic (genealogical and difference) information is rarely available. Nevertheless, arguments for measuring biodiversity value as gene or character richness provide a philosophically and economically defensible starting point as to what is valued in diversity. Accepting that cladistic or taxonomic diversity measures can use the genealogical pattern as a predictor of value also provides a possible key to the problem of finding more practical measures (Williams, 1996). Popular estimates of biodiversity value use either species richness or 'indicator' groups. Fortunately, when dealing with large numbers of species, species richness tends to become a reasonable surrogate for gene or character richness (Williams, 1996). Therefore, identifying genes or characters as the currency of biodiversity value provides an additional justification for using species richness. This is likely to work because (unlike the bumble bee example) as more species are added to the total, at least some representatives of the more divergent, higher groups of organisms co-occur.

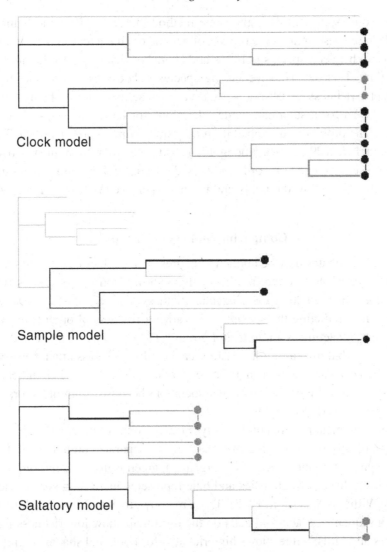

Figure 8.4 Different evolutionary models (clock, sample and saltatory) as applied to the Milkweed butterfly cladogram (see Figure 8.1).

This relationship may be illustrated by joint work with the Royal Botanic Gardens, Kew which examined species richness for five plant families (Dichapetalaceae, Lecythidaceae, Caryocaraceae, Chrysobalanaceae and the genus *Panopsis* of the Proteaceae) from the long-running *Flora Neotropica* project among one-degree grid cells (one degree latitude by one degree longitude) (Williams *et al.*, 1996b). The five families included 729 species, and the best available estimate of character richness had to be derived from the extant taxonomic (non-phylogenetic) classification. Although this is likely to underestimate the degree of character divergence between some species, this is the same method that is used for the bumble bee example in the description of estimates of genetic or character richness and yet the relationship between species richness and character richness for this much larger group is much closer. However, using species richness as a surrogate for gene or character richness is still not a panacea, because there are probably too many species – even for a suburban garden, let alone for large surveys such as South America. The problems are apparent in the same *Flora Neotropica* data. The five families include 729 species among 1751 grid cells, thereby requiring over one million presence/absence records to be established. Such an enormous sampling effort cannot be deployed quickly and is very expensive (Williams *et al.*, 1996b).

Comparing 'indicator' groups

The use of surrogates may be applied to address two problems. First, in the *Flora Neotropica* example for five families of trees, how good is the species richness of tree floras at predicting the character richness of tree floras? Secondly, how good is it for predicting the species and character richness of other groups, or of entire biotas, by acting as an 'indicator group'? They can be predictive under some circumstances but indicator relationships cannot always be assumed, because they can also be weak, absent or even negative, perhaps particularly when indicator and target organisms differ in their habitat associations because different factors govern their distributions (Figure 8.5).

The geographical patterns of diversity for two groups of organisms can be compared graphically by overlaying two maps in two separate colours where black is used to represent no incidence of the organism, green is used to represent increasing species richness of butterflies and blue represents increasing species richness of birds (Williams & Gaston, 1998). These maps are then overlaid in a third digital map. Consequently, black grid cells on the third map show low richness for both butterflies and birds; white shows high richness for both; and shades of grey show intermediate and covarying richness for both. In contrast, areas of the third map with highly saturated green cells show an excess of richness for butterflies over birds, and areas with highly saturated blue show an excess of birds over butterflies (Spearman correlation coefficient, $\rho = 0.25$). The colour classes are arranged to give even frequency distributions of richness scores along both axes (at least within the constraints imposed by tied richness scores).

This technique allows relationships between groups to be compared visually at a broad range of spatial scales. Within Britain, for example, in the first instance any gross differences in the strength of the overall national relationship can be judged from the overall colour saturation of the map; secondly, any regional deviations from

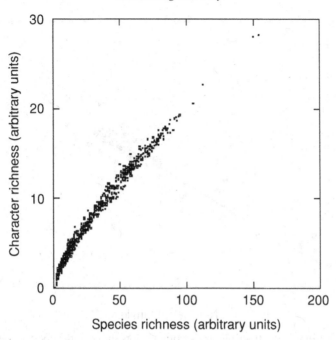

Figure 8.5 Species richness as a surrogate for character richness. A graph showing the good relationship between character richness and species richness for the five families of Neotropical trees.

this national relationship can be seen in regional colour trends; and thirdly, local deviations can be seen as isolated spots of differing colour (Williams & Gaston, 1998).

Practical estimates of biodiversity value using higher taxa as surrogates

Despite the great sampling effort for the five plant families in the *Flora Neotropica* project, these plants still represent less than 1% of the plant species and less than 2% of the plant families recorded within the study region. Consequently some less direct method is needed to survey the broader variety of plants. Higher taxon richness (using genera or families) has been suggested to be useful as a surrogate for species richness, and ultimately as a more remote surrogate than species for gene or character richness (Williams & Gaston, 1994). The particular attraction of this approach is that it may prove suitable for use in more-cost-effective practical surveys, if it allows the taxonomic coverage to be broadened without increasing costs. In comparison with the method of using small indicator groups of species, it should have an advantage of precision for predictions if it permits a broader coverage

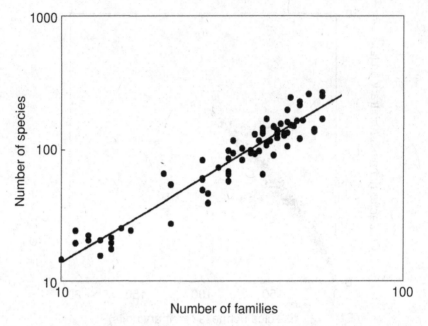

Figure 8.6 A graph of Alwyn Gentry's tropical plots showing the good relationship between species richness and family richness.

of the groups of organisms surveyed, thereby avoiding an enormous extrapolation (mapping 1000 families usually represents a much larger slice of biodiversity than mapping 1000 species). The choice of taxonomic rank to survey must be made with care and there are potential pitfalls (Williams & Gaston, 1994). Nonetheless, the approach shows promise.

Several studies now support the idea of a relationship between the numbers of higher taxa – such as families – and the numbers of species in a given area. Evidence at large spatial scales is very difficult to collect. The largest comparable data set available for plants comes from the work of the late Alwyn Gentry, and this does support a predictive relationship for patches among continents (log-transformed on both axes, $r^2 = 0.91$) (Williams et al., 1994b) (Figure 8.6). In the absence of direct counts of plant species richness, if this kind of relationship can be assumed, then the example can use counts of the numbers of families to represent relative variation in the numbers of species expected among equal-area grid cells.

Comparing taxic and surrogate maps

Other surrogate values include using environmental factors that are likely to govern the distribution of biodiversity, such as the net primary productivity (the amount

Table 8.1. *Scales of taxonomic and environmental surrogates ranging from high to low for both ecosystem richness and genetic richness*

Ecosystem richness high			Genetic richness low
↓	Environmental surrogates	Climate class richness	↑
↓		Terrain class richness	↑
↓		Substrate class richness	↑
↓	Environmental/ assemblage surrogates	Landscape class richness	↑
↓		Habitat class richness	↑
↓	Assemblage surrogates	'Community' class richness	↑
↓		Vegetation class richness	↑
↓	Taxonomic surrogates	Higher taxon richness	↑
↓		Species/subspecies richness	↑
↓		Taxonomic/phylogenetic subtree richness	↑
Ecosystem richness low		Gene/character richness (currency)	Genetic richness high

of growth over one season or year) of the indigenous vegetation. This appears to be useful but a major drawback is that, unlike taxa, there is no critical point information from productivity estimates on the spatial turnover of groups of organisms among the areas being considered. Furthermore, the governing relationships may not always be conveniently linear and therefore are hard to interpret. If greater biodiversity value is associated with richness in a currency of expressible genes (or the characters of organisms for which they code), then when resources for sampling are limited higher levels of biological organisation (or of the environmental factors affecting its distribution) have to be employed for more practical surrogate measures (the term 'assemblages' is used here for non-monophyletic groups of organisms, although it is debatable whether ecosystems – if defined in terms of processes – can be mapped and counted). Choosing a surrogacy level from this scale is a compromise between precision of the measure on the one hand, and availability of data and cost of data acquisition on the other (Williams, 1996) (Table 8.1).

Many of these relationships remain very poorly understood and indeed have never been shown to have an accurate relationship with the basic currency. Therefore, they must be used with a pinch of salt and tested for a close fit whenever possible. This approach provides one possible unified view of the traditional three levels at which biodiversity has been described by the IUCN (the World Conservation Union; the IUCN is an international union to conserve nature and natural resources (IUCN, 2004)) and various non-governmental organisations (NGOs) such as the World Resources Institute (WRI). In effect, the scale of surrogacy uses genetic diversity as a basis for valuing the traditional measures of genetic diversity, species diversity (for their relative richness in different genes) and ecosystem diversity (for the relative richness in the different processes to which the genes ultimately contribute).

Measuring rarity and endemism

As indicated by studies of area selection (e.g. Williams & Humphries, 1994) the boundary limits of a reserve system should include endemic taxa. Areas that contain endemics when left out of a reserve system can never fulfil the goal of representation of wholesale biodiversity (the entire array of organisms, or as much as can be sampled). By the same token then, rarity is important in the conservation of biodiversity for several reasons – including the restrictions imposed by having fewer area options available for narrowly distributed species when attempting to represent as many species as possible for conservation (Williams et al., 1996a). To consider the importance of representation of rarity and endemism the questions to be raised include: what are the taxa; how they relate to occupancy of grid cells on maps; and how they can be measured from these grid maps? Measurement techniques include range-size data, discontinuous measures (thresholds) and continuous measures (weighting).

Measuring rarity or endemism from range-size data

To measure rarity and endemism the two factors can be defined and then combined. Rarity can be described as the condition of occurring infrequently, and takes the form of rarity of occupancy among areas (range-size rarity) as well as rarity of individuals within areas (density rarity). It must be noted though that species with relatively restricted geographical ranges do not always have low local abundances (and vice versa). In contrast, endemism is the condition of being restricted to a particular area with a prescribed extent. Therefore, measures of range-size rarity can be equivalent to measures of narrow endemism, providing that the entire range of the species is encompassed within the study area and that the measures to be compared are expressed relative to the total extent of the survey area. For species-distribution maps that are divided into grid cells, the range sizes of the species can be quantified as counts of the numbers of grid cells occupied. The scores from these measures depend strongly on the spatial scale of the study (both the total survey extent and the grid cell size). For example, bluebells, *Hyacinthoides non-scripta* (L.) Rothm., may have a narrowly restricted range within Europe, but are widespread within the United Kingdom. We have examined two approaches to measuring rarity or narrow endemism: discontinuous measures using artificial cut-off points or thresholds to include only the rarest species; and continuous measures which give individual weights to occurrence in particular grid cells by weighting from a value determined by the entire range size.

Discontinuous measures using thresholds to include only the rarest species

A popular approach to identifying rare or narrowly endemic species has been to use a threshold of range size, measured either as an area figure (for example, $50\,000\,\text{km}^2$ for some worldwide studies (e.g. hotspots) or as a proportion of all species (e.g. the rarest quartile or 25% of species) (using Red data books, etc.; Williams *et al.*, 1996a).

Continuous measures weighting by range size

The threshold approach is essentially arbitrary, because whatever threshold of range size is adopted, it can always miss potentially important species with marginally larger range sizes. An alternative is to use measures of range-size rarity that are continuous functions of range size. This escapes the criticism of arbitrariness concerning the precise threshold used to determine which species are included, because all species make a contribution to the result. Here, rarity is included as a weighting for each species, with the weighting being higher for the more restricted species. However, arbitrariness cannot be avoided entirely, because it remains in the choice of formula for the weighting – for which there is no obvious 'natural' technique to use.

Another treatment that is sometimes used to smooth out the result is to divide the inverse range-size rarity measure by species richness, as a measure of the mean rarity among the species of a grid cell (Williams *et al.*, 1996a; Williams & Gaston, 1998). It might also be more appropriate for scores such as those with a highly skewed frequency distribution to use the median rather than the mean of the inverse range sizes, although the results of comparisons among cells would be broadly similar.

Using biodiversity measures to assess conservation priorities

The whole point of having quantitative measuring methods is to apply them in analyses to realise particular goals of reserve selection. Area-selection methods are identifying priorities for *in situ* conservation and deciding which combinations of available areas could represent the most biodiversity value for the future (Williams *et al.*, 1996a). The difficulty has been how to deal with a problem involving so many species and areas to find efficient solutions, while retaining flexibility and also making the process transparent and accountable to the people for whom conservationists are acting. There are three questions worthy of consideration: What are the aims in assessing areas for conservation priority? How can the most biodiversity be conserved for a given level of investment? How can the assessment process be made repeatable, flexible and – above all – accountable? To answer these questions, it is important to know: the values, goals and priorities; which area-selection methods to use; and how to resolve conflicts. Because choices of areas are so limited these days, most selection methods are used for finding the few options available in gaps within the environmental space.

Values, goals and priorities

Conservation is about ensuring persistence of biodiversity value. Area selection for conservation includes different values, goals and priorities. Values and goals are not universal, but differ among people and among situations. Therefore, they have to be made explicit and have to be agreed upon as broadly as possible at the beginning of any particular area-selection exercise if conflicts are to be minimised. Area-selection methods can then be used to apply rigorous and explicit rules to determine priorities consistent with these values and goals. This procedure should not be viewed dogmatically but rather as a flexible means of exploring the consequences of using different values, rules and data to inform the decision-making process. The result is a process that is explicit, accountable and repeatable. The value chosen for conservation should be broadly shared by the people giving conservationists their mandate and preferably ought to be quantifiable (at least in relative terms) for arguments presented to economists, politicians and their constituencies. Wholesale biodiversity value and its conservation has been the overriding popular choice since the Rio Summit in 1992.

Ensuring the best representation of biodiversity value within a set of areas has been used to approach conservation as a 'proactive' process, as distinct from 'fire-fighting' reactively to the problems of species endangerment or increased vulnerability. Nevertheless, resources for priority action may still be deployed in relation to perceptions of imminent threat. Representativeness simply implies monitoring all valued biota, not just those parts that are currently threatened. Distinguishing areas of higher or lower priority for urgent conservation management, in the context of a particular representation goal, is the purpose of area-selection methods. It does not mean seeking out the minimum set of areas for a particular goal, rather it is more of a ranking system to ensure that all of biodiversity is catered for. The need for priorities is usually unavoidable because competition with incompatible land uses limits the extent of the area that is available for conservation. Intensity of management within priority areas may vary depending upon circumstances, from seeking to exclude some of these land uses (such as certain kinds of agriculture), to being very limited and integrated with other current land uses.

Areas selected as priorities are chosen on the grounds that they are necessary to meet a particular goal, without illusions that they are sufficient for all goals. Analysts begin selection by trying to identify priority areas that meet minimum requirements to satisfy one particular goal. It is clearly important that the goal is explicit and the priority areas are then added to systematically so as to improve prognoses for the biota. However, an acceptance of priorities must recognise that this idea also implies that some areas and biota will be given lower priority. This is not to say that they have no conservation value, rather that in relation to agreed goals the actions are not as urgent. The issues are discussed in terms of the importance of quantitative area-selection methods, the relationship between complementary areas and hotspots of richness or hotspots of rarity, and algorithms for choosing complementary areas.

Quantitative area-selection methods

Area-selection methods are sets of rules (algorithms) – designed to achieve particular goals efficiently and transparently – that aid accountability. They must include: the ability to promote flexibility for the planning process; speed – to facilitate the exploration of alternative values, goals, data and flexible solutions; the ability to deal with incomplete data; and simplicity, to aid communication. Accountability is important to the conservationist mandate so that users can see that their values are being acted upon and that limited resources are not being misused (Williams *et al.*, 1996a, 1996b). Litigation obliges conservationists to defend area choices. Efficiency is important because the area of land available for conservation is usually limited due to competition between conservation and incompatible land uses. If 10 areas are required to represent a goal of 100% biodiversity but it takes 20 areas in the reserve scheme, the reserve scheme is inefficient. Efficiency should always be considered in relation to the goal of ensuring viability and persistence to ensure that effective choices are made. Quantitative methods can be applied to measures of biodiversity value of all the different surrogates described above (Table 8.1) – including cladistic diversity, species richness, higher taxa richness, vegetation or land-classes richness, or most of the other biodiversity surrogates.

The methods used in area selection are similar to those used in modern systematics and historical biogeography. The general form of the data is in areas by attributes (biodiversity surrogates) and matrices (the properties of areas). The general form of the priority problem is one of seeking the maximum representation of the chosen biodiversity surrogates (often species) within the areas selected (i.e. parsimony) – to achieve at least one representation, or multiple representations, within the selected areas set (Williams *et al.*, 1996a, 1996b). Therefore the rarity and narrow endemism of these surrogates are also important factors. Frequently asked questions are of the form: 'What is the minimum set of areas within the United Kingdom required to represent all species of butterflies?' This expresses a simple goal of complete representation, often in the first instance seeking representation of each species in at least one area (Williams *et al.*, 1996a, 1996b). There are many pitfalls with this approach; not only concerning the problems of viability and threat, but also concerning the meaning of 'complete'. It is obvious that a single representation of a species will not represent all intraspecific variation. The problem becomes worse with remote surrogates for biodiversity value. For example, a single representation of every vegetation class from a vegetation classification is unlikely to represent every species, let alone all intraspecific variation. The only reliable solution to representing every difference is to include every area, although more than the minimum can be achieved by the selection of multiple representations. The premise of area-selection methods is that competition with incompatible land uses limits the area available for conservation (conflict resolution). However, if the fundamental value is cumulative richness in different genes or characters (biodiversity value), then minimum-area sets for surrogates at least have the distinct advantage that they can be expected to improve in terms of the chosen currency against undirected selections for the same number of areas. In practice, it may often be more appropriate to pursue goals for maximising biodiversity representation as a maximal-covering set of areas. This approach addresses frequently asked questions such as: 'How can we choose 1% of the total area of Madagascar to represent the greatest number of lemur species?' The principle has long been recognised in approximate techniques for selecting ordered area sequences, for example in prioritising areas by taxonomic diversity (Vane-Wright *et al.*, 1991; Csuti *et al.*, 1997).

Hotspots of richness, hotspots of rarity and complementary areas

A popular approach for selecting priority areas for the conservation of biodiversity has been to select hotspots of diversity. 'Hotspots' is a term often associated with Norman Myers's worldwide review of regions (see UNEP, 1995, p. 142) – combining high richness, narrow endemism and threat. Subsequently, others have considered hotspots in a narrower sense of high scores for species richness within continents or countries (Williams *et al.*, 1996b). With appropriate qualification, hotspots can be used for high-scoring areas on any value scale and on any spatial scale. For example, the consequences might be considered of choosing, as conservation areas, the top 5% of areas (10-km grid cells) within Great Britain by species richness. This method has the appeal of dealing with species-occurrence data with apparent quantitative rigour. It also has the advantage that knowledge of the identity of each species is not required, so it would be possible to use extrapolated richness scores.

Hotspots of rarity or narrow endemism are similar to hotspots of richness, but concentrate on more narrowly distributed species. For example, areas with high richness in just those species with range sizes of less than 50 000 km^2 (discontinuous measures) have been considered as priorities for future protection (see Bibby *et al.*, 1992). This has the advantage of requiring data for only the more restricted species. Furthermore, because the species are necessarily narrowly distributed, it is also likely to select for more highly complementary biotas and so lead to more complete representation of species. A refinement of this technique is to map a continuous function of range size, such as summing the inverse of species range sizes.

Where identities of species or of other surrogates for biodiversity are known, complementarity methods can be applied to seek areas that in combination have the highest representation of diversity. Complementarity refers to the degree to which one or more attributes contributes otherwise unrepresented attributes to one or more other sets of attributes (Vane-Wright *et al.*, 1991). For example, if one area has a fauna consisting of an elephant, cheetah and lion, while another area has a fauna with a lion, leopard and giraffe, then the second-area's fauna complements the first by the leopard and giraffe.

Comparison of hotspots and complementarity methods

Complementary areas are more efficient than hotspots of either richness or rarity. A joint study – between the Natural History Museum, London and the British Trust for Ornithology – of data for British breeding birds, sought a goal of representing as many species as possible within 5% of the 10-km grid cells within Britain (Williams *et al.*, 1996a). While the hotspots of richness and rarity failed to represent all species at least once, the complementary areas represented all of the species at least six times over (or included all representatives for the more narrowly distributed species) within the 5% area limit. Inevitably, hotspots of richness did give the highest total number of species-in-grid-cell records, but most of these were repeat

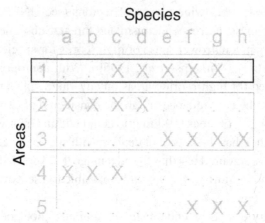

Figure 8.7 Hotspot richness and minimum complementary set for five areas (1 to 5) and eight taxa (a to h). Area 1 is the richest but is not needed to make up a full complement as areas 2 and 3 represent all species at least once and are the most complementary.

representations of more widespread species. In comparison, hotspots of rarity and complementary areas showed a more even representation among species, although the complementary areas increased the number of representations for many of the rarer species, in particular relative to the others (Williams *et al.*, 1996a).

Algorithms for choosing complementary areas

Complementarity can be used to find efficient solutions to area-selection problems. In the simple example of a species-by-areas data matrix (Figure 8.7) (adapted from the work of Underhill (1994)), it is obvious that areas 2 and 3 complement one another perfectly to represent all species a to h between them. In this particular case the hotspot of greatest species richness, area 1, is not needed for a minimum set of complementary representative areas (areas 2 and 3). Areas not needed to attain a particular representation goal may often be identified by including redundancy tests within area-selection procedures. Unfortunately, not all data sets are so simple. There is in fact no fast and general solution to the problem (an 'NP-complete' problem[2]) and the popularity that has come from the efficiency of complementarity has brought a tremendous proliferation of 'brute-force' or heuristic techniques.

---

[2] 'NP' in this case stands for 'non-deterministic polynomial'; the phrase NP-complete refers to non-deterministic polynomial time complete. 'A set or property of computational decision problems which is a subset of NP (i.e. can be solved by a non-deterministic Turing Machine in polynomial time), with the additional property that it is also NP-hard. Thus a solution for one NP-complete problem would solve all problems in NP. Many (but not all) naturally arising problems in class NP are in fact NP-complete.'

One comparative study of many different complementarity-based area-selection techniques was applied to data for vertebrates, in collaboration with the Oregon Gap Analysis project (Csuti *et al.*, 1997). The results illustrate some of the major features of the different groups of algorithms. Exact solutions can be achieved by exhaustive searches of all possible sets, although these are too numerous to be practical for problems with anything but very small numbers of areas. Techniques such as branch-and-bound algorithms can give optimal solutions, but even they can take hours or days to provide results when dealing with just a few hundred species and areas; this renders interactive assessment of priority areas out of the question (Williams *et al.*, 1996a). Consequently, faster but approximate techniques (heuristic algorithms) based on measures of rarity are often employed as an alternative. The practical advantage of speed from heuristic techniques has come to be appreciated (just as in systematics, where comparable optimisation problems occur). Furthermore, the reduction in the efficiency of rarity heuristics that results from incorporating redundancy checks is usually small.

Conclusion

The availability of fast and efficient computers means that over the last 20 years or so the consistent measuring techniques and area-selection methods have come to stay. What has been made apparent is that data sets are still woefully inadequate for completely confident analyses of diversity and conservation. However, the use of data-modelling techniques and mapping means that quantitative diversity measures and area-selection methods can be used and there are a number of different ways that these can be applied and linked to decision analysis. This brief chapter highlights the efforts of the Biogeography and Conservation Laboratory at the Natural History Museum, London (http://www.nhm.ac.uk/research-curation/projects/worldmap/index.html) to mobilise taxonomic data into conservation work so that point information (i.e. a precise locality) and identity data (taxonomic information) improve the quality and accountability of conservation when considering issues of turnover (the amount of change in species composition when comparing one area with the next) and complementarity (the combination of the original set with those taxa not included due to turnover). Rather than being some kind of alternative to the ecologically dominated research that has been used throughout the twentieth century, an examination of the currency of conservation – which we believe basically is best seen as the characters of organisms – has allowed us to put together a surrogacy scheme that can satisfy the usual requirements of NGOs of representing genetic, specific and ecosystem levels of measurement. Indeed our efforts have tried to bring together the entire broad range of surrogates into a single scheme that can form the basis of new developments in the future.

Acknowledgements

I would like to thank particularly my long-term research partner and esteemed colleague, Paul Williams, for allowing me to use many of his ideas; Dick Vane-Wright for continued inspiration and for help putting the Biogeography and Conservation Laboratory together: and the many other colleagues who share interests in biodiversity measurement and area-selection work.

References

Bibby, C. J., Collar, N. J., Crosby, M. J. *et al.* (1992). *Putting Biodiversity on the Map: Priority Areas for Global Conservation*. Cambridge, UK: International Council for Bird Preservation.

Csuti, B., Polasky, S., Williams, P. H. *et al.* (1997). A comparison of reserve selection algorithms using data on terrestrial vertebrates in Oregon. *Biological Conservation*, **80**, 83–97.

Humphries, C. J., Williams, P. H. & Vane-Wright, R. I. (1995). Measuring biodiversity value for conservation. *Annual Reviews of Ecology and Systematics*, **26**, 93–111.

IUCN (The World Conservation Union) (2004). http:www.iucn.org/.

Underhill, L. G. (1994). Optimal and suboptimal reserve selection methods. *Biological Conservation*, **70**, 85–7.

UNEP (1995). *Global Biodiversity Assessment*. Cambridge, UK: Cambridge University Press.

Vane-Wright, R. I., Humphries, C. J. & Williams, P. H. (1991). What to protect? Systematics and the agony of choice. *Biological Conservation*, **55**, 235–54.

Williams, P. H. (1996). Measuring biodiversity value. *World Conservation* (formerly *IUCN Bulletin*), **1**, 12–14.

Williams, P. H. & Gaston, K. J. (1994). Measuring more of biodiversity: can higher-taxon richness predict wholesale species richness? *Biological Conservation*, **67**, 211–17.

(1998). Biodiversity indicators: graphical techniques, smoothing and searching for what makes relationships work. *Ecography*, **21**, 551–60.

Williams, P. H. & Humphries, C. J. (1994). Biodiversity, taxonomic relatedness, and endemism in conservation. In *Systematics and Conservation Evaluation*, eds. P. L. Forey, C. J. Humphries & R. I. Vane-Wright. Oxford, UK: Oxford University Press, pp. 269–87.

(1996). Comparing character diversity among biotas. In *Biodiversity: a Biology of Numbers and Difference*, ed. K. J. Gaston. Oxford, UK: Blackwell Science Ltd, pp. 54–76.

Williams, P. H., Gaston, K. J. & Humphries, C. J. (1994a). Do conservationists and molecular biologists value differences between organisms in the same way? *Biodiversity Letters*, **2**, 67–78.

Williams, P. H., Humphries, C. J. & Gaston, K. J. (1994b). Centres of seed-plant diversity: the family way. *Proceedings of the Royal Society: Biological Sciences*, **256**, 67–70.

Williams, P., Gibbons, D., Margules, C. *et al.* (1996a). A comparison of richness hotspots, rarity hotspots and complementary areas for conserving diversity using British birds. *Conservation Biology*, **10**, 155–74.

Williams, P. H., Prance, G. T., Humphries, C. J. & Edwards, K. S. (1996b). Promise and problems in applying quantitative complementary areas for representing the diversity of some Neotropical plants (families Dichapetalaceae, Lecythidaceae, Caryocaraceae, Chrysobalanaceae and Proteaceae). *Biological Journal of the Linnean Society*, **58**, 125–57.

Williams, P. H., Gaston, K. J. & Humphries, C. J. (1997). Mapping biodiversity value worldwide: combining higher-taxon richness from different groups. *Proceedings of the Royal Society: Biological Sciences*, **264**, 141–8.

9

The need for plant taxonomy in setting priorities for designated areas and conservation management plans: a European perspective

Dominique Richard and Doug Evans

Introduction

Well before the signature of the Convention on Biological Diversity (CBD), which represents one of the most significant components of the 1992 United Nations Conference on Environment and Development (UNCED) in Rio de Janeiro, initiatives had been taken at various decisional levels for protecting plants and their habitats, in particular through the establishment of designated areas. However, biodiversity conservation – including plant species, genetic resources and ecosystems – has been boosted by the launching of the Convention.

Ten years later we are now faced with a critical challenge, stated in the Strategic Plan for the Convention on Biological Diversity (CBD Secretariat, 2002, Decision VI/26) (UNEP, 2002), endorsed at the World Summit on Sustainable Development in Johannesburg in September 2003: 'to achieve by 2010 a significant reduction of the current rate of biodiversity loss at the global, regional and national level, as a contribution to poverty alleviation and to the benefit of all life on earth'. In 2001, the Sixth Environmental Action Programme of the European Union, 'Environment 2010. Our future, our choice' addressed an even more ambitious objective: 'of halting biodiversity decline with the aim to reach this objective by 2010, including prevention and mitigation of impacts of invasive alien species and genotypes' (European Commission, 2001). This objective was also endorsed at pan-European level at the Conference of Ministers held in Kiev in May 2003.

Considering the recognised gaps in knowledge and trends in environmental degradation, such statements are somewhat illusory. However, beyond the difficulty for scientists who now have to take stock of the present knowledge and propose on a very short timescale methods and assessments to monitor the progress towards such targets, as well as priorities for conservation actions, this context provides an opportunity to gather the botanical community – including taxonomists, ecologists, geneticists and conservationists – to work towards a common goal by creating synergies.

Implementing species- and habitats-conservation measures first implies that the species of concern are well defined and recognised: unresolved basic species taxonomy means that action at the species level cannot efficiently address those taxa most in need of conservation, in some cases we do not even know the species exists; protection or restoration of habitats requires knowledge on their species composition (Williams & Humphries, 1994; Ebenhard, 1998; Heywood & Iriondo, 2003; Hollingsworth, 2003).

The present article mainly refers to a European perspective and focuses on the importance of plant taxonomy for biodiversity conservation through the selection of designated areas and management plans.

Designated areas, a tool for biodiversity conservation

Sites of high nature value have been protected from adverse human activities for more than 100 years: in the United States, the Yellowstone National Park was established in 1872; in Europe the Capul Kaliakra National Park was established in Romania in 1890.

Each country has developed its own system of designation types, ranging from very strict nature reserves and national parks to more flexible protection such as landscape parks and areas under specific conservation management. Europe as a whole has some 65 000 sites designated under 600 different national protection systems (national parks, regional parks, nature reserves, protected landscapes, etc.). There has been a huge increase in national designations since the 1970s when most countries started to implement national laws on nature protection (Richard, 2003).

In addition to national initiatives, and given the perceived importance of designated areas for the conservation of the world's natural resources, they have been actively promoted in a wide range of international conventions and cooperative programmes. This results in various networks of sites of international importance, each of them serving specific purposes: protecting targeted species, habitat types, ecosystems; recognising excellence; promoting research and education. Most of the international and European designations overlap with national designations and sometimes among themselves, which in principle ensures stronger protection. Since each designation is made with a specific purpose, a site of particularly high nature value can benefit from several international designations. For instance, Doñana in Spain and the Camargue in France enjoy six overlapping international and European designations (still, risks are always possible, even in highly protected areas!).

Table 9.1. *International, European and regional conservation conventions and legislation with a site-based component (Delbaere & Beltran, 1999)*

Treaty or programme	Date	Geographical coverage	Network of sites
Ramsar Convention	1971	Global	Wetlands of International Importance (Ramsar sites)
World Heritage Convention	1972	Global	World Heritage sites
UNESCO Man and the Biosphere programme	1970	Global	Biosphere Reserves
Council of Europe European Diploma	1965	Pan-Europe	Recipients of the European Diploma
Council of Europe Biogenetic Reserves	1976	Pan-Europe	Biogenetic Reserves
Bern Convention	1979–89	Pan-Europe	Areas of Special Conservation Interest (Emerald network)
Habitats Directive	1992	European Union	Special Areas of Conservation (NATURA 2000 network)
Helsinki Convention	1992	Baltic	Baltic Sea Protected Areas
Barcelona Convention and Protocol	1976	Mediterranean	Specially Protected Areas of Mediterranean Importance
Convention for the Protection of the Marine Environment of the North-East Atlantic (OSPAR)	1992	North East Atlantic	OSPAR Marine Protected Areas

Table 9.1 presents a number of such international and European initiatives of potential relevance for plant and habitat conservation, whether in a direct or an indirect way.

The importance of plant taxonomy is particularly evident for statutory instruments that have defined lists including 'objects' which are of priority for conservation action, such as species and habitats. The designation of sites by countries that implement these international instruments is targeted towards the conservation of the given species and habitats, according to standardised criteria and procedures. This is the case for the Habitats Directive, as presented below, and for the Bern Convention which sets provision for the implementation of a network of sites (Council of Europe, 1999) with 'a list of endangered habitats requiring specific conservation measures' (Resolution 4) and a 'list of species requiring specific habitat-conservation measures' (Resolution 6).

A similar approach has been adopted by other conventions, including OSPAR (protection of the marine environment of the North-East Atlantic), Helsinki (protection of the marine environment of the Baltic Sea area) and Barcelona (protecting the Mediterranean Sea) Conventions which have lists of species and habitats requiring the designation of protected areas.

Conservation through management schemes

Agriculture is recognised as a major driving force influencing biodiversity, particularly in Europe where agricultural land accounts for more than 40% of the total land area (although the proportion varies from less than 10% in Finland, Sweden and Norway to 70% or more in Hungary, Ireland, Ukraine and the United Kingdom) (European Environment Agency, 1995).

The implementation of environmentally friendly farming practices is thus crucial for biodiversity conservation, especially outside protected areas. As part of the Common Agricultural Policy (CAP), a regulation was set up in the European Union (Council Regulation (EEC) 2078/92), to encourage farmers to carry out environmentally beneficial activities on their land. Farmers are paid the costs and income losses for providing the environmental service. These so-called 'agri-environment measures' include such measures as reducing use of pesticides and chemical fertilisers, organic farming, protection of habitats, maintenance of existing sustainable and extensive farming systems, protection of endangered farmed animals and plant varieties, and upkeep of landscapes. By 1998 an average of 13.4% of EU farmers were involved with the programmes and 20% of the total agricultural area in the European Union was covered (European Commission, 1998). Following the enlargement of the European Union in 2004 (with ten new Member States) – including largely rural countries such as Poland, Hungary and Slovakia) such measures will have even more importance.

As in the case of designated areas, a good knowledge of the species involved is needed for efficient management-plan design. Examples of applications of agri-environment measures with direct influence on the wild flora are given below.

1. In Greece
   Conservation of habitats:
   • land-use design at the centre of important habitats;
   • design of land cultivation (perennial vs. annual, legumes vs. cereals, fallow vs. cultivation and conversion of agricultural land into grazing land);
   • low grazing rates or prohibition of grazing;
   • control of sedimentation of water bodies due to erosion.

Protection of agricultural plant diversity:
  • protection of the land-races of agricultural crops and reduction of genetic erosion (Koumas *et al.*, 1998);
2. In the Emilia-Romagna region of Italy
   Conservation of natural habitats such as permanent wetlands, marshy meadows and permanent meadows with scrub patches suitable for wild fauna and flora (G. de Geronimo, personal communication, 2004)
3. In West Fermanagh and Erne Lakeland ESA (Environmentally Sensitive Area) (Northern Ireland, UK)
   Conservation of habitats such as heather moorland, hay meadow, wet pasture and limestone grassland (European Commission, 1998).

Selecting areas of high value for biodiversity

Whether for designating protected areas or areas for conservation management schemes, there is a need to set priorities that target the most valuable or vulnerable areas from a biodiversity point of view. Site selection should be carried out on the basis of specific criteria in relation to clear objectives. Any such selection requires an inventory, which in turn needs taxonomy, which includes that the names of the species to be protected are correct in legal texts (Box 9.2).

Many countries have set up their own national system for site inventory. In the United Kingdom, Sites of Special Scientific Interest (SSSI) under the 1981 Wildlife and Countryside Act are selected according to a series of criteria given for each group of interests (e.g. vegetation, plant species, birds, etc.). For some groups, such as vascular plants, a scoring system is used to select the 'best' sites in each region as worthy for protection giving high value to the presence of Red List species, and less value to rare or nationally scarce species (Nature Conservancy Council, 1989). In France, the ZNIEFF system (Zone naturelle d'intérêt écologique, faunistique et floristique), launched in 1982, aims at informing decision makers of the existence and location of areas of high biodiversity value which should be taken into account in land planning, including for protection. More than 14 000 ZNIEFF ranging from local to international interest have been identified, a large proportion because of their botanical interest (Feraudy *et al.*, 1999).

At the EU level, sites to be designated under the Habitats Directive (see below) are selected following a process described in the Directive, with the Member States asked to propose potential Sites of Community Interest for a given list of habitats and species. These national proposals are compared and assessed within a biogeographic context (six biogeographic zones have been identified for 15 EU countries), which means within a multinational framework. This involves huge harmonisation, standardisation and consultation issues. Only once the list of Sites of Community Interest has been agreed upon at EU level, for each of the six biogeographic regions, can the sites be formally designated by countries as Special Areas of Conservation. So far only the list for the Macaronesia region has been agreed upon (European Commission, 2002).

Very specific to plant conservation is the Important Plant Areas (IPAs) programme led by the non-governmental organisation Plantlife International. This programme aims to quantify and locate the sites that are important for the long-term viability of naturally occurring plant populations in Europe. It is not intended that 'IPA' will become a conservation designation in its own right. The identification of IPAs aims to support, inform and underpin the implementation of the Habitats Directive, the Bern Convention and other relevant statutory instruments in Europe (Anderson, 2002).

Guidelines for selecting IPAs have been developed and refined since 1995. They provide simple, transparent criteria that allow for pan-European and global comparisons, but which also allow for national variation. These are:

- the presence of globally and regionally threatened plant species;
- exceptional botanical richness and diversity (in relation to its biogeographical zone);
- habitat types of global or regional importance.

To qualify, a site has to satisfy one or more of these criteria requiring a sound knowledge of the European flora and habitats at a variety of levels. Guidelines to the identification of Important Plant Areas in Europe have been produced (Anderson & Kusik, 2003).

The identification of IPAs was stated as a global priority of the IUCN Species Survival Commission's Global Plant Conservation Programme adopted at the IUCN World Conservation Congress held in Amman, Jordan in 2000. IPAs also form an essential part of the Global Strategy for Plant Conservation (Target 5) (UNEP, 2002, Decision VI/9) and a major target of the European Plant Conservation Strategy, a joint project between the Planta Europa network and the Council of Europe (Planta Europa and Council of Europe, 2001).

Taxonomy and the Habitats Directive

Council Directive 92/43/EEC of 21 May 1992 on the conservation of natural habitats and of wild fauna and flora, more usually known as the Habitats Directive and the Wild Birds Directive in 1979 (Council Directive 79/409/EEC) constitutes the core of EU legislation to maintain and enhance biodiversity.

Article 3 of the Habitats Directive states:

A coherent European ecological network of special areas of conservation shall be set up under the title Natura 2000. This network, composed of sites hosting the natural habitat types listed in Annex I and habitats of the species listed in Annex II, shall enable the natural habitat types and the species' habitats concerned to be maintained or, where appropriate, restored at a favourable conservation status in their natural range.

The Natura 2000 network shall include the special protection areas classified by the Member States pursuant to Directive 79/409/EEC.

By July 2005 some 20 000 sites had been proposed as Special Areas of Conservation (SACs).

The European Topic Centre on Nature Protection and Biodiversity (ETC/NPB), formerly the European Topic Centre on Nature Conservation, is mandated to provide scientific and technical support to the European Commission in assessing proposals made by Member States for future Natura 2000 sites (Richard, 2002). Some taxonomic problems arose, a major one being the moth *Callimorpha quadripunctaria*. This is a widespread species and is sometimes an agricultural pest; however, it was listed in Annex II as requiring the designation of SACs, instead of the rare and threatened subspecies *Callimorpha quadripunctaria rhodonensis* endemic to the island of Rhodes. Other taxonomic problems related to plant species and habitats also demonstrate the importance of taxonomy for sound assessments of conservation priorities.

Plant species listed in Annex II of the EC Habitats Directive

The list of species in Annex II includes 483 plant species and fortunately the list as published in the *Official Journal* does include authorities (although not for animals). The majority of the listed taxa do not pose any taxonomic problems. However, there are 29 species of bryophytes which pose problems due to the scarcity of bryologists. As a consequence, data on distribution and ecology are poor; this has led to problems in site selection and will pose problems for site management and monitoring. It is also possible that some of these species are not really as rare and threatened as we currently think they are. Rather, the number of people capable of recognising them in the field is very small. Due to the Directive there has been a focus of attention on these species resulting in new localities being found or old records being confirmed. For example, *Buxbaumia viridis* (Moug.) Moug. & Nestl. was only known from one locality in the United Kingdom but has recently been found at two more sites. This indicates that analyses and evaluation of species should remain open-ended to allow for the incorporation of new knowledge and taxonomic revisions.

There are other problematic species in Annex II, for example *Carex panormitana* Guss. is listed as a priority species (higher level of protection than non-priority species). This listing was due to the taxon being known only from very few sites in Greece and Italy, and nowhere else within the European Union. However, in *Flora Europaea* (Tutin *et al.*, 1980) and the new Euro+Med Checklist (www.euromed.org.uk) the name is reduced to a synonym of *Carex acuta* L., a very widespread species whose distribution is 'most of Europe, but rare in the extreme south'.

For site selection the older meaning of *C. panormitana* has been retained to focus attention on the threatened taxon although this can lead to confusion and proposals to classify sites that only host *C. acuta sensu lato*.

In addition to assessment of site proposals by current EU countries, the ETC/NPB has advised the Commission on the inclusion of new species and habitats in respective annexes to take account of the enlargement of the European Union from 15 to 25 Member States. The new Member States were asked to propose species and habitats according to an agreed pro-forma giving information on ecology, distribution, conservation status, etc. These proposals were first assessed against the criteria given in the Directive:

(i) in danger of disappearance in their natural range; or
(ii) have a small natural range following their regression or by reason of their intrinsically restricted area; or
(iii) present outstanding examples of typical characteristics of one or more of the biogeographical regions

However, taxonomic status was used as an additional criterion. For example, three species of *Dactylorhiza* were proposed but as the taxonomy of this and related genera is frequently revised with no agreed treatment at an EU level, we advised that they should not be accepted as there will probably be problems in the future. This work was made harder by the absence of an agreed checklist or Flora of Europe. *Flora Europaea* is not accepted as accurate by many botanists in central Europe and the *Med-Checklist* (Greuter *et al.*, 1984) is only partly complete. There is the Euro+Med Checklist project under way but it was not available at the time.

Habitats listed in Annex I of the EC Habitats Directive

Taxonomic problems are more common with the habitats listed in Annex I. Some of these problems are due to poorly defined, or sometimes overlapping, habitats (see Box 9.1, 'Confusing?'). Many others are due to problems arising from the use of many different systems of classification of vegetation and habitats.

The most widely used system of classifying vegetation in Europe is the phytosociological system developed by Braun-Blanquet and others (Braun-Blanquet, 1928; Westhoff & Van der Maarel, 1978; Mucina, 1997). However, this only covers plant communities and excludes abiotic habitats such as glaciers or lava fields and

---

Box 9.1
Confusing?

Annex I habitat '3110 Oligotrophic waters containing very few minerals of sandy plains (*Littorelletalia uniflorae*)' is defined as 'Shallow oligotrophic waters with few minerals and base poor, with an aquatic to amphibious low perennial vegetation belonging to the *Littorelletalia uniflorae* order, on oligotrophic soils of lake and pond banks (sometimes on peaty soils). This vegetation consists of one or more zones, dominated by *Littorella*, *Lobelia dortmana* or *Isoetes*, although not all zones may be found at a given site.' (European Commission, 1996). However, this is clearly a subset of habitat '3130 Oligotrophic to mesotrophic standing waters with vegetation of the *Littorelletea uniflorae* and/or of the *Isoëto-Nanojuncetea*' as the order *Littorelletalia uniflorae* is within the class *Littorelletea uniflorae*.

There is also habitat '3120 Oligotrophic waters containing very few minerals generally on sandy soils of the West Mediterranean, with *Isoetes* spp.' Defined as 'Dwarf amphibious vegetation of oligotrophic waters with few minerals, mostly on sandy soils of the Mediterranean region and some irradiations in the thermo-Atlantic sector, and belonging to the [class] *Isoeto-Nano-Juncetea*', again a subset of habitat 3130.

This overlapping of habitat descriptions gives rise to many problems in site selection and also to later analysis of the proposed network to ensure a coherent network as demanded by the Directive.

---

the majority of marine habitats, which tend to be classified by fauna or substrate. Also, there has been much dispute between phytosociologists and until recently no agreed checklist at a European level; there have been many problems of synonymy when trying to relate syntaxa in one country with those in another.

Under the CORINE programme (Coordination of Information in Europe), the CORINE biotopes classification (European Communities, 1991) was developed to give a unified system covering marine and terrestrial habitats, both natural and artificial. This was further developed in the more recent Palaearctic habitats classification (Devillers & Devillers-Terschuren, 1996) and the habitat classification developed under the European Nature Information System (EUNIS) (Davies & Moss, 2002; also see http://biodiversity-chm.eea.eu.int/information/database/nature). Only the EUNIS habitat classification has criteria-based keys to allow allocation to a given unit, but these have only been developed to 'level 3' which are broad classes such as B1.2, Sand beaches above the driftline; G1.6, *Fagus* woodland. These are broader units than most Annex I habitats. There is a database giving fuller descriptions of the Palaearctic units, but these are mostly in terms of phytosociological syntaxa (see PHYSIS database, Table 9.2), and contains many nomenclatural errors or misunderstandings.

Table 9.2. *Habitat classification: an extract from the PHYSIS database (Devillers et al. (1996))*

Habcode	NAME	COM	TXT	REF
41.2	Oak-hornbeam forests	[Carpinion betuli]	'Atlantic, medio-European and eastern European forests dominated by [Quercus robur] or [Quercus petraea], on eutrophic or mesotrophic soils, with usually ample and species-rich herb and bush layers. [Carpinus betulus] is generally present. They occur under . . .'	'(Mullenders, 1955; Breton, 1957; Vanden Berghen and Mullenders, 1957b; Ellenberg, 1963, 1988; Izard &, 1963; Tanghe, 1964b, 1967, 1968, 1970; Gaussen, 1964; Dupias and Cabaussel, 1966; Durin &, 1967; Oberdorfer, 1967, 1990; Sougnez, 1967; Noirfalise, 1968 . . .'
41.21	Mixed Atlantic bluebell oak forests	[Carpinion betuli]: [Pulmonario-Carpinenion betuli]: [Endymio-Carpinetum] (incl. [Corylo-Fraxinetum] [p.])	'Atlantic forests of the British Isles, western Belgium and north-western France, mostly on more or less water-retaining soils, characterized by a diverse tree layer, dominated by [Quercus robur] and rich in [Fraxinus excelsior], and by an herb layer rich in . . .'	'(Durin &, 1967; Noirfalise, 1968, 1969; Caron and Géhu, 1976; Noirfalise, 1984: units V.2.3.1, V.2.3.2, V.2.3.3, V.2.3.4; Noirfalise, 1987: D1 [p.], D2 [p.]; Rodwell, 1991a: unit W10; Oberdorfer, 1992b).'

(cont.)

Table 9.2. (*cont.*)

Habcode	NAME	COM	TXT	REF
41.22	Aquitanian ash–oak and oak–hornbeam forests	[Carpinion betuli]: [Hyperico androsaemi-Carpinenion betuli]: [Rusco-Carpinetum], [Saniculo-Carpinetum]	'Forests of [Quercus robur], [Fraxinus excelsior] and [Carpinus betulus] of valley bottoms and cool, damp lower slopes of southwestern France, south to the Pyrenean piedmont, with [Sorbus torminalis], [Ruscus aculeatus] and many thermocline, acidocline and . . .'	'(Izard &, 1963; Gaussen, 1964; Dupias and Cabaussel, 1966; Noirfalise, 1968; Lavergne, 1969; Chastagnol &, 1978; Chastagnol and Vilks, 1982; Bernard, 1983; Botineau and Chastagnol, 1983; Gésan and Plat, 1983; Noirfalise, 1987: E3 [p.]; Oberdorfer, 1992b: . . .'
41.23	Sub-Atlantic oxlip ash–oak forests	[Carpinion betuli]: [Pulmonario-Carpinenion betuli]: [Primulo elatioris-Carpinetum] ([Stellario-Carpinetum stachyetosum] [p.])	'Forests of [Quercus robur], sometimes [Quercus petraea], rich in [Fraxinus excelsior], with [Carpinus betulus], developed on more or less wet, meso-eutrophic soils, in regions of moderate Atlantic influence, from southern Champagne and Lorraine north to l . . .'	'(Ellenberg, 1963, 1988; Noirfalise, 1968; Sougnez, 1973, 1978; Westhoff and den Held, 1975; Rameau and Timbal, 1979; Bournérias, 1979, 1984; Noirfalise, 1984: 102–111; Noirfalise, 1987: D3 [p.] [i.a.]; Pott, 1992: 381; Oberdorfer, 1992b: 159; Schubert & . . .'

Thus the majority of Annex I habitats are defined either directly, or indirectly, by phytosociological syntaxa. Unfortunately, as these often have different interpretations in different countries there has been much discussion, which is still ongoing, about which habitats are present in which country and where. This is a problem particularly in countries where phytosociology has not had a strong tradition, such as Denmark and the United Kingdom.

The European Vegetation Survey, a group of the International Association of Vegetation Scientists, has been working towards a European checklist of syntaxa. Recently a checklist of alliances has been published (Rodwell *et al*, 2002) and a more complete listing, giving authorities and synonyms, is in preparation. However, there is still a need for revision, particularly for associations where there is much synonymy. The European Vegetation Survey (EVS) SynBioSys project (see www.synbiosys.alterra.nl/eu/) will contribute towards this.

There is also a need to know which are the defining and/or key species for optimising site management and for monitoring. Many countries have published guides to their Annex I habitats (e.g. Bensettiti, 2001; Janseen & Schaminée, 2003) but much remains to be done, especially in ensuring similar definitions across the European Union. This work will require expertise in habitat classifications across several classification systems together with a knowledge of the ecology of the habitats and component species.

As for species, new habitats have been proposed for addition to the Annex I of the Habitats Directive, as part of the EU enlargement process. Many of the habitats proposed were, on examination, local variants of habitat types already listed. For example several Tatra or Carpathian alpine habitats were already covered by existing alpine habitats described from the Alps. For several of these we suggested a minor change to the wording in the Interpretation Manual (European Commission, 1996), the official guide to interpreting the habitat names given in Annex I of the Directive, to make it clear that alpine habitats could occur in other mountain ranges.

Conclusion

Ambitious and somewhat unrealistic commitments are given by decision makers for halting biodiversity loss; however, scientists, lawyers and local managers still face a number of practical taxonomic problems in implementing corresponding measures. Even in an area as well studied as Europe, taxonomic skills – both in the field and in the laboratory – remain a key element of any plant conservation measures; both to help frame legislation and for its implementation. For certain groups the lack of taxonomic expertise is causing problems with implementing agreed conservation measures. Furthermore, errors in taxonomic interpretations by lawyers or too-frequent taxonomic revisions lead to mismatches in legal texts and

---

Box 9.2
Much ado about a jurist's blunder!

A French ministerial order dated 20 January 1982 listed plant species to be protected at national level and included the species *Astragalus massiliensis*. But the vernacular name given in the legal text was Astragale de Montpellier instead of Astragale de Marseille. As they believed they had made a double mistake, lawyers working on a revision of the decree, correctly named *Astragalus massiliensis* as Astragale de Marseille. However, they also added *Astragalus monspelliensis* as a reference to Astragale de Montpellier already listed. This plant is widely distributed in southeastern France and does not need priority conservation measures! This led for a while to biased statements concerning the need for protected areas and the specifications therein. The error was corrected in 1995 with a revised version of the ministerial order. Now only the Astragale de Marseille, strictly limited to a few sites in Provence and Corsica, is protected under the scientific name *Astragalus tragacantha* L., including its synonym: *Astragalus massiliensis* (Miller) Lam. (Olivier *et al.*, 1995).

 This example indicates the importance of ensuring that the names of species to be protected are correct in legal texts.

---

necessitates changing the legislation (see Box 9.2). For the Habitats Directive this requires a decision by the Council of Ministers, which is often more difficult than publishing a taxonomic revision. Much effort is still needed to ensure better coordination and to create more synergy between plant taxonomists, conservationists and decision makers. The human factor remains critical.

References

Anderson, S. (2002). *Identifying Important Plant Areas*. London: Plantlife International.

Anderson, S. & Kusik, T. (2003). Eastern promise for plants. *Plant Talk*, **32**, 18–23.

Bensettiti, F. (ed.) (2001). *Cahiers d'habitats Natura 2000: connaissance et gestion des habitats et des espèces d'intérêt communautaire, vol. 1, Habitats forestiers*. Paris: La documentation française.

Braun-Blanquet, J. (1928). *Pflanzensoziologie Grunzüge der Vegetationskunde*. Berlin, Germany: Springer Verlag.

Council of Europe (1999). *The Emerald Network: a Network of Areas of Special Conservation Interest for Europe*. Document T-PVS (99) 36. Strasbourg, France: Council of Europe.

Davies, C. E. & Moss, D. (2002). *EUNIS Habitat Classification*. Final report to the European Topic Centre on Nature Protection and Biodiversity, European Environment Agency, February 2002. Copenhagen, Denmark: European Environment Agency.

Delbaere, B. & Beltran, J. (1999). *Nature Conservation Sites Designated in Application of International Instruments at Pan-European Level*. Nature and Environment Series 95. Strasbourg, France: Council of Europe.

Devillers, P. & Devillers-Terschuren, J. (1996). *A Classification of Palaearctic Habitats*. Nature and Environment Series 78. Strasbourg, France: Council of Europe.

Devillers, P. Devillers-Terschuren, J. & Linden, C. Van der (1996). Palaearctic Habitats. PHYSIS Data Base, Royal Belgian Institute of Natural Sciences website. http://www.kbinirsnb.be/cb/databases/cb_databases_eng.htm (accessed 2005).

Ebenhard, T. (1998). Taxonomy and the Biodiversity Convention. In *Proceedings of the Second European Conference on the Conservation of Wild Plants*, eds. H. Synge & J. Akeroyd. Uppsala: Swedish Threatened Species Unit, Swedish University of Agricultural Sciences and London: Plantlife, pp. 114–18.

European Commission (1996). *Interpretation Manual of European Union Habitats. Version EUR 15*. European Commission DG XI. Brussels, European Commission.

(1998). *State of Application of Regulation (EEC) n° 2078/92, Evaluation of Agri-environmental Programmes*. Report to the Parliament. Brussels: European Commission.

(2001). *Environment 2010. Our Future, our Choice*. Booklet presenting the Sixth Community Environment Action Programme. Brussels: Office for official publications of the European Communities.

(2002). Commission Decision of 28 December 2001 adopting the list of sites of Community importance for the Macaronesian biogeographical region, pursuant to Council Directive 92/43/EEC. *Official Journal of the European Communities*, L5/16 9.1.2002. http://europa.eu.int/comm/environment/nature/natura_biogeographic.htm).

European Communities (1991). *Habitats of the European Community. CORINE Biotopes Manual*. Luxembourg: Commission of the European Communities.

European Environment Agency (1995). *Europe's Environment. The Dobris Assessment*. Copenhagen, Denmark: European Environment Agency.

Feraudy E. de, Hoff, M., Maurin H. & Bardat J. (1999). Contribution des inventaires de zones de grand intérêt écologique et des espaces protégés à la connaissance et la prise en compte de la flore menacée. In *Proceedings of the Workshop on Les plantes menacées de France*, ed. J.-Y. Lesoeuf. *Bulletin de la Société Botanique du Centre-Ouest*, **19**. Saint-Sulpice de Royan, France: Sociéte Botanique du Centre-Ouest, pp. 27–42.

Greuter W., Burdet, H. M. & Long, G. (eds). (1984–) *Med-Checklist: a Critical Inventory of Vascular Plants of the Circum-Mediterranean Countries*. Geneva: Editions des Conservatoire et Jardin botaniques de la Ville de Genève.

Heywood, V. H. & Iriondo, J. M. (2003). Plant conservation: old problems, new perspectives. *Biological Conservation*, **113**, 321–35.

Hollingsworth, P. M. (2003). Taxonomic complexity, population genetics and plant conservation in Scotland. *Botanical Journal of Scotland*, **55**(1), 55–63.

Janseen, J. A. M. & Schaminée, J. H. J. (2003). *Europese Natuur in Nederland: Habitattypen*. Utrecht, the Netherlands: KNNV Uitgeverij.

Koumas, D., Papanikolaou, G. & Thanopoulos, R. (1998). The agri-environmental regulation of the European Union and the management of Biodiversity in Greece. In *Proceedings of the Second European Conference on the Conservation of Wild Plants*, eds. H. Synge & J. Akeroyd. Uppsala: Swedish Threatened Species Unit, Swedish University of Agricultural Sciences and London: Plantlife, pp. 276–80.

Mucina, L. (1997). Classification of vegetation: past, present and future. *Journal of Vegetation Science*, **8**, 751–60.

Nature Conservancy Council (1989). *Guidelines for Selection of Biological SSSIs*. Peterborough, UK: Nature Conservancy Council.

Olivier, L, Galland, J.-P. & Maurin, H. (eds.) (1995). *Livre rouge de la flore menacée, vol. I, Espèces prioritaires*. Institut d'Ecologie et de Gestion de la Biodiversité, Collection Patrimoines Naturels 20, Série patrimoine génétique. Paris: Muséum National d'Histoire Naturelle.

Planta Europa and Council of Europe (2001). *The European Plant Conservation Strategy*. London: Planta Europa Secretariat.

Richard, D. (2002). *European Topic Centre on Nature Conservation Topic Update 2000*. Topic report 3/2002. Copenhagen, Denmark: European Environment Agency.

   (2003). Biological diversity. In *Europe's Environment, the Third Assessment*. Environmental assessment report 10. Copenhagen, Denmark: European Environment Agency, pp. 230–49.

Rodwell, J. S., Schaminée, J. H. J., Mucina, L. *et al.* (2002). *The Diversity of European Vegetation. An Overview of Phytosociological Alliances and Their Relationships to EUNIS Habitats*. Report EC-LNV 2002/054. Wageningen, the Netherlands: EC-LNV.

Tutin, T. G., Heywood, V. H., Burges, N. A. *et al.* (eds.) (1980). *Flora Europaea, vol. 5, Alismataceae to Orchidaceae (Monocotyledones)*. Cambridge, UK: Cambridge University Press.

UNEP (2002). *Strategic Plan for the Convention on Biological Diversity (CBD)*. Decision VI/26, UNEP/CBD/COP/6/20. Montreal, Canada: CBD Secretariat. www.biodiv.org.

Westhoff, V. & Van der Maarel, E. (1978). The Braun-Blanquet approach. In *Classification of Plant Communities*, ed. R. H. Whittaker, 2nd edn. The Hague, the Netherlands: W. Junk, pp. 289–374.

Williams, P. H. & Humphries, C. J. (1994). Biodiversity, taxonomic relatedness and endemism in conservation. In *Systematics and Conservation Evaluation*, eds. P. L. Forey, C. J. Humphries & R. I. Vane-Wright. Oxford, UK: Clarendon Press, pp. 269–87.

10

The identification, conservation and use of wild plants of the Mediterranean region: the Medusa Network – a programme for encouraging the sustainable use of Mediterranean plants

Melpomeni Skoula and Christopher B. Johnson

The current interest in the traditional uses of plants and their natural products represents a complex and fascinating blend of science, tradition and mythology in which scientists endeavour to understand the biological, biochemical and sociological basis underlying the tradition. The word 'mythology is used here in the sense of popular parables handed down through tradition, containing universal truth mingled with fictional embellishments. The myth in this case represents the popular perception of the tradition, but probably also contributes to the tradition as it is handed down through generations. While an understanding of the basis of the traditional knowledge is an important aim for scientists, the economic importance of the wild plants and their perceived value in society is equally dependent on sustaining the myth in the eyes of the public. As we shall see, the reasons for this are not only metaphysical but have a logical rationale that transcends the purely scientific.

The background of a current lively interest in 'natural products' in their broadest sense, and traditional methods of use of natural resources – and especially of plants – provided one of two important stimuli for the initiation of the Medusa Network; in this account of the development of Medusa, both science and tradition can be seen to play major roles in many aspects of the project and in decisions that had to be taken as the development of the Medusa Database proceeded. The other stimulus was provided by the increasing awareness that biodiversity is being destroyed worldwide at unsustainable rates. In a broad sense the aim of the project was the distillation of the essence of the traditional knowledge into a scientifically based database and research programme to encourage the sustainable use of natural resources of the Mediterranean region.

In 1996 the Department of Natural Products of the Mediterranean Agronomic Institute of Chania took the initiative to establish a network among the different Mediterranean countries; this network would collect data on the identification, conservation and exploitation of the wild plants of the Mediterranean region. As defined in 1996 at the start of the project (Skoula *et al.*, 1997), the long-term aim of the

Medusa Network was: 'To propose methods for the economic and social development of rural areas of the Mediterranean region, using ecologically-based management systems that will ensure the sustainable use and conservation of plant resources of the area.' The particular goal of the network in the shorter term was defined as: 'The exploration of possibilities for the sustainable uses of such resources as alternative crops for the diversification of agricultural production.' But what determines the possibilities for the exploitation of such plants as alternative crops? One factor is certainly a need for improved product quality and a prerequisite for that is improved consistency of products. In order to obtain consistency it is necessary to be sure of exactly what plants are being used and what they contain. For the Medusa project, consistency started with consistency of information and in particular of plant nomenclature. Hardly less important was accuracy of information concerning the relevant chemistry of the plants (most notably the range and concentration of important secondary compounds).

The beginnings of Medusa

The Medusa Network was inaugurated in June 1996, funded by the Directorate General I of the European Union, the Centre International de Hautes Etudes Agronomiques Méditerranéennes (CIHEAM) and the Mediterranean Agronomic Institute of Chania (MAICh). At an early stage a steering committee of experts representing international organisations concerned with plant genetic resources was established, with Professor Vernon Heywood (in his capacity as Chairman of the International Council for Medicinal and Aromatic Plants (ICMAP)) as Chairman and Dr Melpomeni Skoula (MAICh) as Executive Secretary. This group, in turn, convened a network of representatives of institutions in the Mediterranean basin (initially Morocco, Algeria, Tunisia, Egypt, Turkey, Greece, Italy, France, Spain and Portugal). It was intended that the Network would eventually include members from all the Mediterranean countries. Although political realities determined that this would never be completely achieved, representatives from Israel, Lebanon and Malta subsequently contributed to the Network.

The Medusa Database

A major project of the Medusa Network has been the development of the Regional Information System, or Medusa Database. At an early stage it was agreed that the database would contain information on:

- plant names;
- conservation status;

- habitat information;
- uses;
- use-related chemistry.

The initial challenge was to develop suitable procedures and field structures enabling standardised input of the data. An initial trial involving information for more than 700 plant species revealed substantial problems related to the standardisation of plant nomenclature, conservation status and chemistry that were to dominate the discussions throughout the development of the database.

Standardising plant names

Consistency of information content starts with consistency of nomenclature and here the role of plant taxonomists is crucial. The chairmanship of Vernon Heywood ensured that a considerable portion of the initial effort was devoted to procedures for ensuring that plant names could be consistently compared regardless of the source of information.

The absence of any single Flora covering the whole Mediterranean region is a problem that remains to this day, despite the efforts of the Euro+Med project (www.euromed.org.uk.) Not all the countries have completed adequate national Floras (e.g. Greece), some others arguably have too many Floras (e.g. Spain). An awareness of the difficulties and requirements is thus not sufficient in itself to ensure taxonomic consistency from a wide range of contributors in different countries, using different Floras and having varying taxonomic philosophies. The Medusa project lacked the resources to develop a comparative analysis of even the main Floras required to cover the region. In addition, it became clear that – even at this level – tradition as well as science may have played a role in nomenclatural decisions taken by some participants, who were sometimes reluctant to lose a perceived distinction between a local species and its standardised scientific name. In all probability, the use of a different name could simply be erroneous; but is it not also possible that, based on a long-standing tradition, it could be a clue that the local plant is different – in some way not obvious from the morphology – from the (usually more common) species named in the reference Floras? For example, the chemical composition of a plant with traditional uses in medicine is of paramount interest. Yet, although chemotaxonomy is recognised as important, classical taxonomy has understandably been reluctant to acknowledge as distinct species, those morphologically similar plants that differ only in chemical profile. Are some of the different names given to plants locally a reflection of different chemical compositions known historically through the successful traditional use of the plants; and is there a risk that in insisting on standardising the information we are losing valuable local knowledge? A full evaluation of this would have required detailed chemical data for each of the local populations in question, information that was only rarely available (see below). An example that illustrates this point concerns *Origanum* distinguished as two species in, for example, the *Flora of Cyprus* (Meikle, 1977–85). *Origanum dubium* Boiss. is a typical 'oregano' plant while *O. majorana* L. is a typical 'majorana' plant. Both are native in Cyprus and these species are separated on minimal morphological difference and almost entirely on their different monoterpenoid composition (carvacrol-rich and sabinine-rich respectively). However, both *Flora Europaea* (Tutin *et al.*, 1964–80) and *Med-Checklist* (Greuter *et al.*, 1984–) consider them both as *O. majorana* but with two chemotypes. There are numerous examples in the Medusa Database of plants of one species (as defined in the standard Floras) with very different uses reported in different countries and it is not impossible that further analysis will reveal differences, especially chemical differences, between these plants that are nominally of the same species. However, the more likely explanation in most cases is informants' lack of information concerning the complementary uses. In order to try to check these we sent all our informants a data sheet indicating uses they had not reported for particular plants, but which had been reported elsewhere. Unfortunately, this initiative met with little tangible success, and we are at present unsure of the extent to which regional chemical variation within species plays a role in the use of the plants.

Against a background of very variable quality of taxonomic accuracy provided in the trial database, the steering committee and an advisory group convened for the purpose concluded that the advantages of following a standardised nomenclature more than compensated for the risk of losing important information (as in the example above) and that only in this way could the integrity of the database be guaranteed.

How should nomenclature be standardised?

The two most substantial Floras covering the region are the *Flora Europaea* (covering the European part of the Mediterranean comprehensively) and the uncompleted *Med-Checklist*, covering the whole Mediterranean but with only partial floristic coverage. It was soon agreed that where a plant could be identified with either of these works the nomenclature should follow one or the other of these, together with an indication of which work had been used in arriving at the name. While these two works provided a solid taxonomic backbone for the project, many species in the non-European countries were excluded and therefore the same system was extended; for species not covered by these two works a short supplementary list of Floras (mostly for north Africa and the Levant) was appended, again correspondents

Table 10.1. *The Floras accepted for the Medusa Database and the extent of use of each*

Flora	Number of species
Flora Europaea (Tutin *et al.*, 1964–80)	530
Med-Checklist (Greuter *et al.*, 1984–9)	127
Flora of Cyprus (Meikle, 1977–85)	67
Nouvelle flore du Liban et de la Syrie (Mouterde, 1966–84)	170
Flora Palestina (Zohary & Feinbrun-Dothan, 1966–86)	126
Flora of Egypt or *Flora of Egypt Checklist* (Boulos, 1995, 1999–)	71
Students' Flora of Egypt (Täckholm, 1974)	120
Flora of Libya (Ali *et al.*, 1976–80)	0
Flore de l'Afrique du Nord (Maire, 1952–)	17
Nouvelle flore de l'Algerie (Quezel & Santa, 1962–3)	275
Flore practique du Maroc (Fennane *et al.*, 1999)	56
Flora of Turkey (Davis, 1965–88)	406

being required to name the Flora they had used in identifying and naming the plant (see Table 10.1).

Although the system we used enables the user to corroborate the plant name used for a record by reference to the cited Flora, it has one serious disadvantage in that not all plants would be found in the database when searched for under their current accepted name. An example: a much-used plant *Salvia fruticosa* (Greek sage) is correctly named in the *Med-Checklist*, as well as some other Floras; however, in *Flora Europaea, S. triloba* is reported as the accepted name, so a correspondent citing this Flora as his reference would use this name – not *S. fruticosa* – and this reference would not be found on a search for *S. fruticosa*. To deal with this problem we should ideally have included a full list of synonyms in our database, but this substantial task was beyond our resources. However, we could not afford to leave synonymy unresolved altogether. Our temporary solution, until the synonymy list being prepared by Euro+Med PlantBase is available, has been a manually implemented check (the hands being those of Professor Heywood) on those species entered, with names that are not currently the accepted names supplemented by the inclusion of the accepted names in these cases.

Standardising conservation-status information

Many of the focal point coordinators were field botanists with detailed knowledge of the plants and where they were growing in their countries. Some of them had strong views on the conservation status of particular plants; views that

Table 10.2. *Percentage of species in the different IUCN*
threat categories. Overall, threatened species constituted
18% of the total (information taken
from Walter and Gillett (1998))

IUCN category	% of species
Extinct	1
Extinct/Endangered	8
Endangered	23
Vulnerable	31
Rare	23
Indeterminate	14

differed from the IUCN data. Should the Medusa Database accommodate the local knowledge of the national focal-point coordinators, or follow the internationally standardised information contained in the *IUCN* Red Lists. After long discussion it was agreed to follow the standardised route and informants were recommended to follow the 1997 *IUCN Red List of Threatened Plants* (Walter & Gillett, 1998). This searchable database (http://www.unep-wcmc.org/species/plants/plants-by-taxon.htm) was developed by UNEP-WCMC (World Conservation Monitoring Centre), formerly WCMC, in collaboration with the Royal Botanic Garden Edinburgh. In order to offset the potential loss of local knowledge entailed in implementing this route, our correspondents were encouraged to inform the IUCN of discrepancies between IUCN records and the situation on the ground. There is no evidence that this approach was consistently followed, and lack of resources prevented the Medusa team from taking on this task in a coordinated way. The percentages of threatened species reported as being used are shown in Table 10.2.

As a side issue there is an interesting and not insignificant consequence for conservation status arising from occasional changes in taxonomic status of plants. The changing of status of a locally endemic or rare species to a subspecies, arising from careful taxonomic studies may have unexpected consequences; for example, changing the status of a previously rare or endemic species. For example, in the recently published *Checklist of the World's Conifers* (Farjon, 1998), two species of *Abies*, *A. marocana* and *A. pinsapo* (the former endemic to a region of Morocco and the latter endemic to a small area in Spain), are merged as *A. pinsapo* (the Moroccan being distinguished only as subsp. *marocana*). These consequences of such apparently neutral taxonomic activity may also contribute to local botanists wishing to preserve artificially species status for some plants that would thereby have the status of endemics or the interest of rarity. Thus conservation issues do bear

Table 10.3. *Endemic species as a percentage*
of the plants reported as being used

Country	% endemics
Algeria	3
Cyprus	13
Egypt	2
France	0
Greece	8
Israel	1
Italy	1
Lebanon	3
Malta	8
Morocco	6
Portugal	0
Spain	24

on the taxonomic ones and this could be seen from time to time while assembling the Medusa Database. Table 10.3 shows, by country, the percentage of endemic species that are reported as being used.

Use information: the TDWG Standard

We were fortunate that the start of the Medusa project coincided with the publication, after much preparatory work involving many experts from the International Working Group on Taxonomic Databases for Plant Sciences (TDWG), of the *Economic Botany Data Collection Standard* (Cook, 1995). The *Economic Botany Data Collection Standard* is a hierarchical structure with three levels to the main part: Level 1 being the most general, Level 3 the most specific. Thirteen Level-1 states cover all uses of plants and these were adopted unchanged in the Medusa Database (see Figure 10.1). According to the Standard, the plant parts used constitute Level 2 for some uses and Level 3 for others, whereas details of use are sometimes Level 3 and sometimes Level 2. Despite the arguments offered in support of this Standard, this inconsistency causes problems both for users and in structuring the database; hence, in the Medusa Database it was decided after some experimentation that we should use the plant part as Level 2 throughout, leaving the detailed descriptions of the use for each plant part as Level 3 throughout. This had the benefit of keeping Levels 1 and 2 identical for all uses, whereas Level-3 content is specific to each use. However, the structure of the database is such that the information can be recovered in the order proposed in Cook (1995) if required. A practical difficulty associated

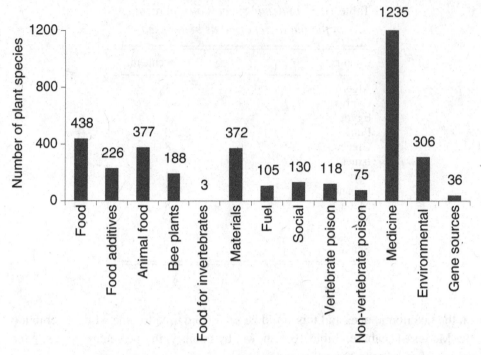

Figure 10.1 The uses of the wild native plants of the Mediterranean as reported to the Medusa Database. All plants are native to the Mediterranean, 87% native in the country concerned, 9% cultivated in the country.

with the TDWG Standard is just how far to go. In practice, a compromise is needed to achieve the best balance between the level of detail set out in the Standard (the ideal record) and a realistic assessment of what might in practice be provided. Especially in complex use categories such as medicine, it is necessary to remember that if the patience of informants is tried too severely they will be less willing to supply useful information – and nothing tries the patience more sorely than being faced with numerous questions that there is little prospect of answering.

Specificity of chemistry information

The detailing of chemistry information posed if anything even greater problems than those associated with plant nomenclature. Much is stressed by taxonomists about the importance of correct plant names in databases. In the case of the Medusa Database, a majority of the records (e.g. Figure 10.2) concern uses, such as food and medicine, where the information concerning the chemical composition of the plants is of an equal level of importance. In an ideal situation it is easy to recognise two criteria of overriding importance. First, the chemical composition should be

PLANT:
ACCEPTED PLANT NAME:
Sideritis syriaca L. syriaca
PLANT NAME IN MEDUSA DATA BASE:
1086 Sideritis syriaca L. syriaca
Greece
Native to the Mediterranean; Native in the country
TAXONOMIC SOURCE:
flora europaea
COMMON NAME(S):
malotira (greek), tsai tou vounou (greek)
CONSERVATION INFORMATION:
IUCN Not threatened (nt) endemic to the country
TRADE INFORMATION:
Traded locally frequently; Traded nationally; Traded inter-
nationally within EU; Traded internationally beyond the EU
HABITAT INFORMATION:
forest; shrubland
SOIL PREFERENCE:
calcareous
LIFE-FORM:
tree
ALTITUDE RANGE:
Less than 1000 metres
GEOGRAPHICAL DISTRIBUTION:
Region:

USES:
food
SPECIFIC USE INFORMATION:
other non alcoholic; infusion with Origanum microphyllum as
a breakfast drink, occasionally added Origanum dictamnus
PLANT PARTS AND RELATED CHEMISTRY:
Leaves
  Fatty Acids
  Flavonoids: Flavones
  Other Phenolics: Phenolic Acids
  Terpenoids: Monoterpenoids, Sesquiterpenoids, Diterpenoids
Inflorescence
  Fatty Acids
  Flavonoids: Flavones
  Other Phenolics: Phenolic Acids
  Terpenoids: Monoterpenoids, Sesquiterpenoids, Diterpenoids
GEOGRAPHY OF USE:
   Region:
EXTENT OF USE: 5
ADDITIONAL USE INFORMATION:
Local: Fragaki E. 1969. Contribution in common naming of
native, naturalised, pharmaceutical, dye, ornamental and
edible plants of Crete. pp. 242. Athens. (In Greek). Havakis I.
1978. Plants and Herbs of Crete. Athens. pp. 315. (In Greek).
Skoula M., Personal data
General:
ADDITIONAL CHEMISTRY INFORMATION:
Local: (beta-caryophyllene, thymol) Aligiannis, N. Kalpoutzakis, E. Chinou, I. B. Mitakou, S. Composition and antimicrobial
activity of the essential oils of five taxa of Sideritis from Greece. Journal of Agricultural & Food Chemistry. 2001. 49: 2, 811-815.
(myrcene) Todorova, M. N. Christov, R. C. Evstatieva, L. N. Essential oil composition of three Sideritis species from Bulgaria.
Journal of Essential Oil Research. 2000. 12: 4, 418-420. Laer, U. Glombitza, K. W. Neugebauer, M. The essential oil of Sideritis
syriaca. Planta Medica. 1996. 62: 1, 81-82. Venturella, P. Bellino, A. Marino, M. L. Three acylated flavone glycosides from
Sideritis syriaca. Phytochemistry. 1995. 38: 2, 527-530. (hypolaetin-4'-methylether-7-O-[6'''-O-acetyl- beta -D-allopyranosyl-(1
right arrow 2)- beta -D-glucopyranoside]) Sattar, A. A. Bankòva, V. Spassov, S. Duddeck, H. Flavonoid glycosides from Sideritis
species. Fitoterapia. 1993. 64: 3, 278-279. Venturella, P. Bellino, A. Marino, M. L. Siderone, a diterpene from Sideritis syriaca.
Phytochemistry. 1983. 22: 11, 2537-2538.
General:

Figure 10.2   Sample record of the Medusa database.

bee plants
    SPECIFIC USE INFORMATION:

    PLANT PARTS AND RELATED CHEMISTRY:
    Exudates: Nectar
    GEOGRAPHY OF USE:
    Region:

    EXTENT OF USE: 0
    ADDITIONAL USE INFORMATION:
    Local: Harizanis, P. The Honey bee and the beekeeping
    techniques. 1993. pp.254 (in Greek).
    General:
    ADDITIONAL CHEMISTRY INFORMATION:
    Local:
    General:

medicine
    SPECIFIC USE INFORMATION:
    Disorders treated: blood system disorders, nutritional disorders,
    mental disorders, injuries and wounds, infections/infestations,
    endocrine system disorders, digestive system disorders.
    digestive system disorders; high cholesterol
    blood system disorders;
    nutritional disorders; obesity
    endocrine system disorders;
    injuries and wounds;
    infections/infestations; fever
    digestive system disorders; lowers cholesterol
    blood system disorders; blood clearing
    nutritional disorders; helps weight loss
    endocrine system disorders; sweating injuries and wounds;
    wound healing infections/infestations; reduce fever
    Vertebrates treated: Mammals; Primates
    PLANT PARTS AND RELATED CHEMISTRY:
    Leaves
        Fatty Acids
        Flavonoids: Flavones
        Terpenoids: Monoterpenoids, Sesquiterpenoids, Diterpenoids
    Inflorescence
        Fatty Acids
        Flavonoids: Flavones
        Terpenoids: Monoterpenoids, Sesquiterpenoids, Diterpenoids
    GEOGRAPHY OF USE:
        Region:
    EXTENT OF USE: 3
    ADDITIONAL USE INFORMATION:
    Local: Fragaki E. 1969. Contribution in common naming of native,
    naturalised, pharmaceutical, dye, ornamental and edible plants of
    Crete. pp. 242. Athens. (In Greek). Havakis I. 1978. Plants and
    Herbs of Crete. Athens. pp. 315. (In Greek). Skoula M., Personal
    data
    General:
    ADDITIONAL CHEMISTRY INFORMATION:
    Local: (beta-caryophyllene, thymol) Aligiannis, N. Kalpoutzakis, E. Chinou, I. B. Mitakou, S. Composition and antimicrobial
    activity of the essential oils of five taxa of Sideritis from Greece. Journal of Agricultural & Food Chemistry. 2001. 49: 2, 811-
    815. (myrcene) Todorova, M. N. Christov, R. C. Evstatieva, L. N. Essential oil composition of three Sideritis species from
    Bulgaria. Journal of Essential Oil Research. 2000. 12: 4, 418-420. Laer, U. Glombitza, K. W. Neugebauer, M. The essential oil
    of Sideritis syriaca. Planta Medica. 1996. 62: 1, 81-82. Venturella, P. Bellino, A. Marino, M. L. Three acylated flavone
    glycosides from Sideritis syriaca. Phytochemistry. 1995. 38: 2, 527-530. (hypolaetin-4'-methylether-7-O-[6'''-O-acetyl- beta -D-
    allopyranosyl-(1 right arrow 2)- beta -D-glucopyranoside]) Sattar, A. A. Bankova, V. Spassov, S. Duddeck, H. Flavonoid
    glycosides from Sideritis species. Fitoterapia. 1993. 64: 3, 278-279. Venturella, P. Bellino, A. Marino, M. L. Siderone, a
    diterpene from Sideritis syriaca. Phytochemistry. 1983. 22: 11, 2537-2538.
    General:

Figure 10.2   (cont.)

that of exactly the plant reported; and secondly, the chemistry reported should be 'use-related' chemistry, rather than simply a list of the chemical composition of the plant. One of the outcomes of the Medusa project has been the realisation that only rarely is information available in a way that satisfies either criterion.

If we had restricted the input of chemical data to satisfy the first requirement, the database would have contained very few records indeed with chemistry information. Some records contain chemical data for plants from the same country, others for the same general region, many only very general information (even from a completely different region of the world) but there are only a few records where the plant identified is directly tied to its chemistry. Even when genuinely local information is available, the situation is not completely clear until it is determined that any variation in 'local' chemical composition and content is genetically rather than environmentally determined. So far, only a few studies have been reported that address this problem, for example by comparing chemical composition of different 'chemotypes' under standard environmental conditions (Skoula *et al.*, 2000, Gotsiou *et al.*, 2002).

The second criterion has also proved difficult. On the face of it the notion of limiting chemical information to 'use-related' chemistry seems clear and sensible. The definition was intended to prevent a full list of every plant constituent from chlorophyll to auxin. However, closer scrutiny (and hindsight) shows that this criterion is usually even more difficult to meet than the first because it implies that we know already, for example, the chemical basis on which a plant is valuable medicinally. For most examples this is simply not the case. Additionally, it is clear that once we do know precisely what compounds produced by a plant are medicinally important (to keep with this example), then the plant soon becomes less important for this purpose: usually the compounds in question can then be chemically synthesised.

Such cases are in any case the rare exceptions. Usually we don't know the chemical information. One reason for this is that plants are perceived as being more effective (in foods, in medicine, as poisons, etc.) than those individual compounds that have been isolated from them and shown to possess appropriate properties. This popular assumption itself consists of two components: one entirely without scientific foundation, the other only partly mythological. The first is the assumption that there is something about a plant or plant extract that is intrinsically greater than the sum of the chemicals it contains. This idea – which has some popular currency – implies that the crucial properties of the plant, handed down by tradition, are not amenable to conventional scientific analysis. Although dealing with such ideas is no part of the Medusa project, they cannot be ignored completely since they contribute to the demand for 'natural' products. The second assumption is based on the notion of synergy between compounds present in plants, leading to effects that are much greater than those of the sum of individual compounds. There are enough known examples of synergistic interactions between compounds to provide fuel for the assertion that plant extracts may be considerably more effective than the individual compounds that have been isolated from them and shown to have active properties.

This assumption is in fact supported by a (small) number of well-founded examples. One of the most striking of these to have recently been elucidated concerns the antimicrobial action of extracts from *Berberis* spp. Several species of *Berberis* show strong antimicrobial properties. Such species produce significant quantities of an alkaloid with demonstrable antimicrobial properties, berberine. However, purified berberine exhibits much less antimicrobial activity than a plant extract containing similar quantities of the compound. It has now been shown (Stermitz *et al.*, 2000) that this is because many bacteria have evolved a mechanism to pump foreign molecules such as berberine out of the cell. These pumps are called multidrug resistance pumps (MDRs) and such pumps greatly reduce the antimicrobial effectiveness of compounds such as berberine. *Berberis* plants, however, contain a compound unrelated to berberine, $5'$ methoxyhydnocarpin ($5'$-MHC). This compound has no antimicrobial activity per se but acts by inhibiting the action of the MDR in the bacterium, with the result that the berberine in the plant extract is several orders of magnitude more effective than berberine administered alone.

Examples like this explain why sometimes plants or plant extracts are more effective than the biologically active compounds isolated from them, and the continuing discovery of such synergistic effects supports those scientists who argue that synergy is a norm rather than an unusual phenomenon. Indeed, the confidence of the public in the value of plant products depends in significant measure upon such generalisations. Although in the case of *Berberis* the use of two compounds will enable the plant-free production of an effective antimicrobial agent (eliminating the need for the plant extract), the assumption that there are many other examples as yet undiscovered feeds the 'synergy myth' and thereby the demand for 'natural products.'

These considerations highlight the difficulties in selecting which chemical information is best included in Medusa records. It is usually impossible to be sure what the 'use-related' compounds are going to be in advance of the sort of discovery described above. For the Medusa Database we were ultimately forced to a pragmatic solution. Almost all those records in the Database that contain chemical data comprise lists of chemicals, mainly secondary products, to provide a guide as to what the possible candidates for 'use-related' chemistry might be. Where these compounds are thought by the correspondent (or by the editors) to be likely to be similar to those of the plants in question (e.g. information from approximately the same region), this information has been classified as 'local'; if the information has been collected from elsewhere, it is classified as 'general'. Where possible the references are given (as in the sample record included as Figure 10.2). In practice, so few of the records are absolutely specific to the plant sample in question that a further category indicating such examples has not so far seemed worthwhile. However, it is clear that as far as those plants whose utility is dependent on their chemical composition are concerned, their description and indeed their taxonomy is incomplete without a chemotaxonomic component. Much work remains to be done before this information will be available for many important wild species. This finding is in itself an important outcome of the Medusa programme.

Conclusions

At the end of the current period of funding, although good progress has been made with the Medusa Database, the ambitious aims of the Medusa project as a whole are some way from being realised. It is clear from the data collected that, not-least through the efforts of Vernon Heywood, the database is on a firm-footing taxonomically; nevertheless, much information needs to be collected – especially relating the chemistry much more closely to taxonomy and use – if we are to distil from mythology and tradition a solid scientific basis for future exploitation, and thereby conservation, through use of these valuable plants. The opportunity for providing this information and developing it now needs to be grasped. The Medusa Database as it stands today can be viewed and searched at http://medusa.maich.gr.

References

Ali, S. I., Jafri, S. M. H. & El-Gadi, A. (1976–80). *Flora of Libya*. Tripoli, Libya: Al Faateh University.

Boulos, L. (1995). *Flora of Egypt Checklist*. Cairo, Egypt: Al Hadara Publishing. (1999–). *Flora of Egypt*. Cairo, Egypt: Al Hadara Publishing.

Cook, F. E. M. (1995). *Economic Botany Data Collection Standard*. Kew, UK: The Royal Botanic Gardens.

Davis, P. H. (ed.) (1965–88). *Flora of Turkey and the East Aegean Islands*, vols. 1 to 10. Edinburgh, UK: University Press.

Farjon, A. (1998). *Checklist of the World's Conifers*. Kew, UK: The Royal Botanic Gardens.

Fennane, M., Ibn Tattou, M., Mathez, J., Ouyahyia, A. & El Oualidi, J. (1999). *Flore practique du Maroc*, vol. 1. Rabat, Morocco: L'Institut Scientifique.

Gotsiou, P., Naxakis, G. & Skoula, M. (2002). Diversity in the composition of monoterpenoids of *Origanum microphyllum* (Labiatae). *Biochemical Systematics and Ecology*, **30**, 865–9.

Greuter, W., Burdet, H. M. & Long, G. (1984–). *Med-Checklist: A Critical Inventory of Vascular Plants of the Circum-Mediterranean Countries*. Geneva: Editions des Conservatoire et Jardin Botaniques de la Ville de Genève.

Maire, R. (ed.) (1952–). *Flore de l'Afrique du Nord*. Paris: Lechevalier.

Meikle, R. D. (1977–85). *Flora of Cyprus*. Kew, UK: Bentham-Moxon Trust, The Royal Botanic Gardens.

Mouterde, P. (1966–84). *Nouvelle flore du Liban et de la Syrie*, eds. A. Charpin & W. Greuter. Beyrouth, Lebanon: Dar El-Machreq.

Quèzel, P. & Santa, S. (1962–1963). *Nouvelle flore de l'Algérie et des régions désertiques méridionales*. Paris: Centre national de la recherche scientifique.

Skoula, M., Griffee, P. & Heywood, V. H. (1997). Identification, conservation and use of wild plants of the Mediterranean region: the Medusa Network. In *Identification of Wild food and Non-Food Plants of the Mediterranean Region*, eds. V. H. Heywood & M. Skoula. *Cahiers Options Méditerranéennes*, **23**, 1–3.

Skoula, M., Abbes, J. E. & Johnson, C. B. (2000). Genetic variation of volatiles and rosmarinic acid in populations of *Salvia fruticosa* Mill. growing in Crete. *Biochemical Systematics and Ecology*, **28**, 551–61.

Stermitz, F. R., Lorenz, P., Tawara, J. N., Zenewicz, L. A. & Lewis, K. (2000). Synergy in a medicinal plant: antimicrobial action of berberine potentiated by 5'-methoxyhydnocarpin, a multidrug pump inhibitor. *Proceedings of the National Academy of Sciences of the USA*, **97**, 1433–7.

Täckholm, V. (1974). *Students' Flora of Egypt*, edn. 2. Cairo, Egypt: Cairo University.

Tutin, T. G., Heywood, V. H., Burges, N. A. *et al.* (1964–80). *Flora Europaea*. Cambridge, UK: Cambridge University Press.

Walter, K. S. & Gillett, H. J. (1998). *1997 IUCN Red List of Threatened Plants*. Gland, Switzerland and Cambridge, UK: IUCN.

Zohary, M. & Feinbrun-Dothan, N. (1966–86). *Flora Palestina*. Jerusalem: The Israel Academy of Sciences and Humanities.

11

Chemosystematics, diversity of plant compounds and plant conservation

Renée J. Grayer

Introduction

During the course of evolution, plants have evolved the ability to synthesise a great variety of molecules to interact with their environment: e.g. to ward off insect pests and microbes, to deter herbivores grazing their leaves and to prevent seeds of competing species germinating in their proximity. Other functions of these compounds include attracting insects and other animals to pollinate their flowers or disperse their seeds, and protection against harmful ultraviolet radiation. Each plant species produces a characteristic set of hundreds of these ecologically important metabolites and this chemical profile can be used as a fingerprint for the species, to distinguish it from related and unrelated species. On the other hand, the production of certain classes of chemicals tends to 'run in families'; e.g. betalains occur exclusively in plant families belonging to the order Caryophyllales. Therefore, from the types of compounds produced by a species it is often possible to identify to which plant family or order it belongs. This is the reason why plant chemicals can provide useful characters for plant systematics and why the study of the diversity of plant constituents and their distribution in the plant kingdom is often called chemosystematics or chemotaxonomy. Biochemical systematics and comparative phytochemistry are other frequently used synonyms for this discipline.

At present the term 'chemosystematics' is usually restricted to the study of compounds of low molecular weight; these are often called 'secondary metabolites' or 'micromolecules'. According to natural-product databases such as the NAPRALERT File (NAtural PRoducts ALERT) (2005), 100 000 to 130 000 different compounds have been reported from natural sources. Approximately 60 000 of these comprise compounds from plants, whereas the remainder comes from microorganisms and animals. As only a small percentage of plant species have been studied thoroughly for secondary compounds, it is estimated that the total number of different secondary substances present in plants is at least ten times higher.

191

This means that the total number of different micromolecules produced by plants may be at least in the same order as the number of extant species of angiosperms, which is currently estimated to be 400 000 (Heywood, 2003). Conservation of plants, therefore, also means conservation of the diversity of plant-derived constituents and all their potential uses. For example, compounds used by plants to ward off fungal infections could be employed as antifungal drugs or as fungicides against diseases of crop plants; constituents present in plants to deter insects could be used as environmentally friendly insecticides etc. On the other hand, the use of a plant substance, e.g. as an anticancer drug, may have nothing to do with the actual function of the bioactive compound in the plant. For instance, the alkaloids vincristine and vinblastine, present in the Madagascar periwinkle *Catharanthus roseus*, perhaps function in the plant to protect its tissues against insect attacks, but by coincidence are very effective in humans against certain cancers. These alkaloids have already been used very successfully for several decades in the treatment of leukaemia and lymphoma (Lewis, 1982). More recently the alkaloid taxol, isolated from *Taxus brevifolia*, has shown activity against ovarian cancer (see Simmonds & Grayer, 1999). But it is not only pure plant compounds that are useful for medicinal purposes. Humans have used crude extracts of plants as medicines since antiquity and these plant preparations still play an important role as medicines in many countries (Padulosi *et al.*, 2002). A mixture of many different plant species is often present in medicines used by Traditional Chinese Medicine (TCM), still practised extensively in China (Farnsworth & Soejarto, 1991). Also in the West, medicinal plant preparations are becoming more and more popular again, not only from oriental species, but also native European plants such as St John's wort (*Hypericum perforatum*). Flower-bud preparations from the latter species are now for sale in pharmacies and health-food shops in many countries and are a popular alternative medicine for treatment of depression. The medicinal qualities of plant constituents provide one more argument to preserve the biodiversity of plants, because when a species becomes extinct the usefulness of its constituents will never be known. Unfortunately, the medicinal properties of plants have more often threatened plant species with extinction than protected them, because of over-collection in the wild. This applies to many medicinal plants used in developing countries, e.g. *Prunus africana* in Africa (Padulosi *et al.*, 2002), but is also happening in the West. Two decades ago *Drosera rotundifolia* was over-collected in the wild by a German manufacturer of pharmaceutical products, even though this insect-eating species was already under threat in many European countries by loss of habitat and by horticultural collectors (Grayer, 1983). More recently, the same has happened to wild populations of *Hypericum perforatum* in Eastern Europe as a result of collectors selling the flower buds to companies dealing in medicinal herbs. Sustainable use of medicinal and aromatic plants is one of the solutions to this problem (Padulosi *et al.*, 2002). If plants of the endangered species were cultivated in quantities sufficient for the demands, they would not have to be collected from the wild. Furthermore, in an ideal world the medicinally or economically important leaves or fruits of threatened wild species would be harvested in a sustainable manner. There are also several ways in which chemosystematic information on the diversity of plant constituents and chemosystematic procedures can be employed to aid plant conservation. This will be discussed below, but first the latest views are presented on why and how plants produce such a great diversity of secondary compounds and why this diversity is important for plant conservation.

Evolution and diversity of plant secondary compounds

As secondary compounds are essential for the survival of plants, these substances have probably played a much more important role in the evolution of plants than most botanists realise. The biosynthesis of each plant constituent involves a multitude of enzymes and these are coded for by a multitude of genes. Recent achievements in molecular biology, such as the *Arabidopsis thaliana* gene project (e.g. Bevan *at al.*, 1998), have given us insight into how new genes evolve and, therefore, how the production of the enormous diversity of plant constituents could have evolved. In addition, the latest classifications based on DNA sequences of slowly evolving genes, such as the rbcL gene of chloroplasts (Angiosperm Phylogeny Group, 1998), have made it possible to trace the evolution of secondary-compound classes in the major plant groups (Kite *et al.*, 2000). It has been estimated that the genome of each plant species contains approximately 20 000 to 60 000 genes, up to a quarter of which encode enzymes for secondary metabolism (Pichersky & Gang, 2000). Some of these genes are almost universally present in the plant kingdom, whereas others are highly specific to certain plant groups or even to just one species. Over 500 million years ago, when the Earth was bombarded with strong ultraviolet irradiation, the only plants on Earth were probably primitive photosynthesising algae; these lived in the sea and, therefore, were protected from this radiation, since the solutes in sea water absorb ultraviolet light. Initially, these algae may have produced almost exclusively primary compounds – such as amino acids, fatty acids, nucleic acids and sugars, needed as building blocks and for the metabolism of the cells. Additionally, they will have produced photosynthetic pigments such as chlorophylls and carotenoids. They must have had a reasonably large genome, encoding all the enzymes required for those products. It is now thought that new genes normally evolve at random by gene duplication and then diverge; the original gene maintains the original function while the copy, after accumulating mutations over a long period of time, could acquire a new function (Pichersky, 1990). This means that enzymes that produce secondary plant compounds probably have evolved – initially by accident – from enzymes used in primary metabolism, but were genetically fixed when they acquired some useful ecological function advantageous to the producing organism. For example, new enzymes that slightly altered the biosynthetic pathway of protein amino acids may have catalysed reactions resulting in the production of various types of secondary compounds – such as non-protein amino acids, alkaloids and cyanogenic glycosides. Some of these compounds happened to protect the plants producing these substances against herbivores or pathogenic microorganisms and, therefore, became genetically fixed. Carotenoids and the phytol side chain of chlorophyll, present in the great majority of plants, are terpenoid in origin and are biosynthesised from isoprenyl units. These isoprenyl precursors became the building blocks of a large range of terpenoid secondary products such as monoterpenoid and sesquiterpenoid essential oils, iridoids and diterpenoids; many of these show a range of biological activities and are potentially useful substances. Only small changes in the amino-acid sequences of the active site in the enzymes may have been sufficient to produce this wide range of products (Pietra, 2002).

New genes for secondary metabolism may not only arise by gene duplication and divergence, but also by allelic divergence without a prior gene duplication (Pichersky & Gang, 2000). Most of the new genes, whether formed by duplication or not, will have evolved very slowly, but during the course of hundreds of millions of years of evolution they resulted in a gigantic number of new secondary products. However, in some cases there may have been abrupt major changes, resulting in very different enzymes and novel types of compounds. I believe that the production of some of these new compound groups were key innovations, which changed the course of evolution. For example, a novel enzyme that combined two biosynthetic pathways used in primary metabolism, that of the amino acid phenylalanine and the acetate/malonate pathway, led to the production of a chalcone, which is the precursor of all flavonoids. This enzyme, chalcone synthase, is present in all terrestrial plants (mosses, liverworts, ferns, gymnosperms and angiosperms). It is also present in a few terrestrial algae such as *Chara*, but is absent from all green aquatic algae. It seems likely that the enzyme evolved hundreds of millions of years ago in a *Chara*-like green alga. Green algae, and especially the Charaphyceae, share a number of biochemical features with terrestrial plants and are thought to be closely related to this group (Kendrick & Crane, 1997). Therefore, perhaps a *Chara*-like green alga in which chalcone synthase evolved, may have been the ancestor of all terrestrial plants. Why was this enzyme so important that it might have initiated the evolution of land plants? The reason may be because flavonoids, the products of chalcone synthase activity, strongly absorb damaging ultraviolet irradiation at a wide range of wavelengths, including ultraviolet-B (Markham *et al.*, 1998), and this acquired protection may have made it possible for plants to come out of the sea and conquer the land. Therefore, the production of flavonoids may have been a major driving force in the initial evolution of terrestrial plants and without flavonoids terrestrial plants might not have evolved the way they have, so that the world would have looked quite different!

The origin of several other compound groups also may represent key innovations, although not such dramatic ones as that of flavonoids. For example, the ability to produce iridoids must have evolved in the ancestor of the Asteridae (*Sensu* The Angiosperm Phylogeny Group (1998)), as these terpenoid constituents are found in all lineages of this species and family-rich superorder, but in only one or two species outside this taxon (Grayer *et al.*, 1999). Therefore, the ability to produce iridoids, which are thought to protect the plants against pests and diseases (Van der Sluys, 1985), may initially have been the driving force behind the evolution of this group of plants, which presently comprises half of all extant angiosperm species. Later, iridoids were replaced in some of the Asteridae families by even more successful compound groups, e.g. sesquiterpene lactones in the Asteraceae, the biosynthesis of which may use the same isoprenoid precursors as iridoids.

It has taken plants hundreds of millions of years to evolve the enormous structural diversity of constituents globally present in extant species at this moment in time. However, perhaps only 10% of the different compounds present in the Plant Kingdom have been structurally elucidated. This means that 90% of plant substances are still waiting to be discovered. A large proportion of these compounds will be present in rare species in areas difficult to access, such as primary rain forests; the very places that are most under threat (Hamann, 1991). Therefore, if these habitats are not preserved, a large proportion of the diversity of plant-derived constituents will be lost before their function in the plant and their potential uses for mankind can be studied.

Chemosystematics and plant conservation

There are a number of ways in which chemosystematic data, or the methods and procedures used in chemosystematics, can help the conservation of plants. Examples are listed below.

1. Showing the economic value of threatened plants.
2. Providing the tools to identify rare and endangered species and to assess their genetic diversity, especially of medicinal or aromatic plants.
3. Indicating alternative plant sources of valuable medicinal drugs that contain the same or similar active ingredients, or their biogenetic precursors, to endangered medicinal plants. Furthermore, methods can be developed to obtain the medicinally active constituents from tissue or cell cultures of those plants.

Chemosystematics and the economic value of threatened plants

A strong argument for the conservation of plants is their value to mankind. Since time immemorial plants have provided not only food, but also the raw materials for shelter, furniture, clothes, cosmetics, perfumes, dyes, weapons, for making boats and fishing nets, and treating many ailments and diseases. In many developing countries, the majority of people cannot afford Western drugs and therefore still use medicinal plants to treat illnesses (Akerele, 1991). More than 9000 species of plants are known to have been used ethnobotanically (Farnsworth & Soejarto, 1991), but the real number is probably much higher than that. Although many plant-based materials have been replaced in developed countries by synthetic ones in the last 50 years, the value of plant products is still enormous and is estimated to be approximately US$ 500 to 880 billion a year (Cordell, 2000). A large percentage of new anticancer medicines is still being developed from plant constituents; for example, 61% of the anticancer agents approved in the 1983–94 period were natural products or their derivatives (Cragg et al., 1997). Additionally, medicinal plant substances have given the chemical blueprints for the development of related synthetic drugs (Hamann, 1991). In the last decade the pharmaceutical industry has attempted to bypass natural products by trying to obtain novel compounds by means of so-called 'combinatorial chemistry'; this is the production of novel synthetic compounds by means of automated systems. However, the products obtained could not match the enormous chemical diversity and bioactivities of the structures developed and selected by plants over hundreds of millions of years of evolution. The latest trend in pharmaceutical research is to use natural plant compounds as starting substrates for a variety of plant enzymes to create a myriad of novel structures, which can be tested for their medicinal potential. This field has been called 'combinatorial biochemistry' (Frick et al., 2001).

It is absolutely essential to create value for the diversity of plants and their products in order to preserve it (Cordell, 2000). For example, the potential value of tree species in primary rain forests can be calculated in terms of their value if they are used to attract tourists interested in the flora and fauna, or if they are employed in a sustainable way to produce medicines and environmentally friendly pesticides. This information could prevent these trees from being cleared for their timber or just for obtaining land for cattle grazing. Furthermore, if useful products such as medicines could be prepared from these trees in a sustainable manner, it could provide jobs for the local population. In addition, medicines, cosmetics and pesticides produced in this way would reduce the need to import expensive products from developed countries. However, before it is possible to establish the potential value of species in areas of rich biodiversity, it is necessary to catalogue the uses of the plant species in those areas and all the chemicals known to be present in them – in addition to taxonomic, distribution and conservation data on those species. A database should be created, in which this information is gathered from existing literature and from new chemosystematic surveys, as many plants in species-rich areas have not yet been investigated chemically. Those chemical studies should preferably be carried out at local universities or research institutes. Projects of this type are already being carried out in special research institutes in developing countries; for example, Iwokrama International Centre in Guyana, Empresa Brasileira de Pesquisa Agropecuária – Embrapa (Brazilian Agricultural Research Corporation) in Brazil and the Instituto Nacional de Biodiversidad (INBio) (National Biodiversity Institute) in Costa Rica. However, it is often difficult to obtain funds for this kind of research.

Chemosystematic tools to identify species and assess the genetic variation within plant populations

One of the most important taxonomic procedures for the purpose of plant conservation is the identification of plant specimens to the species level. The determination of the genetic diversity within endangered species is also very important. Both tasks can be carried out in the first instance by means of morphological characters: complemented where possible by anatomical, palynological and cytological features. Chemical characters such as spot patterns on paper or thin-layer chromatograms of plant extracts have also been used to identify and distinguish species since the 1960s. In the last 15 years, DNA fingerprint systems, such as RAPDs (random amplification of polymorphic DNA), Inter-SSR (inter simple sequence repeat) and AFLP (amplified fragment-length polymorphism) (Bachman *et al.*, 1999; Culham & Grant, 1999), have provided valuable additional evidence, especially to distinguish different genotypes within species. Recently, chemical techniques also have become much more suitable for fingerprinting species and genotypes. Indeed, chemosystematics has come a long way since the 1980s, as sophisticated and accurate chemical techniques have now been developed that make it possible to identify and quantify the compounds present in plants much more accurately. These methods are commonly referred to as 'hyphenated techniques' (Kite *et al.*, 2003), for example: GC-MS and LC-MS (gas chromatography and liquid chromatography, respectively, coupled with mass spectrometry); HPLC-DAD and CE-DAD (high-performance liquid chromatography and capillary electrophoresis, respectively, coupled with diode array detection); and HPLC-NMR (HPLC coupled with nuclear magnetic resonance spectroscopy). Using such techniques, chemical fingerprints can be obtained that distinguish between closely related species and also discriminate between different chemical genotypes (= chemotypes) within species (Vieira *et al.*, 2001, 2003), so that they can be used to assess the genetic diversity of threatened plant species. These chemical methods are generally very fast to carry out (less than one hour's work for each plant extract) and generally need only 50 to 500 mg of dried plant material. They are especially valuable in the case of identifying medicinal or aromatic plant species, as the chemical composition of these plants is very important from an economic point of view. Plant material used in herbal medicine, including Traditional Chinese Medicine, often consists of leaves or roots and does not contain the flowers or fruits that are frequently required for the morphological identification of plant species. The material is often ground, which makes identification by morphological characters almost impossible and, furthermore, medicines used in Traditional Chinese Medicine are often complex mixtures of different plant species, to which non-vegetable matter may have been added, e.g. bear bile. Although DNA techniques can be used for the identification of the species used in these medicines, chemical analysis is the method of choice – especially for quality control. Using chemical techniques, tests can be carried out to determine whether all the required species and their required active ingredients are present in a certain mixture of plants. Furthermore, chemical analysis can show whether the material has been adulterated with undesirable or even toxic species (Kite *et al.*, 2002), and – from a conservation point of view – whether the herbal material contains any protected species, which should not be used. Many medicinal plant species have become rare or endangered because of over-collecting, and in some cases these can be replaced with more common species with similar constituents and medicinal effects. However, a better way, for the purposes of plant conservation, would be the cultivation of medicinal species, since even the common ones eventually may become rare if the demand increases.

Chemical characters should be used in addition to morphological and molecular ones to study the genetic diversity of an endangered medicinal or aromatic plant species and to select genotypes with the right medicinal or essential-oil traits within a species. This is important, because there is not always a correlation between chemical and molecular characters such as RAPDs, or between chemical and morphological features (Grayer *et al.*, 1996; Vieira *et al.*, 2003). Therefore, assessments of genetic diversity of medicinal or aromatic species based only on morphological or molecular assessments may not always give meaningful results.

An example of the use of chemical techniques to protect threatened plant species is the development of chemical fingerprints for species in the genus *Cryptocoryna* Fischer ex Wydler (*Araceae*). This is a genus of aquatic plants that occurs in the tropics from India to Papua, New Guinea. There are around 50 species worldwide (Mabberley, 1987), 10 of which occur in Sri Lanka. Plants belonging to this genus are widely used as ornamentals in aquaria. Some species are common, but the endemic species in Sri Lanka are becoming extremely rare because of over-collection and therefore they are now protected. It is easy to identify and distinguish the common and endangered species using characters of flowers and fruits. The difficulty is, however, that the exported material consists mainly of leaves; these are variable in shape and colour, depending on environmental factors. Therefore, customs officers and even taxonomists cannot easily distinguish leaf material of the common and endangered species. Chemotaxonomic methods have been developed to distinguish between the species and these chemical techniques are now used in addition to morphological and molecular methods (RAPDs) to identify the species and prevent the endangered species being exported (G. I. Seneviratne, personal communication, 2003).

Alternative sources for valuable plant constituents

When a medicinal plant that produces a valuable medicinal drug becomes endangered by over-collection, a chemosystematic approach is worthwhile; i.e. to find out whether there are related species that produce the same constituent. For example, clerodendroside is a flavonoid with antihypertensive activity, which until recently was known from only one plant source, *Clerodendrum trichotomum* (Lamiaceae, formerly Verbenaceae) (Morita *et al.*, 1977). Recently, a chemotaxonomic survey of species in related genera has shown that this compound also occurs in very high concentrations in the garden plant *Tripora divaricata* (Maxim.) P. D. Cantino (= *Caryopteris divaricata* Maxim.) (Grayer *et al.*, 2002); therefore, this species could be used for the production of clerodendroside if necessary. In principle, it is possible to look up alternative sources for valuable plant constituents in natural-product databases. However, many phytochemical journals have made it their policy to publish only papers describing new compounds, and not plant sources of known constituents (even if they are very rare), so that in practice it may be difficult to find the required information. An excellent source of reference on the occurrence of certain natural constituents in the Plant Kingdom is the 11 volumes of the *Chemo-taxonomie der Pflanzen* by Hegnauer (1962–2001). The literature quoted in this work extends from the most recent to nineteenth-century references, which is very useful as most commercial databases do not go back further than 10 to 20 years.

As mentioned before, the alkaloids vincristine and vinblastine from the leaves of *Catharanthus roseus* (Apocynaceae) are very valuable agents against certain forms of leukaemia. However, the amounts of these alkaloids produced in the leaves are extremely low, so that tens of kilograms of plant material are needed to produce 1 mg of the drugs. Therefore, much research has been carried out to elucidate the biosynthesis of these compounds, to find precursors that could be used as starting points for synthesising these drugs. One of the main precursors in the biosynthesis of these alkaloids is secologanin, a seco-iridoid. This is a well-known chemotaxonomic marker, characteristic of many species of Apocynaceae and also of related families such as Rubiaceae and Loganiaceae (Hegnauer, 1989, 1990). So there is no shortage of this biogenetic precursor!

Another alternative method to obtain a medicinally active constituent in cases where the plant source is endangered, is to develop cell-suspension cultures of those plants that produce the required compounds (Zenk, 1991). When successful, the threatened species no longer has to be collected from the wild. Other advantages of cell-suspension cultures are that with the addition of the right elicitors they can produce much higher levels of the required metabolites than the whole plant, and that the compounds can be produced all year round rather than seasonally. However, this and other technologies for the production of plant drugs, such as semi- or total synthesis, are still very expensive. Therefore, in the short run, the sustainable utilisation of medicinal plants should be the primary option to prevent medicinal species becoming endangered. Additionally, their cultivation and processing are likely to benefit local people by providing employment and cheaper medicines, especially in developing countries.

References

Akerele, O. (1991). Medicinal plants: policies and priorities. In *The Conservation of Medicinal Plants,* eds. O. Akerele, V. Heywood & H. Synge. Cambridge, UK: Cambridge University Press, pp. 3–11.

Angiosperm Phylogeny Group (1998). An ordinal classification for the families of flowering plants. *Annals of the Missouri Botanical Garden*, **85**, 531–53.

Bachman, K., Blattner, F. & Dehmer, K. (1999). Molecular markers for characterisation and identification of genebank holdings. In *Taxonomy of Cultivated Plants: 3rd International Symposium*, eds. S. Andrews, A. Leslie & C. Alexander. Kew, UK: The Royal Botanic Gardens, pp. 239–52.

Bevan, M., Bancroft, I., Bent, E. *et al.* (The EU Arabidopsis Genome Project) (1998). Analysis of 1.9 Mb of contiguous sequence from chromosome 4 of *Arabidopsis thaliana. Nature*, **391**, 485–8.

Cordell, G. A. (2000). Biodiversity and drug discovery: a symbiotic relationship. *Phytochemistry*, **55**, 463–80.

Cragg, G. M., Newman, D. J. & Snader, K. M. (1997). Natural products in drug discovery and development. *Journal of Natural Products*, **60**, 52–60.

Culham, A. & Grant, M. L. (1999). DNA markers for cultivar identification and classification. In *Taxonomy of Cultivated Plants: 3rd International Symposium*, eds. S. Andrews, A. Leslie & C. Alexander. Kew, UK: The Royal Botanic Gardens, pp. 183–98.

Farnsworth, N. R. & Soejarto, D. D. (1991). Global importance of medicinal plants. In *The Conservation of Medicinal Plants*, eds. O. Akerele, V. Heywood & H. Synge. Cambridge, UK: Cambridge University Press, pp. 25–51.

Frick, S., Ounaroon, A. & Kutchan, T. M. (2001). Combinatorial biochemistry in plants: the case of *O*-methyltransferases. *Phytochemistry*, **56**, 1–4.

Grayer, R. J. (1983). *Medicinal Plants*. WWF Plants Campaign 1983–4. Godalming, UK: WWF.

Grayer, R. J., Kite, G. C., Goldstone, F. J. *et al.* (1996). Infraspecific taxonomy and essential oil chemotypes in sweet basil, *Ocimum basilicum*. *Pytochemistry*, **43**, 1033–9.

Grayer, R. J., Chase, M. W. & Simmonds, M. J. S. (1999). A comparison between chemical and molecular characters for the determination of phylogenetic relationships among plant families: an appreciation of Hegnauer's *Chemotaxonomie der Pflanzen*. *Biochemical Systematics and Ecology*, **27**, 369–93.

Grayer, R. J., Veitch, N. C., Kite, G. C. & Garnock-Jones, P. J. (2002). Scutellarein 4′-methyl ether glycosides as taxonomic markers in *Teucridium* and *Tripora* (Lamiaceae, Ajugoideae). *Phytochemistry*, **60**, 727–31.

Hamann, O. (1991). The joint IUCN–WWF plant conservation programme and its interest in medicinal plants. In *The Conservation of Medicinal Plants*, eds. O. Akerele, V. Heywood & H. Synge. Cambridge, UK: Cambridge University Press, pp. 13–22.

Hegnauer, R. (1962–2001). *Chemotaxonomie der Pflanzen*, vols. 1 to 11. Basel, Switzerland: Birkhäuser Verlag.

(1989). *Chemotaxonomie der Pflanzen*, vol. 8. Basel, Switzerland: Birkhäuser Verlag.

(1990). *Chemotaxonomie der Pflanzen*, vol. 9. Basel, Switzerland: Birkhäuser Verlag.

Heywood, V. H. (2003). Red Listing: too clever by half? *Plant Talk*, **31**, 5.

Kendrick, P. & Crane, P. R. (1997). *The Origin and Early Diversification of Land Plants*. Washington, DC and London: Smithsonian Institution Press.

Kite, G. C., Grayer, R. J., Rudall, P. J. & Simmonds, M. S. J. (2000). The potential for chemical characters in monocotyledon systematics. In *Monocots: Systematics and Evolution*, eds. K. L. Wilson & D. A. Morrison. Melbourne, Australia: CSIRO, pp. 101–13.

Kite, G. C., Yule, M. A., Leon, C. & Simmonds, M. S. J. (2002). Detecting aristolochic acids in herbal remedies by liquid chromatography/serial mass spectrometry. *Rapid Communications in Mass Spectrometry*, **16**, 585–90.

Kite, G. C., Veitch, N. C., Grayer, R. J. & Simmonds, M. S. J. (2003). The use of hyphenated techniques in comparative phytochemical studies of legumes. *Biochemical Systematics and Ecology*, **31**, 813–43.

Lewis, W. H. (1982). Plants for man: their potential in human health. *Canadian Journal of Botany*, **60**, 310–15.

Mabberley, D. J. (1987). *The Plant Book*. Cambridge, UK: Cambridge University Press.

Markham, K. R., Tanner, G. J., Caasi-Lit, M. *et al.* (1998). Possible protective role for 3′,4′-dihydroxyflavones induced by enhanced UV-B in a UV-tolerant rice cultivar. *Phytochemistry*, **49**, 1913–19.

Morita, N., Arisawa, M., Ozawa, H., Chen, C.-S. & Kan, W.-S. (1977). Clerodendroside, a new glycoside from the leaves of *Clerodendron trichotomum* Thunb. var. *fargesii* (Verbenaceae). *Yakugaku Zasshi*, **97**, 976–9.

NAPRALERT File (NAtural PRoducts ALERT) 2005. http://www.cas.org/ONLINE/DBSS/napralertss.html.

Padulosi, S., Leaman, D. & Quek, P. (2002). Challenges and opportunities in enhancing the conservation and use of medicinal and aromatic plants. In *Breeding Research on Aromatic and Medicinal Plants*, eds. C. B. Johnson & C. Franz. Binghampton, NY: Haworth Herbal Press, pp. 243–67.

Pichersky, E. (1990). Nomad DNA: a model for movement and duplication of DNA sequences in plant genomes. *Plant Molecular Biology*, **15**, 437–48.

Pichersky, E. & Gang, D. R. (2000). Genetics and biochemistry of secondary metabolites in plants: an evolutionary perspective. *Trends in Plant Science*, **5**, 439–45.

Pietra, F. (2002). *Biodiversity and Natural Product Diversity*. Amsterdam, the Netherlands: Pergamon.

Simmonds, M. S. J. & Grayer, R. J. (1999). Plant drug discovery and development. In *Chemicals from Plants*, eds. N. J. Walton & D. E. Brown. London: Imperial College Press, pp. 215–49.

Van der Sluys, W. G. (1985). *Secoiridoids and Xanthones in the Genus Centaurium Hill (Gentianaceae)*. Ph.D. Thesis, University of Utrecht.

Vieira, R., Grayer, R. J., Simon, J. E. & Paton, A. (2001). Genetic diversity of *Ocimum gratissimum* L. based on volatile oil contents, flavonoids and RAPD markers. *Biochemical Systematics and Ecology*, **29**, 287–304.

Vieira, R. F., Grayer, R. J. & Paton, A. J. (2003). Chemical profiling of *Ocimum americanum* using external flavonoids. *Phytochemistry*, **63**, 555–67.

Zenk, M. H. (1991). Chasing the enzymes of secondary metabolism: plant cell cultures as a pot of gold. *Phytochemistry*, **30**, 3861–3.

Part IV

Conservation strategies: taxonomy in the practice and measurement of effective conservation action

12

'The business of a poet': taxonomy and the conservation of island floras

David Bramwell

The business of a poet, said Imlac, is to examine, not the individual but the species; to remark general properties and large appearances. He does not number the streaks of the tulip

Samuel Johnson, *The Prince of Abissinia. A Tale*, 1759

Ever since Darwin developed his theory of evolution by natural selection from his observations of the finch species of the Galapagos Islands, biologists have paid particular attention to the biota and ecosystems of oceanic islands. Although they are generally small in size and strongly geographically isolated, islands tend to have rich floras with a high degree of endemism (e.g. Hawaii, New Zealand, Canary Islands). Mueller-Dombois (1981) argued that the insular conditions of small size, isolation and relatively short time scales have resulted in a 'different' evolution in island ecosystems: 'Thus on islands, more so perhaps than in biogeographically different continental ecosystems, the effect of a unique evolution becomes evident.' On the other hand 'different evolution' may simply be the result of a different focus from the people studying island organisms; island biologists do, at times, tend to be more obsessed with differences due to the smaller scale of island populations.

Evolution proceeds by means of adaptation and speciation (Darwin, 1859; Carson, 1981) and one of the fascinating evolutionary features of islands is that some organisms speciate extensively (e.g. *Asparagus and Ceropegia* in the Canary Islands) and others almost not at all (e.g. *Dracaena and Caralluma* in the Canary Islands) even though both types of organism are subject to similar environmental conditions. A question arises, therefore, about the results of evolution in both cases; are the species arising as the result of active speciation equivalent to those organisms that Carson describes as non-speciating organisms? The question of the nature and delimitation of species and the numbers of existing species is the subject of a continuing debate, which is extremely significant in the light of current trends towards species-orientated biodiversity conservation and the utilisation of species-based information systems, such as Red Lists etc., as a basis for taking major conservation decisions.

'Short of the death penalty, there will always be splitters and lumpers' (Davis & Heywood, 1963)

As a result of having to handle the products of species radiation, taxonomists dealing with island organisms are often accused of excessive splitting and describing too many species. On the other hand, however, attempts at lumping island species are usually unsatisfactory and often obscure the understanding of insular biodiversity; as in the case, for example, of recent revisions of the genera *Ceropegia* (Bruyns, 1986) and *Monanthes* (Nyffeler, 1992) in the Canary Islands. Unwarranted lumping in such cases can have a negative effect on biodiversity conservation by leaving unique elements of diversity unrecognised.

This leads to the broader question of the delimitation of species in general. Despite over 200 years of debate, species concepts and definitions are – as Hey (2001a) points out – still difficult and controversial. One of the most lucid and relevant discussions of species is still to be found in *The Principles of Angiosperm Taxonomy* by Davis and Heywood (1963), to the extent that on reading it again after many years, the phrase *'plus ça change, plus c'est la même chose'* came immediately to mind. The same discussions about the nature of species are still taking place. We may have more information and new classes of data such as molecular and micromorphological characters but even perfect data sets will not resolve the species problem because what we delimit as individual species are not necessarily comparable and are not necessarily the results of the same processes or even similar evolutionary parameters. In fact, the more we know about species the more complex the situation seems to become. As Davis and Heywood (1963) pointed out, the concepts covering our notions of species are 'constructions of the human mind and cannot be defined'. Indeed, taxonomy itself is a science in which each species is simply an individual hypothesis put forward by the taxonomist to be tested by research and observation.

In order to communicate with each other, humans, however, need to give things names; in this case, the entities as defined in the original species hypothesis by the taxonomist. Further, under the International Code of Botanical Nomenclature (Greuter *et al.*, 2000) all plants belong to a species for the purpose of being named. Though it would be most convenient if the units in the taxonomic and nomenclatural hierarchy that are treated as species for any particular purpose were as similar as possible whether they be taxonomic, biological or conservation species (those species classified as threatened in Red Lists), this would seem to be an impossible task (cf. Hey, 2001b). This is not only because there are important psychological differences involved in placing the emphasis on shared characteristics (lumping) or emphasizing differences (splitting). It is also because species result from a spectrum of processes, both metaphysical and biological.

'Fuzzy logic is a natural language for modelling biology' (Sokhansanj *et al.*, 2001)

A few years ago I was a member of a scientific visiting group at a major taxonomic institution and I remember getting the following reply on asking a member of staff responsible for a number of important monographs which species concept she used: 'I don't use a species concept, I know what a species is when I see one.' A subtler version is quoted by Davis and Heywood (1963) as 'any systematic unit classified as a species by a competent systematist' and in this case it becomes an alternative definition of a taxonomic species.

Both these quotations, of course, refer to the classical taxonomic species concept where the units are generally delimited on morphological grounds, the one we use to comply with the rules of the Code of Nomenclature. In the majority of cases of plant species recognised at the present time, this is the species concept employed. This is principally because the plant taxonomist does not usually have enough (if any) information available about reproductive isolation or gene exchange and has to rely on morphological boundaries to delimit species. This is the *'force majeure'* noted by Davis (1978) of morphological characters usually being the only ones available. The plant taxonomist relies on a concept of species derived basically from Aristotelian logic, that of mutually exclusive groups, A or not A. He looks for boundaries that include or exclude individuals, but are such boundaries real in nature? Or is nature a fuzzy continuum where degrees of discontinuity exist which permit values between A and not A, between the completely true and the completely false, making well-defined species a subset of fuzzy species? Fuzzy logic systems are designed to handle complex systems where an inexact model exists or where ambiguity and vagueness is common. As Wilkins (2002), in an Internet review of Hey (2001a), indicates, 'species are fuzzy from the working of evolution (i.e. variation, speciation and local selection) and so it is clear to biologists that there should be a problem counting and delineating them, and yet biologists persist in trying to count them to resolve questions of biodiversity and conservation'.

'Everything is vague to a degree that you do not realize until you have tried to make it precise' (Russell, 1956)

My own analogy of what is a species is as follows: sitting on a boat out on a choppy sea, waves are not well-defined entities, for in the ebb and flow it is hard to distinguish boundaries or troughs where one ends and another begins. Out at sea, waves do not move regularly but go in all directions; they are ephemeral and are subjected to the selection pressures of tidal forces, wind, etc., rather like evolving species. They can only be counted when they reach the shore and break, in effect becoming extinct in the process. In this analogy the taxonomist, when he defines a species instantaneously freezes the ebb and flow and describes what he sees at that moment in time. He describes the frozen waves but they only exist in the described form for that short instant. Furthermore, in the same way that movement in the water does not always generate a wave, the speciation process does not always lead to species production. Mayr (1988) commented, '*because species speciate and occasionally merge, their borders are sometimes "fuzzy" and the point at which one species leaves off and another begins is often arbitrary*'. Thus species need not be entirely mutually exclusive and perhaps we can accept degrees of membership of a species, that is, accept an inexact fuzzy species model in the Zadehian-logic sense. This is in fact probably what we have in nature, a series of varying and short-lived wave crests that we call 'good' species surrounded by indefinable swirls and troughs that may or may not participate in or contribute to any particular wave-crest. What we understand as 'species' are simply wave peaks in a sea of continuous and continuing evolution. Furthermore, ocean waves, when seen from high above, from a plane window or from a high mountainside, appear to be static or fixed. It is only when we get near to them that we detect their ephemeral, constantly changing nature. Species are the same; it is only when we research them closely that we begin to appreciate the complex nature of the concept of 'species'.

More recently, Cracraft (1989) has introduced an additional phylogenetic species concept in which species are defined as: 'an irreducible cluster of organisms, diagnostically distinct from other such clusters, and within which there is a parental pattern of ancestry and descent'. The validity of such a concept depends on knowing the history of the organisms to be included within a species, and there is some difficulty in reconciling putative phylogenies with the inherent reticulate nature of the speciation process in many organisms. We have not yet found ways of constructing phylogenies, of understanding the water beneath the waves, that actually reflect the history of species and this makes the practical application of such a concept difficult. Davis' presentation (1995) at a symposium on Species Concepts and Phylogenetic Analysis – and the subsequent papers in that symposium – provide an excellent discussion of phylogenetic species concepts; in view of the fact that they are still difficult to apply to practical exercises such as the production of checklists and local Floras, I will not consider them further here.

Islands, species, taxonomy and conservation

Whatever the processes involved in island speciation, they have resulted in large numbers of unique taxa, high endemism, and many rare and threatened species. Most of the species have been described historically using the taxonomic species concept based on morphological similarities and discontinuities. In a survey of species concepts used by monographers, McDade (1995) found that: 'the vast majority [of authors] indicated that they adopted a morphological or taxonomic concept of species and used morphological differences to distinguish them'. In support, perhaps, of Davis' *force majeure*, McDade goes on to express the opinion that the monographic work of systematics thus appears to have been largely unaffected by recent contributions to the species controversy.

On occasions over the last 30 years or so, biosystematic information has also been used to help delimit species. In some cases revisions have taken into account the fact that new data on reproductive biology and population genetics have become available, facilitating the possibility of applying a biological species concept. In some cases a 'pattern based species concept' (Nelson & Platnick, 1981) has been adopted, for example by Moylan *et al.* (2002) in their account of *Hemigraphis* from the Philippines. Under this concept, species are defined as: 'the smallest detected samples of self-perpetuating organisms that have a unique set of characters'; this differs very little from the phylogenetic species of Cracraft. In the *Hemigraphis* revision, however, no information on the 'self-perpetuation' of the samples of organisms delimited as species is presented and, in fact, species delimitation is based on morphological characters and their discontinuities and thereby employs what is simply a classical taxonomic species concept.

In fact, in general terms, almost all island species are still defined on the basis of a classical taxonomic species concept. One of the major problems in applying any other concept is that island species tend to be morphologically distinct as both adaptive radiation and vicariance, the driving forces behind insular evolution, are both divergent processes; however, the formation of reproductive barriers lags behind morphological differentiation. As Davis (1978) says, 'in the higher plants every graduation occurs from full intersterility between taxonomic species to full (at least potential) interfertility'. The examples cited in the relevant chapters of Stuessy and Ono (1998, Chapters 3 and 4) serve to validate Davis' comment.

However, in recent years a further, important factor has crept, almost undetected, into the practical application of species definitions. This is the requirement for taxonomy and systematics to serve the multiple needs of conservation. Conservation legislators require checklists, inventories, Floras, species lists with Red List Categories included, etc. On the other hand, conservation managers dealing with species survival need to know about their biology, phenology, reproductive biology, pollinators, population size, fluctuations and behaviour, etc. Thus, we have a dichotomy; on the one hand, the needs of legislation can be served using a taxonomic species concept and on the other a biologically based species concept would seem to be more appropriate to field management. Ghiselin (1997) suggests that, as taxonomy is pluralistic, we can consider these different species concepts as complementary provided that they are not applied as an incoherent mixture of the two and that we do not mix the purposes. In conservation there seems to be considerable justification for taking such an approach, especially if we consider both concepts as subsets of a fuzzy concept of species. Such subsets will have a major degree of overlap in almost all cases so that there should be a high degree of coincidence between the species listed in the legislation process, i.e. the units of taxonomic convenience, and those biological units requiring conservation management. Furthermore, the legislation process of arriving at a full list of the world's threatened plant species has been considerably handicapped in the past by the lack of appreciation of the inherently 'fuzzy' nature of the IUCN Red List Categories. This is especially so because the continual revision of the categories or moving of the goalposts has tended to make the exercise a long-term, process-orientated rather than product-orientated one; a problem which is still to be addressed by the conservation community in general.

The conservation of island floras with their high levels of biodiversity and endemism is a major issue on a world scale. Recent estimates (Bramwell, 2002) suggest that about 50 000 species are insular endemics and that some 20 000 of them are threatened. Many of these species are poorly known and in an alpha-taxonomy state where no information is available about their biology or even their rarity or abundance. The floras of Indonesia, Madagascar, New Guinea, Cuba, etc. yield many new species annually, mainly described from a few herbarium specimens. The only species concept we can apply to them is the taxonomic/morphological one. However, in an era of disappearing forests, major fragmentation of insular ecosystems leading to patch and edge effects, increasing population isolation and decline, loss of genetic diversity, disruption of species interactions (predators, dispersers and pollinators, etc.), permanent change in landscapes and loss of niches for regeneration, their real conservation is a matter for population biology and a biological species concept.

References

Reading is sometimes an ingenious device for avoiding thought
Arthur Helps, *Friends in Council*, 1847

Bramwell, D. (2002). How many plant species are there? *Plant Talk*, **28**, 32–34.
Bruyns, P. (1986). The genus *Ceropegia* on the Canary Islands (Asclepiadaceae-Ceropegieae). A morphological and taxonomic account. *Beiträge zur Biologie der Pflanzen*, **60**(3), 427–58.
Carson, H. L. (1981). Microevolution in insular ecosystems. In *Island Ecosystems: Biological Organization in Selected Hawaiian Communities*, eds. D. Mueller-Dombois, K. W. Bridges & H. L. Carson. US/IBP Synthesis Series 15. Stroudsburg, PA: Hutchinson Ross.
Cracraft, J. (1989). Speciation and its ontology: the empirical consequences of alternative species concepts for understanding patterns and processes. In *Speciation and its Consequences*, eds. D. Otte & J. A. Endler. Sunderland, MA: Sinauer Associates.

Darwin, C. (1859). *The Origin of Species by Means of Natural Selection or the Preservation of Favoured races in the Struggle for Life*. London: John Murray.

Davis, J. I. (1995). Species concepts and phylogenetic analysis: introduction. *Systematic Botany*, **20**, 555–9.

Davis, P. H. (1978). The moving staircase: a discussion on taxonomic rank and affinity. *Notes from the Royal Botanic Garden Edinburgh*, **36**, 325–40.

Davis, P. H. & Heywood, V. H. (1963). *The Principles of Angiosperm Taxonomy*. Edinburgh, UK: Oliver and Boyd.

Ghiselin, M. T. (1997). *Metaphysics and the Origin of Species*. New York: Albany State University Press.

Greuter, W., McNeill, J., Barrie, F. R. *et al.* (2000). *International Code of Botanical Nomenclature (Saint Louis Code) Adopted by the Sixteenth International Botanical Congress, St Louis, Missouri, July – August 1999. Regnum Vegetabile* **131**. Königstein: Koeltz Scientific Books.

Hey, J. (2001a). *Genes, Concepts and Species: the Evolutionary and Cognitive Causes of the Species Problem*. New York: Oxford University Press.

  (2001b). The mind of the species problem. *Trends in Ecology and Evolution*, **16**(7), 326–9.

Mayr, E. (1988). *Toward a New Philosophy of Biology*. Cambridge, MA: Belknap Press of Harvard University Press.

McDade, L. A. (1995). Species concepts and problems in practice: insight from botanical monographs. *Systematic Botany*, **20**, 606–22.

Moylan, E. C., Pennington, R. T. & Scotland, R. W. (2002). Taxonomic account of *Hemigraphis* Nees (Strobilanthinae-Acanthaceae) from the Phillipines. *Kew Bulletin*, **57**, 769–825.

Mueller-Dombois, D. (1981) Some environmental conditions and the general design of IBP research in Hawaii. In *Island Ecosystems: Biological Organization in Selected Hawaiian Communities*, eds. D. Mueller-Dombois, K. W. Bridges & H. L. Carson. US/IBP Synthesis Series 15. Stroudsburg, PA: Hutchinson Ross, pp. 3–32.

Nelson, G. & Platnick, N. (1981). *Systematics and Biogeography: Cladistics and Vicariance*. New York: Columbia University Press.

Nyffeler, R. (1992). A taxonomic revision of the genus *Monanthes* Haw. (Crassulaceae). *Bradleyia*, **10**, 49–82.

Russell, B. (1956). *Logic and Knowledge: Essays, 1901–1950*. London: George Allen and Unwin.

Sokhansanj, B. A., Rodrigue, G. H. & Fitch, J. P. (2001). Applying URC fuzzy logic to model complex biological systems in the language of biologists. In 2nd International Congress on Systems in Biology, California Institute of Technology, Pasadena. Website (2004)
  www.icsb2001.org/Posters/092_sokhansanj.pdf

Stuessy, T. F. & Ono, M. (eds.) (1998). *Evolution and Speciation of Island Plants*. New York: Cambridge University Press.

Wilkins, J. S. (2002). Species as a psychological problem, or, What the Hey are species. Internet Review.
  www.members.dodo.com.au/~wilkinsjandp/papers/GCSreview.htm (accessed 2004).

13

The role of the taxonomist in conservation of critical vascular plants

T. C. G. Rich

Introduction

Critical plant groups can be broadly defined as those that are difficult to identify. They include a range of taxa such as apomicts (e.g. *Hieracium* L.), infraspecific taxa (e.g. *Anthyllis vulneraria* L. subspecies) or closely related groups of out-breeding species which require careful examination (e.g. *Cochlearia* L.), and a range of taxonomic problems (e.g. Valentine, 1960). In practice, of course, they form one end of a spectrum; at the opposite end of this spectrum are those groups that are easy to identify (e.g. *Bellis perennis* L.).

There are many reasons why critical groups may be difficult to identify.

1. There are real taxonomic problems. Taxonomic treatments may differ between experts, may require updating to incorporate recent work, may be published in an inaccessible form or be in the process of being published, or may simply be horrendously difficult.
2. The taxa are apomictic or in-breeding, with a large number of morphotypes that are difficult to distinguish from one another (e.g. *Taraxacum* F. H. Wigg.).
3. Hybridisation may blur the limits between taxa, making individuals difficult to identify (e.g. *Rosa* L.).
4. Some taxa are morphologically reduced with few characters available to distinguish species (e.g. *Lemna* L.).
5. Some taxa can only be identified at specific times of year restricting collection of information (e.g. *Salicornia* L.), or require material from more than one time of year (e.g. flowers and fruit for *Ulmus* L.).
6. Growth forms may be heavily modified by the environment (e.g. *Rubus* L. shaded in woodlands).

Historically, conservation of the critical taxa, and especially apomicts, in Britain has taken second place to conservation of 'ordinary' species (i.e. classic Linnean sexuals) for two main reasons. First there is an attitude that apomicts are clonal and not really worth conserving as they are not equivalent to 'real' species. This point has been addressed recently by Stace (1998) who argued cogently that apomicts should be treated taxonomically in exactly the same way as other species, and there is now increasing evidence that genetic variation is present in most apomictic groups and they are not simply clonal (e.g. Stace *et al.*, 1997).

Secondly, the quantity and quality of information available about critical taxa is poor. Floristic mapping programmes, which provide most of the basic information used to assess conservation status, are usually carried out by amateur volunteers – who vary enormously in their botanical skill and knowledge; this is the major factor causing variation in recording (Rich & Smith, 1996). As identification of critical taxa usually requires more time and skill – and access to resources such as herbaria, literature or microscopes – fewer botanists study them; this results in distribution information that is poor, and often patchy. Data for some critical taxa may be considered so poor that they are not even presented (e.g. segregates of *Utricularia intermedia* Hayne in Preston *et al.*, 2002). The recent *Vascular Plant* volume of the *British Red Data Books* (Wigginton, 1999) gave IUCN threat categories for small critical apomictic genera such as *Sorbus* and *Alchemilla*, but for *Hieracium* and *Taraxacum* only provided simple lists of rare species using different thresholds without IUCN criteria, and did not cover *Rubus* at all. This was attributed to lack of information (yet they were not categorised as 'Data Deficient'), which belied the more fundamental problem that they were simply too large and daunting to be dealt with by the conservationists.

However, there has been an increasing acceptance of the importance of the critical genera to biodiversity and that they require conservation. Again, this is for several reasons.

1. They form a significant part of the biodiversity. In the British Isles, there are currently about 2682 species of native flowering plants and ferns (Kent, 1992; Stace, 1997 and additional data) of which about one-third can be regarded as 'critical taxa' (Table 13.1).
2. They often have high rates of endemism. Table 13.1 shows that at least 541 taxa (*c.* 20% of the flora) in critical groups are endemic to the British Isles, and consequently the duty to protect them lies with both the British and Irish governments.
3. Many of these taxa are rare, are restricted to a few sites and are probably significantly threatened. Unfortunately, the current status of the majority of them, and therefore their priorities for conservation, is poorly known. Some are already extinct, e.g. the only known site of *Hieracium hethlandiae* (F. J. Hanb.) Pugsl. was quarried away in 1976 (Scott & Palmer, 1987) and *H. snowdoniense* P. D. Sell & C. West was thought to be extinct until one plant was rediscovered in 2002 (Rich & Hand, 2003).
4. The Convention on Biological Diversity, signed by the United Kingdom at the Rio Summit in 1992 and ratified in 1994, sets out a requirement to conserve biodiversity at genetic, species and ecosystem level; critical taxa contribute to all of these.
5. They may be valuable for showing on-going evolutionary processes and phytogeographical patterns (e.g. Tyler, 2000).

Table 13.1. *Number of native taxa and number of endemics in the major critical groups in Britain. Number of endemics updated from Rich* et al. *(1999)*

	Number of native taxa	Number of endemics (%)
Critical group		
Alchemilla L.	12	1 (8%)
Euphrasia L.	21	9 (43%)
Hieracium L.*	*c.* 400	250? (63%)
Limonium Mill.*	14	7 (50%)
Ranunculus auricomus L.	(100+)	?
Rosa L.	12	0 (0%)
Rubus L.	*c.* 325	*c.* 210 (65%)
Salicornia L.	7	0 (0%)
Sorbus L.	21	15 (71%)
Taraxacum F. H. Wigg.	*c.* 230	39 (17%)
Ulmus L.*	(70+)	?
Other flowering plants and ferns	1470	10 (0.7%)
Totals	*c.* 2682	*c.* 541 (*c.* 20%)

*Genera currently undergoing significant revisions in Britain.

The problem of lack of information about critical taxa is a practical one; taxonomists can have an important contribution to make here, due to their unique position in being able to provide quality information and – above everything else – a correct name. I will illustrate how amateur and professional taxonomists alike can contribute with examples mainly drawn from my recent work on conservation of rare Welsh critical taxa at the National Museums and Galleries of Wales; this work is being carried out in collaboration with the Countryside Council for Wales and the National Botanic Garden of Wales.

Provision of names

Taxonomists should be able to provide names (including synonymy) for critical taxa, the cornerstone of species information retrieval (Heywood, 1963). Without knowing the correct name, much information might be irretrievable or misapplied.

With access to recently revised herbarium material and the literature, taxonomists are in a strong position to name appropriately collected material sent for identification by other botanists. For instance, I have checked voucher material for *Sorbus* surveys carried out on cliffs in the Wye Valley by Mrs L. Houston and others between 2000 and 2002; this has resulted in many interesting and previously unknown records, and much excellent material for the herbarium. However, I struggle to name *Hieracium* species other than those I am familiar with, and also some material I collect in the field; the assistance of other experts such as P. D. Sell, D. J. Tennant and D. McCosh has been particularly helpful in this respect.

Taxonomic evaluation

For work on any critical taxon, we must be certain that the taxon is a good one and that its rank is appropriate. Such information may already be available, but some groups may have to be studied from scratch which can take several or more years. There is no need for a taxonomist to be a specialist in a particular group to start with, although it is often quicker if they are! There is also no guarantee that different taxonomists will agree, and opinions may change with time as more information becomes available.

Historically there has been much uncertainty about whether *Asparagus prostratus* Dumort. should be a species in its own right, or a subspecies or variety of *A. officinalis* L. A study by Kay *et al.* (2001) found that it merited species rank as it differed morphologically with distinctive characters that were retained in cultivation, that it was tetraploid compared to the diploid *A. officinalis*, that the two taxa were reproductively isolated, and that *A. prostratus* was a northwest European maritime endemic while *A. officinalis* was of Eastern European or western Asian origin. The clarified status now allows it to be treated equally across Europe for conservation.

I came to the opposite conclusion with *Anthyllis vulneraria* L. subsp. *corbierei* (Salmon & Travis) Cullen, which was described as differing from subsp. *vulneraria* in having fleshy leaves, spreading hairs on the stem, green calyx tips and yellow flowers (Cullen, 1986). As a rare British endemic it was included in the third edition of the *Vascular Plant* volume of the *British Red Data Books* recorded from Cornwall, Anglesey and Caithness, although little was known about it (Wigginton, 1999). After a request for information about the Welsh populations, I carried out field surveys to assess its status in Anglesey including its type locality, but found that plants there graded into subsp. *vulneraria* in varying combinations of characters. After further field and herbarium studies of all British and some northwestern French material, I was unable to find any correlation between the characters and concluded that subsp. *corbierei* was simply one part of the polymorphic variation in subsp. *vulneraria*. As such, it did not merit recognition at subspecific rank and I recommended that it should be deleted from the *Red Data Book* (Rich, 2001).

During one field survey, I was excited to find a tiny population of a very distinctive leafy *Hieracium* in Cwm Haffes, Brecon that I eventually identified as *H. trinitatis*

Pugsl. which is restricted to this site (Pugsley, 1948). I consulted the *Hieracium* expert P. D. Sell, who explained that during his revision of the British species he had come to the conclusion that this was best included as part of the many-faceted variation in *H. uiginskyense* Pugsl. which Pugsley had interpreted more strictly (P. D. Sell, personal communication, 2000). Consequently, instead of being the only population of one species, it was one of many populations of a more widespread species (albeit still a British endemic), and its priority for conservation was a lot lower.

Training and communication

Taxonomists can also play an important role through training non-specialist botanists to identify groups of critical taxa, or just the selected target species. I have had some success training other botanists where only a small number of *Hieracium* species were present, but anything more than eight species in one day rapidly becomes too complicated. The provision of identification manuals, photographs, annotated illustrations and keys to the taxa at a particular site is another approach that has proved popular with the local site managers (e.g. keys to *Hieracium* in Cheddar Gorge and at Craig y Cilau were provided by Rich & McDonnell, 2001). We have also tried to produce simple summary guides to identification of many smaller critical plant groups (Rich & Jermy, 1998), which has undoubtedly improved their accessibility and recording by amateur botanists.

Taxonomists must also communicate to non-specialists in a non-technical way. I have often wondered how to conserve *Taraxacum cherwellense* A. J. Richards, a little known, rare endemic of grassland in central southern England (Dudman & Richards, 1997). It is best known from its type locality in the parks of Oxford, and persuading the park keepers not to cut the grass so that a dandelion can set seed would surely require communication skills of the highest order!

Collating existing data

Herbaria and taxonomic literature are key reference sources for critical groups. These are standard tools of the trade for taxonomists, who are used to using them and are aware of the synonymy and nomenclatural problems, the limitations and potential pitfalls.

Herbaria provide the key reference collections where the identification of particular records can be checked, and from which locality details etc. can be abstracted. Taxonomists are usually aware of which herbaria are worth looking at for collections of particular taxa or collectors, and have experience of handling collections with the occasional mislabelling, reading difficult handwriting and interpreting missing or partial information from a wider knowledge of collections ('meta-data').

Tracing cited material can be difficult, especially when some may have been rede-termined (for instance, vouchers for two of the three 10-km squares reported for *Hieracium linguans* (Zahn) Roffey had been redetermined or were errors; Rich & Motley, 2001), but may become easier with the ongoing compilation of comput-erised herbarium databases.

The literature provides another valuable source of information, which may often give different and more expansive information than is included on herbarium sheets. Records in the literature cannot be checked without vouchers, but knowledge of the collector and his/her expertise can help with their interpretation. Information in print often differs from that on herbarium sheets, and the two require careful integration.

National databases such as those held by the Biological Records Centre at Monks Wood or the Botanical Society of the British Isles (BSBI) Threatened Plants Database can also be useful sources of information, although the original data have often been reinterpreted and standardised for computer entry and should be treated cautiously. Wherever possible, the original sources should be consulted.

These sources are invaluable in accumulating and interpreting information and should be researched before field surveys are carried out. I have found them an invaluable way of getting to know critical taxa before seeing them in the field, and try to do it routinely for all species. Frequently the archive data also require reinterpretation in light of information from the field surveys.

Field surveys

Once the background information is drawn together, the taxonomist can be usefully applied in carrying out field surveys, providing the key expert identification in the field that negates the need to collect large numbers of vouchers of potentially rare taxa as well as the expertise to pick out the right species. However, pressed specimens in the herbarium often look quite different to plants growing in the field, and initially careful checks of diagnostic characters may be needed before the field 'jizz' can be picked up. Collection of voucher specimens is always worthwhile, but depending on rarity or population size, photographs including characteristic features may suffice.

I have now carried out specific field surveys for eight *Hieracium*, two *Rubus* and three *Sorbus* species, often in collaboration with other experts. Surveying popula-tions of rare critical plants in the field can be good discipline for several reasons. Some sites may have many taxa in a critical group (for instance, Corrie Etchachan in Scotland has 12 members of *Hieracium* Section *Alpina* alone! D. J. Tennant, personal communication, 2000) and having to put a name to each and every spec-imen tests both the taxonomy and the taxonomist. If carrying out surveys for one particular taxon, not everything else needs to be fully named at the same time, but as one rare species often grows with another, several taxa can be covered at once. It is surprising how much more variation may be noted in the field than is written down in descriptions. The surveys have also provided material for further study and raised many new questions, and in the last six years we have found at least four new *Sorbus* or *Hieracium* taxa.

If the surveys are carried out in collaboration with ecologists and site managers, there is often profitable exchange of information and ideas about the practicalities of conserving the taxa *in situ*. Seed or cuttings can be collected at the same time for *ex situ* conservation in botanic gardens, as can DNA samples for genetic studies. Such liaison can also help to overcome the prejudice against critical taxa held by many non-specialists.

Monitoring

Monitoring is essential in conservation to demonstrate the success or failure of conservation practice but it is a complex subject. For monitoring critical plant groups, taxonomists are best employed to provide critical identification and population counts.

It is important to be clear about what is required from the outset and to adapt the method to the objectives – broadly based, poorly defined monitoring projects rarely get repeated or give answers. The procedures are best kept simple. First ensure that the organism can be identified readily at the survey time, or from material brought back to the laboratory (e.g. flowering time, fruiting time – vegetative rosettes may be impossible to identify in *Hieracium*). Secondly, if systematic whole-population censuses are impractical and samples are required, take statistical advice on sampling methods before commencing work. Thirdly, and perhaps most important, test and develop the census method and then get someone else to repeat it immediately and independently to give an indication of the observer variation and errors involved (observer bias is especially important). Fourthly, ensure the data are written up and documented.

Craig y Cilau is a stunning National Nature Reserve (NNR) in the Brecon Beacons in South Wales. A series of carboniferous limestone cliffs, capped by a millstone grit plateau, support at least 7 rare *Sorbus* species and 14 *Hieracium* species, 5 of which are endemic and are either virtually confined to this site or have the bulk of their populations there. The populations were first assessed by Lynne Farrell and Peter Sell in one day in June 1975 (Table 13.2), a phenomenal achievement given the physical scale of the site and the number of critical taxa, and the Countryside Council for Wales (CCW) required updated counts to inform conservation management. I repeated the surveys, *Hieracium* in June 2000 and *Sorbus* in September 2002, spending at least 6 days in the field. For *Sorbus*, I found approximately double

Table 13.2. *Number of* Hieracium *and* Sorbus *plants reported for Craig y Cilau NNR in 1975 (P. D. Sell & L. Farrell) and 2000/2002 (T. Rich). Data from Rich and McDonnell (2001) and Rich (2002, 2003)*

Taxon	1975	2000/2002	Change
Hieracium asteridiophyllum P. D. Sell & C. West	253	458	81% increase
Hieracium cillense Pugls.	396	296	25% decrease
Hieracium cyathis (Ley) W. R. Linton	857	690	19% decrease
Sorbus anglica Hedl.	61	103	69% increase
Sorbus leptophylla E. F. Warb.	28	45	61% increase
Sorbus minima (Ley) Hedl.	305	730	139% increase
Sorbus porrigentiformis E. F. Warb. s.l.	95	107	13% increase
Sorbus rupicola (Syme) Hedl.	69	101	46% increase

the number of plants, perhaps not surprising in light of the more detailed coverage, but the results for *Hieracium* were more mixed indicating that some real changes in populations may have occurred (Table 13.2).

Conclusions

Taxonomists thus have important roles to play in conservation of critical taxa – through provision of names, evaluation of taxa, collation of information, field surveys and monitoring. The correct name is the key to information and consequently conservation of any species, and especially so for critical species; taxonomists are uniquely able to provide this correct name.

Although it is not primarily the role of the taxonomist to provide conservation assessments of taxa (e.g. IUCN threat categories), they can usually readily do so from the historical and field-survey information (e.g. Rich, 2002; Tennant & Rich, 2002). Some taxonomists have even provided assessments of new taxa when they have described them (e.g. for *Taraxacum* sect. *Palustria* species by Kirschner & Štepànek, 1998).

Acknowledgement

I would like to thank Andy Jones of the Countryside Council for Wales for his lively debate and support.

References

Cullen, J. (1986). *Anthyllis* in the British Isles. *Notes from the Royal Botanic Garden Edinburgh*, **43**, 277–81.
Dudman, A. & Richards, A. J. (1997). *Dandelions of Great Britain and Ireland*. London: BSBI.

Heywood, V. H. (1963). The 'species aggregate' in theory and practice. *Regnum Vegetabile*, **27**, 26–37.

Kay, Q. O. N., Davies, E. W. & Rich, T. C. G. (2001). Taxonomy of the Western European endemic *Asparagus prostratus* (*A. officinalis* subsp. *prostratus*) (Asparagaceae). *Botanical Journal of the Linnean Society*, **137**, 127–37.

Kent, D. H. (1992). *List of Vascular Plants of the British Isles*. London: BSBI.

Kirschner, J. & Štepànek, J. (1998). *A monograph of Taraxacum sèct. Palustria*. Pruhonice: Institute of Botany, Academy of Sciences of Czech Republic.

Preston, C. D., Pearman, D. A. & Dines, T. D. (eds.) (2002). *New Atlas of the British and Irish Flora*. Oxford, UK: Oxford University Press.

Pugsley, H. W. (1948). A prodromus of the British *Hieracia*. *Journal of the Linnean Society of London (Botany)*, **54**, 1–356.

Rich, T. C. G. (2001). What is *Anthyllis vulneraria* L. subsp. *corbierei* (Salmon & Travis) Cullen (Fabaceae)? *Watsonia*, **23**, 469–80.

   (2002). Conservation of Britain's biodiversity: *Hieracium asteridiophyllum* and *H. cillense* (Asteraceae). *Watsonia*, **24**, 101–6.

   (2003). Establishment of a monitoring regime for *Sorbus* species at Craig y Cilau NNR. Unpublished contract report to Countryside Council for Wales, January 2003.

Rich, T. C. G. & Hand, S. O. (2003). Conservation of Britain's biodiversity: *Hieracium snowdoniense* (Asteraceae), Snowdonia Hawkweed. *Watsonia*, **24**, 513–18.

Rich, T. C. G. & Jermy, A. C. (eds.) (1998). *Plant Crib 1998*. London: BSBI.

Rich, T. C. G. & McDonnell, E. J. (2001). Distribution and conservation of *Hieracium cyathis*, Challice Hawkweed. Unpublished report from National Museum of Wales to Countryside Council for Wales.

Rich, T. C. G. & Motley, G. S. (2001). Conservation of Britain's biodiversity: *Hieracium linguans* (Asteraceae), Tongue Hawkweed. *Watsonia*, **23**, 517–23.

Rich, T. C. G. & Smith, P. A. (1996). Botanical recording, distribution maps and species frequency. *Watsonia*, **21**, 155–67.

Rich, T. C. G., Hutchinson, G., Randall, R. D. & Ellis, R. G. (1999). List of plants endemic to the British Isles. *BSBI News*, **80**, 23–7.

Scott, W. & Palmer, R. (1987). *The Flowering Plants and Ferns of the Shetland Islands*. Lerwick, UK: The Shetland Times Ltd.

Stace, C. A. (1997). *New Flora of the British Isles*, 2nd edn. Cambridge, UK: Cambridge University Press.

   (1998). Species recognition in agamosperms: the need for a pragmatic approach. *Folia Geobotanica*, **33**, 319–26.

Stace, C. A., Gornall, R. J. & Shi, Y. (1997). Cytological and molecular variation in apomictic *Hieracium* sect. *Alpina*. *Opera Botanica*, **132**, 39–51.

Tennant, D. J. & Rich, T. C. G. (2002). Distribution maps and IUCN threat categories for *Hieracium* section *Alpina* in Britain. *Edinburgh Journal of Botany*, **59**, 351–72.

Tyler, T. (2000). Detecting migration routes and barriers by examining the distribution of species in an apomictic species complex. *Journal of Biogeography*, **27**, 979–88.

Valentine, D. H. (1960). The treatment of apomictic groups in *Flora Europaea*. In *Problems of Taxonomy and Distribution in the European Flora. Proceedings of the Flora Europaea Round Table Conference, Vienna 1–7 April 1959*, ed. V. H., Heywood. Berlin: Akademie-Verlag, pp. 119–27.

Wigginton, M. J. (ed.) (1999). *British Red Data Books*, vol. 1, *Vascular Plants*, 3rd edn. Peterborough, UK: JNCC.

14

Plant taxonomy and reintroduction

John R. Akeroyd

Introduction

Sound taxonomy and accurate identification of plant material provide firm and necessary foundations for the implementation of strategies such as Species Recovery Programmes to reintroduce, restock or restore wild plant populations. The reintroduction of rare plant species and restoration programmes that enhance, restore or recreate native or semi-natural habitats are now essential weapons in conservation's armoury (Akeroyd & Wyse Jackson, 1995). The old distinction between *ex situ* – sometimes admittedly the last-ditch defence for rare or threatened species – and *in situ* conservation projects has largely disappeared. Today, the trend is to combine *ex situ* and *in situ* species-conservation measures within an integrated conservation programme (Falk, 1990; Maunder 1992; Falk *et al.*, 1996), involving collaboration between botanic gardens and other centres of scientific or horticultural expertise and conservation or land agencies and private owners. The procedures are almost inevitably long term, time consuming and expensive; they require strict guidelines and controls.

Rigorously sourced and identified plant material is a primary prerequisite for success. The integrated conservation strategy required for a Species Recovery Programme combines ecological research on an individual species with the scientific management of the communities and ecosystems in which it occurs. It is important that the programme should study and, for practical reasons, capture and maintain the genetic variation of the populations of the species. This may include any ecotypic and taxonomic variants that are characteristic at biome, regional or local level, and usually requires the input of a high level of taxonomic expertise.

The principal aim of a successful reintroduction is to re-establish a viable population of a species, subspecies or 'race' in the wild that will survive long term (IUCN/Species Survival Commission (SSC) Re-introduction Specialist Group, 2003). The SSC's own recommendations state: 'An assessment should be made of the taxonomic status of individuals to be reintroduced. These should preferably be of the same subspécies or race as those which were extirpated, unless adequate numbers are not available . . . A study of genetic variation within and between populations of this and related taxa can also be helpful.' In the case of some rare plants this taxonomic assessment may be necessary to fulfil legal requirements. For example, *Rhododendron ponticum* subsp. *baeticum*, endemic to streamsides and sheltered gullies in southern Spain and Portugal, is protected in Europe under the terms of the EU Habitats Directive. The other populations of *R. ponticum* in Western Europe, where this species is introduced and frequently an invasive weed of woodland and heathland, enjoy no such protection and clearly would have no value as a source of material for reintroduction (although some relict native populations of *R. ponticum* subsp. *ponticum* do occur in Turkey-in-Europe and adjacent Bulgaria).

Taxonomic variation within species

Since the pioneering work of Professor A. D. Bradshaw and his co-workers in the 1950s to 1970s, a considerable body of research has been published that demonstrates physiological variation in natural populations of plants in relation to a range of environmental factors, notably tolerance to salts of heavy metals such as copper, lead and zinc (Antonovics, 1971; Antonovics *et al.*, 1971). Despite extensive gene flow, powerful disruptive selection across ecological boundaries has been demonstrated even over distances of a few centimetres. A particularly illuminating example was demonstrated in populations of Annual Meadow-grass (*Poa annua*) investigated by Warwick & Briggs (1978a, 1978b), where selection across a sharp lawn–flower-bed boundary favoured distinct genetic variants of contrasting life history. Short-lived perennial plants of Annual Meadow-grass adapted to survive in lawns, a modified grassland habitat, were replaced in adjacent flower beds by rapidly growing and reproducing annual plants.

Most widespread plant species exhibit obvious phenotypic morphological variation. Although the taxonomic treatment of this intraspecific differentiation does not always receive the attention it deserves, botanists have long identified and classified much phenotypic variation in natural populations, for example coastal ecotypes along the western seaboard of Europe (Akeroyd, 1997). One of the most valuable features of *Flora Europaea* (Tutin *et al.*, 1964–80, 1993) was the extensive use of subspecies as a taxonomic category to denote cytological, geographical and ecological species variants. This has proved to be a practical means of recognising and cataloguing significant patterns of variation, especially within highly polymorphic species such as *Biscutella laevigata* (E. Guinea & V. H. Heywood in Tutin *et al.*,

1964, 1993) and *Anthyllis vulneraria* (J. Cullen in Tutin *et al.*, 1968) with a complex mosaic of variation across Europe and the Mediterranean region.

The subspecies has also served as a device to reconcile what are sometimes profound national differences in taxonomic treatment and a regrettably high level of parochial inconsistency (Akeroyd & Walters, 1987; Akeroyd & Jury, 1991). The establishment of a pan-continental overview of species variation was one of the triumphs of *Flora Europaea* (another was that the project was completed!). *Flora Europaea* and the longer-term mapping project *Atlas Florae Europaeae* (Jalas & Suominen, 1971–94; Jalas *et al.*, 1996, 1999), which has now covered all the families in *Flora Europaea* volume 1, exhibit some differences in taxonomic recognition of species; but these are frequently mere differences of opinion over rank. One should also accept that authors in different regions frequently raise varieties familiar to them up to subspecific rank.

Failure to use properly identified plant material

Species Recovery Plans are but a small element in plant reintroduction and habitat restoration. In much of North America and Europe the loss, degradation and fragmentation of biodiversity-rich native and semi-natural habitats has greatly impeded the natural regeneration and colonisation of many plants that were formerly more widespread. Restoration ecology, increasingly applied, and executed with ever-greater levels of precision, offers the best practical solution to re-establish these species for posterity (Dobson *et al.*, 1997). A much more extensive use of these techniques occurs in the landscaping and restoration of sometime disturbed ground – either waste land or former industrial sites, or land associated with the construction of new roads and buildings (Department of Transport, 1993), to create amenity grassland (Wells *et al.*, 1981, 1989), woodland or wetland. In the United Kingdom in recent years, there has been considerable discussion about the provenance of the plant and seed material used in such ecological restoration projects.

Particular interest has focused on the provenance of 'native' wildflower seed, much of which has apparently been of foreign origin (Akeroyd, 1994a, 1994b). Not only are native plants better adapted to local conditions and are more authentic elements of habitats in the landscape, but also in many cases they represent genetically relict populations themselves in need of conservation. This is particularly true of habitats in Britain that have been profoundly damaged by intensive agriculture and other large-scale habitat loss. The disruption of natural patterns of plant distribution makes the interpretation of geographical and ecological data increasingly difficult, thus undermining an effective biological tool at a time of climate change. Apart from the problem of potential damage to the integrity of native biodiversity, the use of non-native wildflower seed may be detrimental to the spirit of the landscape itself, creating a facsimile of countryside that is more park than living countryside.

Due to the large quantity of agricultural seed on the market for forage crops, the several clovers and other legumes widely used in amenity landscaping or ecological restoration too often derive from non-native sources (Akeroyd, 1997): notably seed of Kidney Vetch (*Anthyllis vulneraria*), Bird's-foot Trefoil (*Lotus corniculatus*), Black Medick (*Medicago lupulina*), Lesser Trefoil (*Trifolium dubium*), Red Clover (*Trifolium pratensis*) and Common Vetch (*Vicia sativa*). Alsike Clover (*Trifolium hybridum*), which is introduced, and White Clover (*T. repens*), which has few or very restricted truly wild populations, are sometimes present as gross agricultural variants with fistulose stems and huge leaves and flower-heads. All these leguminous plants are tall, erect, leafy variants formerly selected as forage crops that can be readily cut during hay-making. Other widely sown species appear to derive from horticultural stocks: for example *Achillea millefolium*, *Leucanthemum vulgare* and *Sanguisorba minor*.

Anthyllis vulneraria is commonly sown on to new road verges, and flowering material upon inspection is almost invariably found to belong to subsp. *carpatica* (Pant.) Nyman var. *pseudovulneraria* (Sag.) Cullen. This plant, intermediate between subsp. *vulneraria* of the lowlands of northwestern Europe (the common wild variant in lowland Britain) and the montane subsp. *carpatica*, is widespread in western and west-central Europe (Cullen, 1976), and is typical of hay meadows in valleys in the Alps. It has undoubtedly spread with cultivation and the seed appears to be widely available commercially. During the 1950s and 1960s another variant, subsp. *polyphylla* (DC.) Nyman from central and Eastern Europe, was widely sown, including along the verges of an embryonic motorway network (Akeroyd, 1991). However, there are few recent records of this subspecies and presumably subsp. *carpatica* var. *pseudovulneraria* has largely replaced it in commercial seed mixtures.

Many landscaping practitioners as well as conservationists have expressed concern about the presence of non-native material in planting schemes. Nevertheless, little experimental taxonomic or genetic research has been carried out on these alien taxa and variants, apart from a study of variation in *Lotus corniculatus* (e.g. Bonnemaison & Jones, 1986), a species that has attracted the attention of geneticists since the late 1960s. In particular we have few data on how alien variants, often more robust than native plants, may out-compete or hybridise with native populations. Some recent experimental evidence from common weeds used to plant field margins for game and other birds (Keller *et al.*, 2000) does, however, suggest that these sort of hybridisation events can lead to reduction in fitness of native populations, with considerable levels of outbreeding depression. If hybridisation between native and non-native variants does break up adaptive polygene complexes, then the repeated introduction (as in Britain) of non-native stocks of wildflower seed into often small and fragmented native populations is likely to have considerable deleterious effects.

Applying taxonomy

Taxonomists have an important role, and responsibility, to check the identity of plants employed in restoration and amenity plantings, and to alert those concerned to the presence of inappropriate taxa or variants. Unfortunately, in the United Kingdom at least, two impediments exist to the sensible application of taxonomic and genetic knowledge to the selection of plant material for use in habitat restoration and amenity plantings. One is a growing lack of expertise in traditional plant taxonomy, identification and 'whole-plant' science, a subject that has been debated elsewhere (e.g. Hedberg, 1998). The other is an inherent political correctness at the heart of the conservation movement. There has been considerable debate in the popular and semi-popular wildlife and conservation literature about the merits or otherwise of allowing populations of alien plants to persist and expand. In general scientists favour the rigorous monitoring and control of non-natives, whereas those who have come into conservation from the arts, journalism or politics see no problem with aliens (at least in a UK context), and indeed are even unhappy with the use of the word alien.

In view of the devastating and frequently unpredictable effects of emigrant plants and animals on ecosystems worldwide (Williamson, 1996), and the little that we know about how processes such as hybridisation and natural selection involving native and introduced species and variants operate in natural populations, it is surely best to err on the side of caution. Above all, we need a factual basis from which to proceed. A sound body of taxonomic knowledge, not least involving the thorough assessment of intraspecific variation, is a valuable tool for all those who wish to restore, transplant or reintroduce seeds and plants for whatever reason.

References

Akeroyd, J. R. (1991). *Anthyllis vulneraria* subsp. *polyphylla* (DC.) Nyman, an alien kidney-vetch in Britain. *Watsonia*, **18**, 401–3.

(1994a). *Seeds of Destruction? Non-native Wildflower Seed and British Floral Biodiversity*. London: Plantlife.

(1994b). Some problems with introduced plants in the wild. In *The Common Ground of wild and Cultivated Plants*, eds. A. R. Perry & R. G. Ellis. Cardiff, Wales: National Museum of Wales, pp. 31–40.

(1997). Intraspecific variation in European coastal plant species. In *Ecosystems of the World, Vol. 2C, Dry Coastal Ecosystems*, ed. E. van der Maarel. Amsterdam, the Netherlands: Elsevier, pp. 154–62.

Akeroyd, J. R. & Jury, S. L. (1991). Updating '*Flora Europaea*'. *Botanika Chronika*, **10**, 49–54.

Akeroyd, J. R. & Walters, S. M. (1987). *Flora Europaea*: background to the revision of volume 1. *Botanical Journal of the Linnean Society*, **95**, 223–6.

Akeroyd, J. R. & Wyse Jackson, P. S. (1995). *A Handbook for Botanic Gardens on the Reintroduction of Plants to the Wild*. Kew, UK: Botanic Gardens Conservation International.

Antonovics, J. (1971). The effects of a heterogeneous environment on the genetics of natural populations. *American Scientist*, **59**, 593–9.

Antonovics, J., Bradshaw, A. D. & Turner, R. G. (1971). Heavy metal tolerance in plants. *Advances in Ecological Research*, **7**, 1–85.

Bonnemaison, F. & Jones, D. A. (1986). Variation in alien *Lotus corniculatus* L. 1. Morphological differences between alien and native British plants. *Heredity*, **56**, 129–38.

Cullen, J. (1976). The *Anthyllis vulneraria* complex: a résumé. *Notes from the Royal Botanic Garden Edinburgh*, **35**, 1–38.

Department of Transport (1993). *The Wildflower Handbook*. London: Department of Transport.

Dobson, A. P., Bradshaw, A. D. & Baker, A. J. M. (1997). Hopes for the future: restoration ecology and conservation biology. *Science*, **277**, 515–22.

Falk, D. A. (1990). Integrated strategies for conserving plant diversity. *Annals of the Missouri Botanical Garden*, **77**, 38–47.

Falk, D. A., Millar, C. I. & Olwell, M. (eds.) (1996). *Restoring Diversity: Strategies for Reintroduction of Endangered Plants*. Washington, DC: Island Press.

Hedberg, I. (1998). Teaching crisis for whole-plant biology. *Plant Talk*, **15**, 4.

IUCN/SSC Re-introduction Specialist Group (2003). www:iucn.org/themes/ssc/programs/rsg.htm (accessed 2003).

Jalas, J. & Suominen, J. (1971–94). *Atlas Florae Europaeae*, vols. 1–10. Helsinki, Finland: Committee for Mapping the Flora of Europe and Societas Biologica Fennica Vanamo.

Jalas, J., Suominen, J. & Lampinen, R. (1996). *Atlas Florae Europaeae*, vol. 11. Helsinki, Finland: Committee for Mapping the Flora of Europe and Societas Biologica Fennica Vanamo.

Jalas, J., Suominen, J., Lampinen, R. & Kurtto, A. (1999). *Atlas Florae Europaeae*, vol. 12. Helsinki, Finland: Committee for Mapping the Flora of Europe and Societas Biologica Fennica Vanamo.

Keller, M. Kollmann & Edwards, P. J. (2000). Genetic introgression from distant provenances reduces fitness in local weed populations. *Journal of Applied Ecology*, **37**, 647–59.

Maunder, M. (1992). Plant reintroduction: an overview. *Biodiversity and Conservation*, **1**, 51–61.

Tutin, T. G., Heywood, V. H., Burges, N. A. *et al.* (1964–80). *Flora Europaea*, vols. 1 to 5, vol. 1 (1964), vol. 2 (1968). Cambridge, UK: Cambridge University Press.

Tutin, T. G., Burges, N. A., Chater, A. O. *et al.* (1993). *Flora Europaea*, vol. 1, 2nd edn. Cambridge, UK: Cambridge University Press.

Warwick, S. I. & Briggs, D. (1978a). The genecology of lawn weeds, I. Population differentiation in *Poa annua* L. in a mosaic environment of bowling green lawns and flower beds. *New Phytologist*, **81**, 711–23.

(1978b). The genecology of lawn weeds, II. Evidence for disruptive selection in *Poa annua* L. in a mosaic environment of bowling green lawns and flower beds. *New Phytologist*, **81**, 725–37.

Wells, T., Bell, S. & Frost, R. (1981). *Creating Attractive Grasslands Using Native Plants*. Shrewsbury, UK: Nature Conservancy Council.

Wells, T., Cox, C. & Frost, R. (1989). *The Establishment and Management of Wildflower Meadows*. Peterborough, UK: Nature Conservancy Council.

Williamson, M. (1996). *Biological Invasions*. London: Chapman and Hall.

15

Rattans, taxonomy and development

John Dransfield

Introduction

As practising plant taxonomists, we need no convincing of the fundamental importance of our discipline to the rest of plant sciences; but even within plant sciences and more so outside, taxonomy is often viewed as an irrelevance and plant taxonomists as self-serving. Yet nothing could be further from the truth. Robust stable taxonomies provide an essential framework for the dissemination of information about plants. In this chapter I aim to demonstrate the fundamental importance of taxonomy in an applied development arena, the development of sustainable management models for rattan palms.

Rattans are conspicuous components of many wet forest types in the Old World tropics. They are spiny climbing palms belonging to subfamily Calamoideae (Uhl & Dransfield, 1987). There are about 550 species in all (Govaerts & Dransfield, 2003), with some local floras being very diverse. For example the Malay Peninsula has approximately 106 species belonging to 9 genera (Dransfield, 1979 and additions) and the whole of Borneo has 150 species in 8 genera (Dransfield & Patel, 2005).

Although there is a long history of important local use of rattans in everyday life within the southeast Asian region for basketry, matting, general tying and other purposes, the major use that has impacted on rattan populations is as a source of cane for the cane-furniture industry. In much of the region, rattan is quite simply the most important non-timber forest product – and the harvesting of this forest product involves some of the poorest people in the community. Annually trade is worth about US$6.5 billion (ITTO, 1997), so it is a significant trade and much of the cane entering world trade comes from the wild. With forest loss and over-harvesting, rattan stocks have become seriously depleted, with grave consequences for sustainability and indeed the survival of some species.

Rattans were attractive to agencies interested in social forestry, and as serious shortages began to be manifest in the 1970s and 1980s, research was started on many aspects of rattans – from taxonomy through to processing and marketing – with the overall stated aim of safeguarding the rattan industry. A better understanding of the biology of the wild resource, potential for cultivation, improvement of processing and manufacturing, and a better understanding of the sociology of rattan gatherers, for example, could all contribute to a better future for the rattan industry. What was immediately obvious was that with 550 species, there is great variation in the species between one geographical area and the next: their ecology varies greatly, their morphology varies greatly and they are varied in cane size and quality. General statements about rattan must, therefore, inevitably be qualified according to species; it becomes imperative that we know what species we are dealing with. That should be no problem because local taxonomists – i.e. village people living near the forest, the people most involved with the harvesting of the wild resource – already know their rattans and have local names for them. The trouble is that these local names are not consistently applied and ultimately the only sure basic reference is the scientific name that is rooted in a type specimen. In other words, there has to be a properly researched basic taxonomy to provide the firm reference points for future research and development. Lest it be thought that taxonomic surveys are of mere academic interest, it is worth emphasising that discoveries made during such surveys may show otherwise poorly known species to be of commercial significance. For example, until 1979, *Calamus subinermis* was known as a single herbarium collection in the Kew Herbarium, collected by Hugh Low in the early 1900s in Sabah. The species was 'rediscovered' during fieldwork in the late 1970s and was shown to be widespread in northern coastal Sabah and to have great commercial potential. It was later shown to occur in Palawan and North Sulawesi. Ecological data collected during field surveys were used in developing planting methods and the species has now proven to be one of the most important large-diameter canes of Borneo, with excellent silvicultural potential (Dransfield & Manokaran, 1993).

The problem of chauvinistic taxonomies

Most rattan taxonomic studies have been country-based. This is not surprising as with 550 species to deal with the taxonomist has to devise achievable short-term goals for projects usually supported by funding agencies that specify particular countries or parts of countries. Such an approach allows the building up of an intimate knowledge of local variation and ecology of the rattan species but in concentrating on a defined political area, there is a tendency for unidentified species to be described as new local endemics when they may well be species described and well known in neighbouring areas. Problems occur, for example in Indochina, where species' distributions transgress political boundaries – particularly where diverse early-European influence occurred. This is an area of complex interdigitation of vegetation types dependent on different climatic influences and superimposed on this are the political boundaries of Burma, Bangladesh, India, China, Thailand, Lao PDR, Vietnam and Cambodia. Ex-colonial powers Britain and France are involved along with two countries that have remained independent throughout (Thailand and China). At the end of the nineteenth century and the early part of the twentieth century, palms from the entire region were described mostly by the great Italian botanist Beccari, who had material from throughout the region at his disposal. Not only was he able to resolve political over-description, but he tended not to describe the same thing twice. After his death in the first part of the twentieth century, no one continued this overall approach and Burmese, Bangladeshi and Indian plants tended to be described by British botanists, whereas those of Vietnam, Cambodia and Lao PDR were described by French botanists. More recently this politically based approach has continued. The account of the rattans in the Flora of China (Pei & Chen, 1991) illustrates the problem. There has been a tendency for the description of new taxa from China without sufficient comparison with material from over the borders in neighbouring Asian countries to see if the species have not already been described. This has been coupled with what I regard as a rather narrow species concept, perhaps influenced also by chauvinism, so there has been a proliferation of new names. Resolving some of the taxonomic and nomenclatural problems related to this has been difficult, given language barriers and the difficulty of exchanging material. In particular there are several rattans where new Chinese varieties of extra-Chinese taxa have been described. *Calamus giganteus* Becc. (Malayan) is now regarded as a synonym of *C. manan* Miq. (Dransfield, 1977) that is not known further north than the Isthmus of Kra in Thailand. What then is *Calamus giganteus* var. *robustus* Pei & Chen, recently described (Pei & Chen, 1991)? Evans, taking a regional approach, has been able to resolve this and several other problems of over-description in his joint project on the rattans of Laos (Evans *et al.*, 2002). So little was known of Lao rattans at the beginning of the Darwin Initiative for the Survival of Species (UK Government)-funded rattan project that he had perforce to look at the rattans of the neighbouring countries. Painstaking work in regional and European herbaria has allowed him to resolve almost all the major taxonomic problems of Indochinese rattans, and, incidentally, showing that *Calamus giganteus* var. *robustus* is synonymous with *Calamus platyacanthoides*, originally described from neighbouring Vietnam.

Calamus caesius in the Philippines

There seems to be a common perception that taxonomy is self-serving and irrelevant to the rest of botanical science and conservation. Yet those who may hold these attitudes continue to use plant names, whether scientific or vernacular, and are thus implicitly using a taxonomy. Taxonomy is unavoidable if scientific results are to be communicated. It provides the basic reference tool to which other information can be attached. It seems to be a lack of understanding of this point that has had such a deleterious effect on rattan research. A fine example is provided by the case of 'Sika' in the Philippines.

For at least a century and a half the slender rattan *Calamus caesius* has been recognised as the premier small-diameter cane in Borneo for weaving and mat-making. It is also used for the manufacture of chair-cane that is woven into the webbing panels that form the backs of chairs in either hardwood or rattan. Canes of this species have been traded from Borneo and Sumatra, and to a lesser extent from Peninsular Malaysia, for many centuries. In Borneo, particularly, *C. caesius* received rudimentary cultivation around long-houses, and has been the subject of silvicultural trials by forest departments since the 1920s. Its beautiful lustrous canes are strong and flexible and of the very highest quality. Growth rates in cultivation can be spectacular – as much as 7 m a year (see Dransfield and Manokaran (1993) for a summary of silvicultural aspects of this species). The superb quality, coupled with the tradition of small-scale cultivation, make this a very obvious candidate for trials on a commercial scale in perhumid areas of the tropics. Records of silvicultural trials of this species date back to the 1920s. However, it was not until the 1970s that experiments with *C. caesius* became both extensive and intensive. By the late 1970s there were already sufficient experimental data and experience for commercial estates of this species to be established in Sabah, East Malaysia. There was thus a great deal of published information attributed to the name *Calamus caesius*.

In the Philippines, interest in rattan was also becoming quite intense during the 1970s as forestry agencies realised that the harvesting of rattan was occurring at unsustainable rates and that there was a real danger that the trade might collapse, with consequent serious effects on employment in rural areas. With the depletion of rattan stocks in Luzon and Mindoro the rattan traders looked increasingly to areas more remote from Manila and, in particular, Palawan became pre-eminent as a source of cane at that time. Among the canes was a good small-diameter species with the local name of 'Sika'. In the Philippines there has always been heavy reliance on Merrill's *An Enumeration of Philippine Flowering Plants* (1922) for identification, particularly when trying to equate local names with Latin equivalents. Foresters, looking up 'Sika' in Merrill's *Enumeration*, found as Latin equivalent *Calamus spinifolius*, and so in forestry reports *Calamus spinifolius* was often cited as the best small-diameter cane from Palawan. In fact there are no herbarium records of *Calamus spinifolius* from Palawan, this species being a rare rattan confined to Luzon. When I visited Palawan in 1979, I was able to identify immediately rattans shown to me as 'Sika' as *Calamus caesius* (Dransfield, 1980). In fact *C. caesius* was quite common in Palawan but no herbarium collections until then had ever been made. Why should it be so important to know the correct identity of Sika? After all, people in Palawan knew very well that Sika was a good-quality cane and surely it does not matter whether the scientific name is *C. spinifolius* or *C. caesius* – the quality and potential uses of the plant remain the same. In fact, correctly identifying this rattan and placing it on a firm taxonomic basis has immediate significance and importance. All the legacy of research and cultivation experience attached to the name *Calamus caesius* can be applied to a Philippine species. In other words, instead of going through the long process of trials, forestry authorities in Philippines would immediately be able to use the results of experimentation elsewhere to establish plantations (from their own source material), this at a time when export of rattan seed was already being controlled. In this way a robust reliable taxonomy allows the transference of technology.

The inventory of the wild resource

With 106 species native to Peninsular Malaysia, the Malaysian Forestry Department had at its disposal an astonishingly broad gene pool of material for development. In the forest itself, Malay aborigines have long used rattans, and ascribe uses to almost all the rattans growing in the local area; they have also developed robust folk taxonomies that are used fairly consistently within each small area where they live. Rattans were harvested for thatch; for house construction; for weaving baskets; for medicine, dye and toys; and at the same time the canes of a few elite species were harvested for trading with rattan merchants as a source of income. Of the 106 species, perhaps 20 consistently enter trade and of these perhaps 4 or 5 are truly elite canes, commanding high prices. Faced with diminishing forest cover and intense pressure on remaining forest areas, an inventory of Malaysian forest areas that estimates standing crops of timber becomes an essential tool for land-use planning and economic-development forecasts. National forest inventories have been conducted on a regular basis and there are well-accepted methods for sampling and drawing up conclusions. However, if an inventory is to be reliable, the items inventoried have to be correctly identified. Rattans were included in the *Third National Forest Inventory* (Forest Department Malaysia, 1990–3). Given that only a few of the 106 species are commercially significant, 4 rattans only were included in the inventory and were referred to by their vernacular names: Sega (*Calamus caesius*), Semambu (*Calamus scipionum*), Manau (*Calamus manan*) and Dahan (*Korthalsia* spp. various). All individuals, from seedlings to adults, identified as these 4 types of rattan were recorded. Herein lies the root of the problem in interpreting the results. Seedlings are notoriously difficult to identify to species. The results of the inventory seemed to suggest that there was an abundance of commercial rattan in Peninsular Malaysia and that far from there being a shortage, there was, for example, sufficient Manau to sustain the industry until 2020. This result was at complete variance with the experience of rattan traders. They were in a state of crisis with most rattan middlemen complaining about severe shortages of Manau. This led to both the substitution of canes of inferior quality (such as Rotan Dok or Mantang – *Calamus ornatus*) and the closure of many rattan trading companies. In furniture design it also led to the substitution of Manau by braided Sega, *Calamus caesius*. How could the inventory results be believed when the industry was in crisis? Unreliable primary data are almost certainly the source of this discrepancy.

In fact, the few rattan inventories that can be relied upon (those rooted in a firm taxonomic base with species properly vouchered by herbarium specimens) all seem to indicate that stocks of commercially valuable cane are thinly spread and coupled with growth-rate observations, the studies suggest that harvesting would need to be on a very long cycle (sometimes as long as 40 years) if the exploitation were to be truly sustainable (see, for example, Bøgh (1996) and Supardi *et al.* (1999)).

Confusion between local taxonomies and the scientific taxonomy

If rattan research is to be rigorous and scientific it must be repeatable. This is an accepted principle of scientific research. To be repeatable, the research needs to have verifiable reference points. One of these is a robust taxonomic and nomenclatural framework, and the identities of any species mentioned need to be verifiable. The only reliable way of providing verification is by the citation of voucher specimens. In the case of rattans, with large numbers of sympatric species, it is essential that we know what species are being referred to and specimens need to be cited in case doubts over the identity of the subjects arise. All too often research has lacked these verifiable reference points. Numerous studies have been carried out on the physical properties of cane, citing trade names for the subject material without any vouchering. Trade names often cover several different species recognised by the scientist – species with different properties, different ecological requirements and different silvicultural potential – and some trade names may be applied even to different size classes in different parts of the rattan-growing region. Cane traded as 'Rotan Batu' in Peninsular Malaysia may belong to *Calamus insignis* var. *insignis* with a diameter of approximately 10 mm, whereas in East Malaysia 'Rotan Batu' is a large-diameter cane (at least 20 mm diameter) and usually refers to *Calamus subinermis* – with a completely different end use. An even more complex case is provided by the local name 'Tohiti' in Sulawesi. Tohiti is one of the names given to canes of good quality for furniture construction. Heyne (1922) published a Beccari manuscript name, *Calamus inops*, for a rattan specimen carrying this local name. Ever since, the name *Calamus inops* has been provided as the scientific name for a traded cane bearing this local name. The anatomy, bending properties and standing stock of Tohiti (= *Calamus inops*) have all been published without there ever being a herbarium specimen to voucher the observations. Why is this a problem? *Calamus inops*, based on its type specimen, is a rather-small-diameter cane with stems about 15 to 20 mm in diameter (Kramadibrata & Dransfield, 1992) whereas what is usually traded as 'Tohiti' has a cane diameter of at least 22 mm. In fact, 'Tohiti' has been applied to at least four different species, including *Calamus subinermis* mentioned above under Rotan Batu and which also occurs in northern Sulawesi. Much of the research published on Tohiti is valueless as the identity of the rattan studied is unverifiable.

Conclusions

It remains astonishing that some researchers still appear unable to grasp the fact that research on rattans (and for that matter any plant group) demands a firm reference base – the taxonomy – and that journals continue to accept for publication papers where the taxonomic base is not established and species are not verifiable by specimens. Clearly there is a communication gap somewhere. The taxonomist has a duty to make data available in an easily used form. This is not to compromise on taxonomic method or the publication of results – it is about the need to prepare products from taxonomic research that are properly targeted at the end user and are designed to be attractive and easy to use. A good example is provided by the *Field Guide to the Rattans of Lao PDR*, authored by Evans *et al.* (2001). This small, attractively designed book was based on a full reliable taxonomic monograph of the rattans of the Indochinese region (Evans *et al.*, 2002), but was specifically targeted at the non-taxonomists who need to use the taxonomic information – such as foresters and conservationists conducting inventories. A similar approach has been taken in designing the interactive key to the 150 species of rattan in Borneo (Dransfield & Patel, 2005). Here an electronic key has been devised that allows the easy identification of rattans, even if the material is incomplete. Eschewing abstruse terminology and profusely illustrated with scanned images of rattans and abundant interactive glossaries, the key should appeal to a wide range of end users including those normally put off by dichotomous keys and obscure descriptions.

References

Bøgh, A. (1996). Abundance and growth of rattans in Khao Chong National Park, Thailand. *Forest Ecology and Management*, **84**, 71–80.
Dransfield, J. (1977). The identity of 'rotan manau' in the Malay Peninsula. *Malayan Forester*, **40**(4), 197–9.

(1979). *A Manual of the Rattans of the Malay Peninsula.* Malayan Forest Records 29. Malaysia: Forest Department.

(1980). The identity of 'sika' in Palawan, Philippines. *Kalikasan*, **9**(1), 43–8.

Dransfield, J. & Manokaran, N. (eds.) (1993). *Plant Resources of South East Asia (PROSEA)*, vol. 6, Rattans. Wageningen, the Netherlands: Pudoc.

Dransfield, J. & Patel, M. (2005). *An Interactive Key to the Rattans of Borneo*, CD. Kew, UK: The Royal Botanic Gardens.

Evans, T. D., Sengdala, K., Viengkham, O. V. & Thammavong, B. (2001). *A Field Guide to the Rattans of Lao PDR.* Kew, UK: The Royal Botanic Gardens.

Evans, T. D., Sengdala, K., Thammavong, B., Viengkham, O. V. & Dransfield, J. (2002). A synopsis of the rattans (Arecaceae: Calamoideae) of Laos and neighbouring parts of Indochina. *Kew Bulletin*, **57**, 1–84.

Forest Department Malaysia (1990–3). *Third National Forest Inventory.* Kuala Lumpur, Malaysia: Forest Department.

Govaerts, R. & Dransfield, J. (2003). World Palm Checklist On-line. http://www.rbgkew.org.uk/data/monocots/palm`all.pdf

Heyne, L. (1922). *De Nuttige Planten van Nederlandsch Indië.* Batavia: Ruygrok.

ITTO (International Tropical Timber Organisation) (1997). Bamboo and rattan: resources for the 21st century? *Tropical Forest Update*, **7**(4), 13.

Kramadibrata, P & Dransfield, J. (1992). *Calamus inops* (Palmae: Calamoideae) and its relatives. *Kew Bulletin*, **47**, 581–93.

Merrill, E. D. (1922). Palmae. In *An Enumeration of Philippine Flowering Plants*, vol. 1. Manila, Philippines: Bureau of Printing, pp. 142–72.

Pei, S. J. & Chen, S. Y. (1991). *Flora Reipublicae Popularis Sinicae*, Tomus 13(1). Beijing, People's Republic of China: Science Press.

Supardi, N., Dransfield, J. & Pickersgill, B. (1999). The species diversity of rattans and other palms in the unlogged lowland forest of Pasoh Forest Reserve, Negeri Sembilan, Malaysia. In *Rattan Cultivation: Achievements, Problems and Prospects*, eds. R. Bacilieri & S. Appanah. Malaysia: CIRAD, pp. 22–37.

Uhl, N. W. & Dransfield, J. (1987). *Genera Palmarum. A Classification of Palms Based on the Work of Harold E. Moore Jr. L. H. Bailey Hortorium & International Palm Society.* Kansas: Allen Press.

16

Molecular systematics: measuring and monitoring diversity

Alastair Culham

The discipline of molecular systematics now focuses largely on the use of DNA technologies in the assessment of diversity and the understanding of its genetic basis. The focus on measures of genetic diversity, whether by genomic fingerprinting or by sampling genomic sequences allows a new estimate of diversity among and within species. Molecular systematics offers many features of interest to the conservation biologist both in terms of understanding biodiversity patterns and in the interpretation of specific conservation issues. The understanding of biodiversity is promoted through:

   (i) phylogenetic analysis of plants based on sequence divergence giving measures of genetic divergence among lineages;
  (ii) relative measures of species genomic diversity within genera;
 (iii) measures of genetic diversity within species;
  (iv) a critical re-evaluation of taxon limits from species level upwards;
   (v) evaluation of species concepts using molecular measures;
  (vi) estimates of the rate of establishment of lineages over time through ultrametric trees;
 (vii) incorporation of recently extinct species into modern genetic comparisons;
(viii) measures of geographical distribution of genetic diversity;
  (ix) identification of cryptic hybridisation;
   (x) molecular markers for identification of biological fragments.

The development of molecular systematics

Molecular technology has grown in two decades to the state where diversity can be studied at the level of the Kingdom down to somoclonal variation among parts of an individual. Change in technology and methodology has been astoundingly rapid. Crawford (1990) gave a review of molecular systematics that offered a mixture predominantly of serology, allozyme studies and DNA restriction analysis (with a little on sequencing), accurately summarizing the subject at that time. Few laboratories had used the techniques and few taxa had been sampled. The development of the polymerase chain reaction (PCR) in the mid 1980s (Saiki *et al.*, 1985; Mullis & Faloona, 1987; Mullis, 1990) opened the application of molecular technologies to tiny fragments of material and minimally destructive sampling.

By 1990, DNA fingerprinting using mini- and microsatellites had become an established set of techniques, largely used in the study of population variability in vertebrates (Jeffreys & Wilson, 1985; Jeffreys *et al.*, 1985a, 1985b; Litt & Luty, 1989; Tautz, 1989). Within a further five years, three new techniques – RAPD (random amplification of polymorphic DNA), ISSR (inter simple sequence repeat) and AFLP (amplified fragment-length polymorphism) – were introduced to allow genetic fingerprinting in plants (Williams *et al.*, 1990, Zietkiewicz *et al.*, 1994; Vos *et al.*, 1995). In parallel to this, DNA sequencing was developing from a laborious manual task where radioisotopes generated fuzzy radiographs that were read and manually entered into a computer one base at a time, to a fully automated process of simultaneous electrophoresis, data gathering and recording onto computer. Computing power has increased rapidly but the accumulation of sequence data and the need for complex analysis has demanded all the processor speed that is available. As a measure of the rapid growth of molecular techniques in diversity studies, a survey of the Thomson-ISI Web of Knowledge (2004) database shows a steady increase in papers using the terms 'molecular taxonomy', 'molecular systematics' or 'molecular phylogenetics' in their titles, keywords or abstracts (Figure 16.1). Obviously these numbers are a major underestimate of the actual number of publications because many papers in this field use more restrictive keywords and the database does not include all relevant journals. However, largely post-1990, growth in the field is seen clearly. From 1981 to 1990 only 43 papers are listed that use the term 'molecular phylogeny' but from 1991 to 2000 a further 1972 had been published, an increase of 460%.

The development of DNA technologies and computing power seems to have acted synergistically in the development of modern molecular systematics. The qualitative sequences of DNA suited the application of computationally intensive cladistic and maximum-likelihood analyses that were available. The ability to record DNA sequence as an objective series of letters, compared with discursive morphological descriptions, allowed laboratories to share data more easily. The establishment of internationally recognised data repositories for DNA sequence data (EMBL (European Molecular Biology Laboratory) Nucleotide Sequence Database Collaboration (2004) between Genbank (United States), EMBL Database (Europe) and DDBJ (DNA Databank of Japan)) provide a medium for storage and retrieval of sequence data. The database at the European Bioinformatics Institute (EBI) now contains DNA sequences from 178 000 species, of which 35 000 are flowering plants. This gives coverage of around 10% of flowering plant species. There are 46 extinct

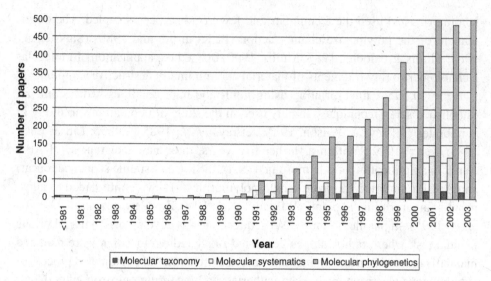

Figure 16.1 Occurrence of papers using the terms 'molecular systematics', 'molecular phylogeny' or 'molecular taxonomy' in the Thomson-ISI Web of Knowledge (2004) database until December 2003. Search limits allow a maximum of 500 returns to any query.

species represented in the database, of which only 4 are plants. At Genbank (2004) at the National Center for Biotechnology Information, USA, a taxonomy browser allows the user to research within a hierarchical taxonomic framework so that data can be explored species by species.

The ten items listed at the beginning of this chapter effectively fall into two broad areas: the techniques used to classify and name species (i.e. the framework on which all biology stands), and the techniques used to understand genetic variation within an existing taxonomic framework (i.e. the understanding of local variation of species and their component populations).

Applications

Phylogenetic analysis of plants based on sequence divergence to give measures of genetic divergence among lineages has required large concerted team efforts to generate the necessary data. The ground-breaking paper by Chase *et al.* (1993) is an example of collaboration of 41 authors in the construction of a phylogenetic tree for 500 species of flowering plants. Only in major floristic projects have similar numbers of taxonomists worked together on a single publication. This paper led to the establishment of the Angiosperm Phylogeny Group who have now established a single gene (rbcL) phylogeny for >2500 plant species and have proposed a new ordinal classification of flowering plants (Bremer *et al.*, 2003). The phylogeny of green plants as a whole is the remit of the Green Plant Phylogeny Group (2004) whose sampling and analysis is often at higher ranks than those of interest to most conservationists. The value of such work to conservation is in the highlighting of isolated taxa that may be the sole representative of ancient lineages. Such phylogenies can be incorporated into specialist software such as WorldMap (Vane-Wright *et al.*, 1991), which allows a cladistic measure of taxic diversity that could be used to prioritise lineage rather than taxon conservation.

The critical re-evaluation of taxon limits from species level upwards is tightly linked to the construction of phylogenies. Studies in molecular systematics commonly result in three transferable outcomes: new DNA sequence data deposited in public databases, new phylogenetic hypotheses available through publications and TreeBase (2004), and new taxonomies available through published papers. The exemplar study by Compton *et al.* (1998) used a nuclear and a plastid DNA region in combination with morphological data to construct species-level cladograms of *Actaea* and *Cimicifuga* in the Ranunculaceae with the aim of establishing classifications of the genera. Analysis of individual datasets and data in combination showed clearly that the generic delimitation was not congruent with phylogenetic relationships. The two morphological characters used to separate the genera did not correlate perfectly leaving one species in limbo having conflicting diagnostic characters of fruit and pedicel. The combination of molecular evidence with morphological gave compelling support for combining the two genera into one and establishing a more robust infrageneric classification that better represented the genetic diversity. The broadening of the generic concept of *Actaea* changed the genus from one containing a group of few very widely distributed species to one largely consisting of numerous narrowly distributed species. A study of the phylogeny shows distinct geographic and morphological groups recognisable as sections. Taking the North American section Oligocarpae as an example, it is notable that of the three species it contains: *Actaea arizonica* is within the 'Highly safeguarded' category of the Arizona Game and Fish Department due to its narrow distribution and endemic status, *Actaea elata* is rare in Canada, while *A. cordifolia* is widespread and not under threat. While two species may be of interest to conservationists, the lineage they are part of is secure. There are many such examples of the re-evaluation of generic limits based on combined molecular and morphological data. Molecular systematics does not work in isolation from morphological or other data. The increase in objectivity of taxonomy by utilising molecular data is as much a product of the quantity of data available and changes in approaches to analysis as the fact that DNA sequence has been used. While new classifications per se are not a direct part of conservation action, much conservation is based around the recognition of species as defining units of conservation policy.

Figure 16.2 Isotype of *Ulmus plotii* Druce at the Reading University Herbarium (photo: Alistair Culham).

Taxonomic limits within species complexes have also been investigated using molecular techniques. The investigation of the elm *Ulmus plotii* (Coleman *et al.*, 2000), recognised by a distinct unilateral growth of the crown, is a case where species evaluation had direct consequences to conservation evaluation. A study of RAPD banding patterns in the *U. minor* complex was used to establish whether there was one or several species in the group. Several elms recognised as rare and narrowly distributed species by some authors have been treated as parts of a much broader species that is consequently commoner and in little need of urgent conservation. The RAPD data supported recognition of two species, *Ulmus glabra* and *U. minor*, but demonstrated clearly that the rare endemic *U. plotii* (Figure 16.2) is a single genotype (i.e. a clone). Some feature of this genotype may have led to its widespread use in agricultural landscapes of the midlands of the United Kingdom. Coleman *et al.* (2000) suggested that cultivar status is the most appropriate.

Phylogenies can also be used to establish patterns of spread and endemism of species relative to each other. A study of *Pelargonium* by Bakker *et al.* (1998, 1999a, 1999b, 2000a, 2004) used DNA sequences to establish the relative rates of diversification in the genus and identified a series of species radiations nested one within another. The centre of species diversity for the genus is in the South African Cape flora but most of the earliest lineage splits occur in species from East Africa (Figure 16.3). This implies that the major diversification, at least in terms of surviving species, occurred many millions of years after the genus first appeared. The split in distribution between Africa and Australia could have indicated an ancient division due to continental drift (vicariance) but the DNA-sequence-based phylogeny showed the Australian species to be recent relatives of a weedy group of South African species. The distribution was due to long-distance dispersal. The single most speciose section of *Pelargonium* is sect. *Horaea* which has a great diversity of floral morphology. The 70+ species in this section show very little DNA sequence variation telling us that the diversification has been recent and rapid. This raises the question of the comparative value of species in conservation. Should every species be treated as an equal or are some more important than others?

Estimates of the rate of establishment of lineages over time through ultrametric trees have recently become a focus of systematic research. The Deep Time Project (2004) has established a network of scientists largely in the United States who are combining rate-smoothed phylogenetic trees with fossil and morphological data to calibrate the time at which lineages have diverged. An understanding of the palaeohistorical patterns of evolutionary diversification may give an improved insight into how lineages will respond to future changes in ecology due to climate change. Estimates of the age of lineages also give an insight into the rate of spread

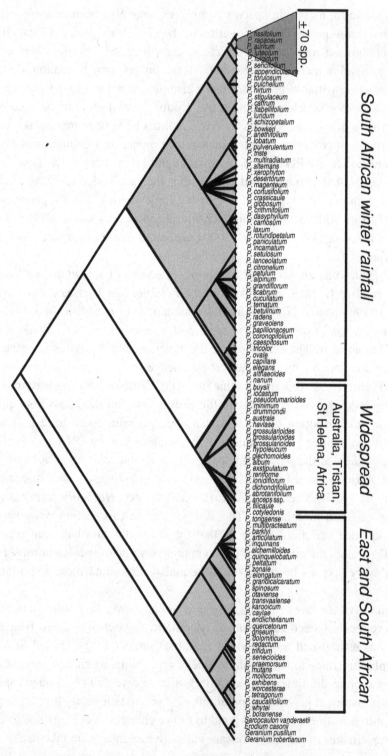

Figure 16.3 Diversification and distribution of *Pelargonium*.

of lineages and species and the rate of speciation. In angiosperms, Wickstrom *et al.* (2001) calibrated the rbcL-based phylogeny using critical fossils to establish ages of nodes, providing an age framework for diversification of the major lineages of flowering plants. There is considerable controversy surrounding the techniques of rate smoothing and time estimation in phylogenetic trees (Sanderson, 1997; Graur & Martin, 2004). Despite the legitimate concerns about calibration and dating of nodes in phylogenies, the technique has so much to offer in the understanding of rates of biological change that even these far-from-perfect models are beginning to challenge biological assumptions about plant species migration and establishment around the globe. The BiodiversityWorld Project (2004) is currently developing the application of these techniques to allow studies to be integrated with climate modelling to better understand the impact of climate change on species distribution and diversity. Phylogenetic analyses can be used to address other issues than organismal evolution. Dunwell *et al.* (2001) is one example among many where these techniques were used to understand the evolution of a group of proteins, while Bakker *et al.* (2000b) investigated the patterns of substitution seen in a plastid region to evaluate the assumptions made about the nature of the sequence change in systematic analysis.

The relative measure of species genomic diversity is a more complex issue that has not yet been resolved. Although largely complete nuclear genome sequences are available for *Arabidopsis thaliana*, *Solanum esculentum*, *Medicago sativa* and *Oryza sativa*, the technology to produce these large datasets is too expensive to apply to anything but an elite set of commercially important species at present. The use of expressed sequence tags (ESTs) is a more economically viable approach to giving an insight into differences in genomes and particularly the genes being expressed, but even that is beyond the budget of wide sampling among angiosperms. Estimates of diversity are currently based on DNA sequences from a few favoured DNA regions (rbcL, 18SrDNA and atpB for the Angiosperm Phylogeny Group for instance) that have to be taken as representative of the rest of the genome. In time, genomic studies will become important as data quantity increases and technology cheapens. The increasing quantity of data will eventually lead to an improved, if rather patchy, dataset for phylogenetic reconstruction of flowering plants.

The identification of cryptic hybridisation has been revolutionised by molecular systematics. The concept of chloroplast capture did not exist until molecular analyses found conflict between the morphological appearance of individuals and their chloroplast genome. Hybridising populations show gene flow that can be measured both by DNA-sequencing and DNA-fingerprinting approaches. There are many studies in this field that show cryptic hybridisation with no obvious macromorphological features to allow hybridisation to be detected (e.g. Harris & Ingram, 1991; Riesberg & Soltis, 1991). It has become evident that the level of hybridisation in plant populations is much higher than morphology-based estimates would suggest. Levin *et al.* (1996) review this topic and conclude that much of this hybridisation is usually deleterious to conservation efforts, particularly where non-hybrid populations of the species of concern are extant.

Evaluation of species limits and concepts based on DNA sequences has become an active and somewhat acrimonious debate. The increase in DNA-sequence availability and the storage of these sequences on searchable web databases has led some biologists to propose a new systematics where species are delimited on sequence information (Tautz *et al.*, 2002, 2003) and the current system abandoned. There has been strong resistance to this proposal (e.g. Seberg *et al.*, 2003) on the grounds that the resulting classification will be unusable due to: lack of any ready measure of where species boundaries will lie, potential lack of correlating morphological characters, lack of a biological basis for species recognition and – most compellingly – the fact that the current system of classification has continued to work through some 250 years of use, more than can be said for any computerised system yet. There can be no doubt that DNA techniques are being used to discover new areas of biodiversity that can not be studied by any other means. Analysis of DNA sequences from environmental samples of sea water have generated measures of biodiversity based purely on DNA-sequence diversity with no associated data on morphological diversity for practical reasons (Venter *et al.*, 2004). Thus we are already at a stage where some measures of local biodiversity are observable only by molecular means. Dealing with the management of conservation of organisms that are not known, let alone named, will provide a major challenge for conservation biology in the next few decades.

Incorporation of recently extinct species into modern genetic comparisons is evidenced by the sequencing of only a (large) handful of species. The 46 extinct species currently released on the EMBL database include the Dodo and several species of New Zealand Moas. While it is too late to conserve species that are already extinct, Jurassic Park aside, herbarium material can provide evidence of past genetic diversity and identity in species with greatly reduced living populations (Maunder *et al.*, 1999). A study of *Sophora toromiro* from Easter Island by Maunder *et al.* (1999, 2000) incorporated herbarium material into a survey of genetic diversity using DNA-fingerprinting techniques. Comparison of the historic fingerprint with those of the few remaining survivors of this species showed that much genetic diversity had already been lost permanently.

Measures of genetic diversity within species fall at the junction of systematics and population biology. The systematic aspects are focused on species delimitation and the identification of genetic discontinuities that represent species or infraspecies boundaries. The identification of species-specific RAPD markers for endangered

Figure 16.4   Young plants of *Pyrus cordata* being planted at the reintroduction site (photo: Alistair Culham).

UK populations of *Pyrus cordata* allowed possible introgression with *Pyrus communis* to be investigated (Jackson *et al.*, 1997). The same fingerprint data also allowed the assessment of the number of remaining genotypes in this clonally reproducing species. The lack of natural seed set was a result of the remaining UK sites for the species each being a single clone. The species is self-incompatible. The discovery that only two genotypes were present allowed a planned breeding programme to be established and a new population to be planted based on sound genetic principles (Figure 16.4). There are many such examples of fingerprinting techniques that allow both species limits and overall genetic diversity to be evaluated.

Measures of geographical distribution of genetic diversity have advanced rapidly. A recent study combined geographical information systems (GIS) technology with diversity of 13 sequenced loci in *Arabidopsis thaliana* (Hoffmann *et al.*, 2003). A distinct geographic pattern in variation was found showing most of the genetic variation to occur along the Atlantic coast of Iberia and southern United Kingdom. This study was able to employ phylogenetic measures to identify diversity hotspots that had been impossible to interpret with anonymous fingerprint markers. Such new approaches will be of fundamental importance in establishing not just what should be conserved but where the most valuable populations are.

Figure 16.5 Detail of young growth of *Olea europaea* (photo: Alistair Culham).

Olea europaea, the olive, is a major crop in the Mediterranean basin but it has several wild relatives that are much less common. Conservation of wild relatives of crop species is often considered important. Morphology alone can not distinguish some of the taxa reliably. The study by Médail *et al.* (2001) combined a range of data, including both morphological and molecular data, to characterise a rare Moroccan element of the olive, *O. europaea* subsp. *maroccana*. Data were gathered from chloroplast and mitochondrial DNA using a combination of PCR and RFLP to establish organellar haplotypes for 76 individuals in *O. europaea*. In addition these individuals were studied using 79 RAPD markers to give an estimate of the overall genetic diversity within the species. Correspondence analysis identified a core area of genetic diversity corresponding to *O. europaea* subsp. *europaea* and cultivars, *O. europaea* subsp. *laperrinei* (Figure 16.5) and *O. europaea* subsp. *cuspidata* from Yemen. Satellite to this were four separate foci representing three different subspecies: *O. europaea* subsp. *guanchica*, *O. europaea* subsp. *maroccana*, *O. europaea* subsp. *cuspidata* (Iran) and *O. europaea* subsp. *cuspidata* (Kenya). The combination of morphological and diverse molecular evidence allowed formal recognition of *O. europaea* subsp. *maroccana* and a consequent evaluation of its conservation status. Three main risks were identified to the subspecies: small populations due to limited range; damage by humans and livestock; and hybridisation between *O. europaea* subsp. *maroccana* and cultivated *O. europaea* subsp. *europaea*. It is estimated that this is one of the ten most threatened tree taxa in the Mediterranean basin (Barbero *et al.*, 2001).

The endangered filmy fern *Trichomanes speciosum* (Figure 16.6) has been studied using a combination of allozyme analysis and chloroplast DNA haplotyping (Rumsey *et al.*, 1999). Fieldwork showed that while the sporophyte generation of this species is very rare, the gametophyte is much more widespread. Genetic analysis identified a strong geographic partitioning of genetic variation. Many localities showed only a single genotype to be present. The consequence to conservation is that the loss of any individual site for the species will result in the loss of genetic variation for the species as a whole. Conservation must be approached for the whole species across its distribution and genetic diversity will not be preserved by maintaining just a few sites for the species.

Molecular markers for identification of morphologically cryptic biological fragments can be developed readily if the nature of sequence variation is already known. Application of coloured fluorescent markers linked to species-specific primers allowed Mishra *et al.* (2003) to identify fragments of *Fusarium* species without the need to culture them. DNA profiling is widely used in forensic science to link biological remains to particular individuals or taxa. In conservation, reliable DNA markers can allow detection and reliable identification of rare species that might be the subject of illegal smuggling, where the diagnostic morphological characters are missing.

Techniques

An introduction to techniques and their limits will be valuable, if not essential to understand the areas where molecular systematics can be applied to conservation. The plant material used for molecular systematics is varied. In most cases the ideal material is fresh leaf tissue collected and extracted with minimum delay. DNA can be retrieved from a much wider range of material. The rapid drying of plant material with silica gel is a common technique while air-drying and pressing are also sometimes used. The ability to retrieve DNA from herbarium specimens has given

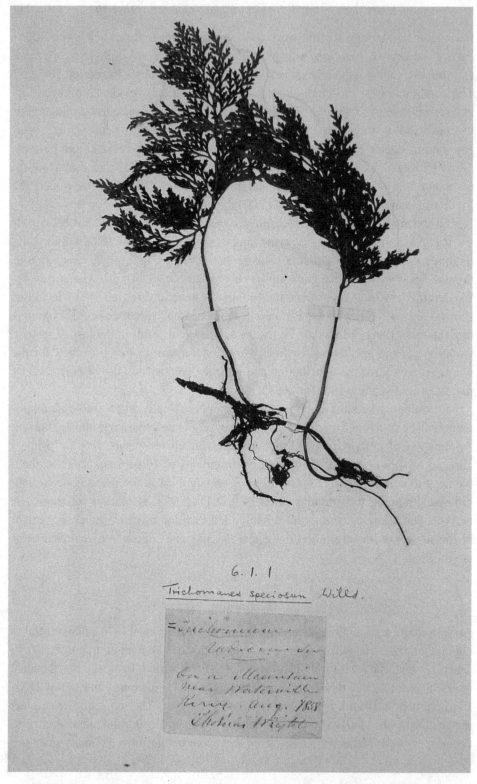

Figure 16.6 A specimen of *Trichomanes speciosum* from Ireland (photo: Alistair Culham).

a new role to herbaria where vast repositories of well-documented and carefully filed plant material can be found. Fresh material can be preserved by grinding in CTAB (hexadecyl trimethyl ammonium bromide) buffer in the field and later stored at room temperature, or it can be taken back to the laboratory for storage as frozen tissue. Each method has pros and cons and often only empirical testing can give information on the best way to collect material from any individual plant group. In a study of Lardizabalaceae, Hoot *et al.* (1995a, 1995b) used herbarium material of the South American genus *Boquila* due to lack of available living material, while recently DeCastro *et al.* (2002) used material that was largely of herbarium origin in their study of *Genista*. In conservation, the use of herbarium specimens can give access to genetic variation in populations and individuals now extinct. For instance, Maunder *et al.* (1999) used ISSR fingerprints from a herbarium specimen of what was probably the last wild plant of *Sophora toromiro* in a study of genetic diversity in *ex situ* stocks of the species. There are risks of contamination to be considered before material is extracted. Some plants have fungal endophytes, especially in the root systems, and fungal DNA will co-extract with plant DNA if the sample is mixed. Quality of starting material is essential to the validity of any DNA-based study.

The extraction of DNA for plant systematics has largely followed the protocol of Doyle and Doyle (1987) based on CTAB. This extraction process is popular among plant systematists because it has proved reliable among a wide range of genera. The yields of DNA are moderately good and the quality is sufficient for most subsequent work. The CTAB technique is slow, usually taking several hours from beginning to end, but is amenable to modification for use with very small quantities of source material and for coaxing out small quantities of DNA from old material such as herbarium specimens. In recent years a number of DNA extraction kits have been marketed that offer much more rapid and simple extraction of small quantities of DNA from small amounts of plant tissue. The use of these is increasing, especially where DNA content and quality is high in the source material. Extracted DNA can generally be stored for many years with minimal degradation in a suitable buffer at low temperatures ($-80\,°C$ commonly) and some projects have been set up to store DNA for long periods in DNA banks, such as one at the Jodrell Laboratory, Kew. For long-term banking, DNA usually undergoes additional purification by caesium chloride ultracentrifugation. Even after this treatment, some DNAs do not store well but the reasons for this are not yet clear.

The development of the PCR technique revolutionised molecular systematics. The gradual decrease in price of the components of the reaction and the hardware to run the reactions allowed the PCR technique to become generally available by 1990. Without the PCR the low yields of DNA from herbarium specimens would not be useful and sampling of living plants would be far more destructive. This reaction is used both to amplify specific regions of DNA for later sequencing and to generate DNA fingerprints of various kinds. The ability of the PCR to amplify millions of copies of a specific region of DNA from a tiny amount of starting material is both a strength and a weakness. Tiny quantities of contaminating DNA can cause amplification of a product that is not from the target individual. This is a particular risk in non-specific fingerprinting techniques but much less of a risk for specific primers such as those for parts of plant chloroplast DNA where fungal or human contamination will not cause a PCR product.

There are a group of technically different protocols that generate DNA fingerprints. The widely used RAPD (Williams *et al.*, 1990), ISSR (Ziekiewicz *et al.*, 1994) and AFLP (Vos *et al.*, 1995) techniques all produce banding patterns based on chance proximity of primer sites (or restriction sites) in genomic DNA. The patterns can be complex and the numbers generated by a single reaction vary. These bands are usually dominant markers, which change by the loss or gain of a primer site. Diversity is measured as the proportion of shared bands among individuals using one of several similar algorithms. Since 2000, the microsatellite technique has become more widely used in plants. The technique has been available for longer than the three previously mentioned approaches (Litt & Luty, 1989) but has much higher initial development costs. Recent modification to the protocol and the increasing ease of high-throughput sequencing have made the identificaion of microsatellite markers more easy (Zane *et al.*, 2002). While the non-specific techniques are relatively cheap they are, by their very nature, prone to sensitivity to contamination from any other DNA source and sometimes are difficult to reproduce except under very tightly controlled experimental conditions. The specific microsatellites suffer the opposite problem in that they are rarely transferable to even closely related species.

DNA-sequence-based studies rely on the sequencing of one or, usually, more regions of DNA and use these to construct phylogenetic relationships. It has been estimated that about 6000 base pairs of phylogenetically informative sequence are needed per sample to construct a stable phylogeny (Hillis, 1991). Few studies come close to achieving this. Despite the lack of data there are now thousands of DNA-based phylogenies available. While the tests for support of these phylogenies show that some areas are weak, other areas – usually those represented in existing taxonomies – remain stable. There are a number of textbooks detailing aspects of the data collection and analysis such as those of Hillis *et al.* (1996) and Felsenstein (2002). The use of DNA sequencing in conservation is usually through the establishment of phylogenetic trees to demonstrate the relationships of rare species, by the development of species-specific markers and in the detection and evaluation of introgression.

The primary aim of molecular systematics is to establish a stable measure of biotic diversity at a global level, from evaluation of the genetic nature of species through to relationships among the kingdoms of life. While these aims are not the same as those of conservation, the two are tightly linked; systematics defines the units of biology, conservation ensures those units remain a biological reality.

References

Bakker, F. T., Helbrugge, D, Culham, A. & Gibby, M. (1998). Phylogenetic relationships within *Pelargonium* sect. *Peristera* (Geraniaceae) inferred from nrDNA and cpDNA sequence comparisons. *Plant Systematics and Evolution*, **211**, 273–87.

Bakker, F. T., Culham, A., Daugherty, L. C. & Gibby, M. (1999a). A trnL-F based phylogeny for species of *Pelargonium* (Geraniaceae) with small chromosomes. *Plant Systematics and Evolution*, **216**, 309–24.

Bakker, F. T., Culham, A. & Gibby, M. (1999b). Phylogenetics and biogeography of *Pelargonium* (Geraniaceae). In *Molecular Systematics and Plant Evolution*, eds. P. M. Hollingsworth, R. M. Bateman and R. J. Gornall Systematics Association special volume 57. London: Taylor & Francis, pp. 353–740.

Bakker, F. T., Culham, A., Pankhurst, C. E. & Gibby, M. (2000a). Mitochondrial and chloroplast DNA-based phylogeny of *Pelargonium* (Geraniaceae). *American Journal of Botany*, **87**, 727–34.

Bakker, F. T., Culham, A., Gomez-Martinez, R. *et al.* (2000b). Patterns of nucleotide substitution in angiosperm cpDNA trnL(UAA)-trnF(GAA) regions. *Molecular Biology and Evolution*, **17**, 1146–55.

Bakker, F. T., Culham, A., Hettiarachi, P., Toloumenidou, T. & Gibby, M. (2004) Phylogeny of *Pelargonium* (Geraniaceae) based on DNA sequences from three genomes. *Taxon*, **53**, 17–28.

Barbero, M., Loisel, R., Médail, F. & Quézel, P. (2001). Signification biogeographique et biodiversite des forets du bassin mediterranean. *Bocconea*, **13**, 11–25.

BiodiversityWorld Project (2004). http://www.bdworld.org.

Bremer, B., Bremer, K., Chase, M. W. *et al.* (2003). An update of the Angiosperm Phylogeny Group classification for the orders and families of flowering plants: APG II. *Botanical Journal of the Linnean Society*, **141**, 399–436.

Chase, M. W., Soltis, D. E., Olmstead, R. G. *et al.* (1993). Phylogenetics of seed plants: an analysis of nucleotide-sequences from the plastid gene rbcL. *Annals of the Missouri Botanical Garden*, **80**, 528–80.

Coleman, M., Hollingsworth, M. & Hollingsworth, P. M. (2000). Application of RAPDs to the critical taxonomy of the English endemic elm *Ulmus plotii* Druce. *Botanical Journal of the Linnean Society*, **133**(3), 241–62.

Compton, J. A., Culham, A. & Jury, S. L. (1998). Reclassification of *Actaea* to include *Cimicifuga* and *Souliea* (Ranunculaceae): phylogeny inferred from morphology, nrDNA ITS and cpDNA TrnL-F sequence variation. *Taxon*, **47**, 593–634.

Crawford, D. J. (1990). *Plant Molecular Systematics: Macromolecular Approaches*. New York and Chichester: John Wiley.

De Castro, O., Cozzolino, S., Jury, S. L. & Caputo, P. (2002). Molecular relationships in *Genista* L. Sect. *Spartocarpus* Spach (Fabaceae). *Plant Systematics and Evolution*, **231**, 91–108.

Deep Time Project (2004).
    http://www.flmnh.ufl.edu/deeptime/.
Doyle, J. J. & Doyle, J. L. (1987). A rapid DNA isolation procedure for small quantities of
    fresh leaf tissue. *Phytochemical Bulletin*, **19**, 11–15.
Dunwell, J. M., Culham, A., Carter, C. E., Sosa-Aguirre, C. R. & Goodenough, P. W.
    (2001). Evolution of functional diversity in the cupin superfamily. *Trends in
    Biochemical Sciences*, **26**, 740–6.
EMBL (European Molecular Biology Laboratory) Nucleotide Sequence Database
    Collaboration (2004).
    http://www.ebi.ac.uk.
Felsenstein, J. (2002). *Inferring Phylogenies*. Sunderland, MA: Sinauer Associates.
Genbank (2004).
    http://www.ncbi.nlm.nih.gov/Taxonomy/taxonomyhome.html/.
    Graur, D. & Martin, W. (2004). Reading the entrails of chickens: molecular
    timescales of evolution and the illusion of precision. *Trends in Genetics*, **20**,
    80–6.
Green Plant Phylogeny Group (2004).
    http://ucjeps.berkeley.edu/bryolab/GPphylo/.
Harris, S. A. & Ingram, R. (1991). Chloroplast DNA and systematics: the effects of
    intraspecific diversity and plastid transmission. *Taxon*, **40**, 393–412.
Hillis, D. M. (1991). Discriminating between phylogenetic signal and random noise in
    DNA sequences. In *Phylogenetic Analysis of DNA Sequences*, eds. M. M. Miyamoto
    and J. Cracraft. New York: Oxford University Press, pp. 278–94.
Hillis, M., Moritz, C. & Mable, B. K. (1996). *Molecular Systematics*, 2nd edn.
    Sunderland, MA: Sinauer Associates.
Hoffmann, M. H., Glaß, A. S., Tomiuk, J. *el al.* (2003). Analysis of molecular data of
    Arabidopsis thaliana (L.) Heynh. (Brassicaceae) with Geographical Information
    Systems (GIS). *Molecular Ecology*, **12**, 1007–19.
Hoot, S. B., Culham, A. & Crane, P. R. (1995a). The utility of atpB gene chloroplast DNA
    gggs in resolving phylogenetic relationships in the Lardizabalaceae, including
    comparisons with *rbc*L and 18S ribosomal DNA sequences. *Annals of the Missouri
    Botanical Garden*, **82**, 194–207.
    (1995b). Phylogenetic relationships of the Lardizabalaceae and Sargetodoxaceae:
    chloroplast and nuclear DNA sequence evidence. *Plant Systematics and Evolution*
    (Supplement), **9**, 195–9.
Jackson, A., Erry, B. & Culham, A. (1997). Genetic aspects of the species recovery
    programme for the plymouth pear (*Pyrus cordata* Desv.). In *The Role of
    Genetics in Conserving Small Populations*, eds. T. J. Crawford, J. Spencer,
    D. Stevens *et al.* Peterborough, UK: Joint Nature Conservation Committee,
    pp. 112–21.
Jeffreys, A. J. & Wilson, V. (1985). Hypervariable regions in human DNA. *Genetical
    Research*, **45**, 213.
Jeffreys, A. J., Wilson, V. & Thein, S. L. (1985a). Hypervariable minisatellite regions in
    human DNA. *Nature*, **314**, 67–73.
    (1985b). Individual-specific fingerprints of human DNA. *Nature*, **316**, 76–9.
Levin, D. A., Francisco-Ortega, J. & Jansen, R. K. (1996). Hybridization and the
    extinction of rare plant species. *Conservation Biology*, **10**, 10–16.
Litt, M. & Luty, J. A. (1989). A hypervariable microsatellite revealed by *in vitro*
    amplification of a dinucleotide repeat within the cardiac-muscle actin gene.
    American Journal of Human Genetics, **44**, 397–401.

Maunder, M., Culham, A., Bordeu, A., Allainguillaume, J. & Wilkinson, M. (1999). Genetic diversity and pedigree for *Sophora toromiro* (Leguminosae): a tree extinct in the wild. *Molecular Ecology*, **8**, 725–38.

Maunder, M., Culham, A., Alden, B. *et al.* (2000). Conservation of the toromiro tree: case study in the management of a plant extinct in the wild. *Conservation Biology*, **14**, 1341–50.

Médail, F., Quezel, P., Besnard, G. & Khadari, B. (2001). Systematics, ecology and phylogeographic significance of *Olea europaea* L. ssp *maroccana* (Greuter & Burdet) P. Vargas *et al.*, a relictual olive tree in south-west Morocco. *Botanical Journal of the Linnean Society*, **137**(3), 249–66.

Mishra, P. K., Fox, R. T. V. & Culham, A. (2003). Development of a PCR-based assay for rapid and reliable identification of pathogenic *Fusaria*. *FEMS Microbiology Letters*, **218**, 329–32.

Mullis, K. B. (1990). The unusual origin of the polymerase chain reaction. *Scientific American*, **262**, 56–65.

Mullis, K. & Faloona, F. (1987). Specific synthesis of DNA *in vitro* via a polymerase catalyzed chain reaction. *Methods in Enzymology*, **55**, 335–50.

Riesberg, L. H. & Soltis, D. E. (1991). Phylogenetic consequences of cytoplasmic gene flow in plants. *Evolutionary Trends in Plants*, **5**, 65–84.

Rumsey, F. J., Vogel, J. C., Russel, S. J., Barrett, J. A. & Gibby, M. (1999). Population structure and conservation biology of the endangered fern *Trichomanes speciosum* Willd. (Hymenophyllaceae) at its northern distributional limit. *Biological Journal of the Linnean Society*, **66**, 333–44.

Saiki, R. K. S., Scharf, S., Faloona, F. *et al.* (1985). Enzymatic amplification of beta globin sequences and restriction site analysis for diagnosis of sickle cell anemia. *Science*, **230**, 1350–4.

Sanderson, M. J. (1997). A nonparametric approach to estimating divergence times in the absence of rate constancy. *Molecular Biology and Evolution*, **14**, 1218–31.

Seberg, O., Humphries, C. J., Knapp, S. *et al.* (2003). Shortcuts in systematics? A commentary on DNA-based taxonomy. *Trends in Ecology and Evolution*, **18**, 63–5.

Tautz, D. (1989). Hypervariability of simple sequences as a general source for polymorphic DNA markers. *Nucleic Acids Research*, **17**(16), 6463–71.

Tautz, D., Arctander, P., Minelli, A., Thomas, R. H. & Vogler, A. P. (2002). DNA points the way ahead in taxonomy. *Nature*, **418**, 479.

(2003). A plea for DNA taxonomy. *Trends in Ecology and Evolution*, **18**, 70–4.

Thomson-ISI Web of Knowledge (2004).
http://www.isinet.com.

TreeBase (2004).
http://www/treebase.org/treebase.

Vané-Wright, R. I., Humphries, C. J. & Williams, P. H. (1991). What to protect: systematics and the agony of choice. *Biological Conservation*, **55**, 235–54.

Venter, J. C., Remington, K., Heidelberg, J. F. *et al.* (2004). Environmental genome shotgun sequencing of the Sargasso Sea. *Science*, **304**(5667), 66–74.

Vos, P., Hogers, R., Bleeker, M. *et al.* (1995). AFLP: a new technique for DNA-fingerprinting. *Nucleic Acids Research*, **23**(21), 4407–14.

Wickstrom, N., Savolainen, V. & Chase, M. W. (2001). Evolution of the angiosperms; calibrating the family tree. *Proceedings of the Royal Society of London. Series B*, **268**, 2211–20.

A. Culham

Williams, J. G. K., Kubelik, A. R., Livak, K. J., Rafalski, J. A. & Tingey, S. V. (1990). DNA polymorphisms amplified by arbitrary primers are useful as genetic-markers. *Nucleic Acids Research*, **18**, 6531–5.

Zane, L., Bargelloni, L. & Patarnello, T. (2002). Strategies for microsatellite isolation: a review. *Molecular Ecology*, **11**, 1–16.

Zietkiewicz, E., Rafalski, A. & Labuda, D. (1994). Genome fingerprinting by simple sequence repeat (SSR)-anchored polymerase chain-reaction amplification. *Genomics*, **20**, 176–83.

17

Legislation: a key user of taxonomy for plant conservation and sustainable use

H. Noel McGough

Taxonomy is the service industry of botanical science – it supplies the classification that is used in conservation, government policy, trade and horticulture. In terms of legislation, taxonomy is at the core of species-based national legislation, lists of protected species and national action plans. Examples of regional and international conservation legislation that are wholly or partly species based include the European Union Habitats Directive and the Convention on International Trade in Endangered Species (CITES). All of these initiatives, which are key to the global conservation and sustainable use of plant and animal resources, need to be firmly anchored by good taxonomy. The importance of taxonomy and taxonomists in the genesis and implementation of one major international convention, CITES, is examined.

CITES

The Convention was established to control and monitor international trade in plants and animals threatened or potentially threatened by commercial exploitation (Royal Botanic Gardens, Kew, 2002). The Convention controls trade in listed plants and animals by means of a permit system. The Convention came into force in 1975 and over 160 countries are now party to it. The list of taxa controlled by the Convention is revised every two years when all member countries (Parties) meet to review its implementation at a meeting of the Conference of the Parties (COP).

This list of CITES-controlled taxa consists of three appendices. Commercial trade in species listed on Appendix I is banned. The Convention states (Wijnsteckers, 2002) that:

Appendix I shall include all species threatened with extinction, which are or may be affected by trade. Appendix II shall include all species which although not necessarily now threatened with extinction may become so unless trade . . . is subject to strict regulation to avoid utilization incompatible with their survival. Appendix III shall include all species which any Party identifies as being subject to regulation within its jurisdiction for the purpose of preventing or restricting exploitation, and as needing the co-operation of other Parties in the control of trade.

At the core of the Convention are these three lists of species.

CITES monitors trade by means of a globally accepted permit system. The CITES Conference of the Parties have established criteria, through debate and negotiation, for listing species on the Appendices. The Convention has also established a series of committees to service the needs of the Parties. The Standing Committee is the formal administrative Committee which takes decisions between meetings of the CITES Conferences of the Parties. Three technical committees advise the Standing Committee and the Conference of the Parties. These are the Animals Committee, the Plants Committee and the Nomenclature Committee. The Nomenclature Committee has the remit to prepare or recommend for adoption by CITES standard nomenclatorial references, to review the appendices with regard to the correct nomenclature, vet listing proposals to ensure correct names are used and generally advise on nomenclature. In effect, the Nomenclature Committee is the reference point to ensure that good taxonomy is at the core of the Convention with its work extending beyond strict issues of nomenclature. The Conference of the Parties appoints two individuals to carry out this task: one for plants and one for animals. The two individuals then use a global network of taxonomic experts and institutions to consolidate advice for CITES.

The species concept in CITES

In conservation legislation the definition of species, or indeed the concept of what is covered by the terms 'fauna' and 'flora', may differ from that used in the scientific world. In CITES, for example, Article 1(a) defines 'species' as any 'species, subspecies or geographically separate population thereof'. This tinkering with the species concept allows, for example, split listing of populations in different countries. It does not take away the requirement that to ensure effective implementation of a listing the species concerned must be 'good' (taxonomically sound). The CITES listing criteria asks proponents of species-listing proposals to: 'provide sufficient information to allow the Conference of the Parties to identify clearly the taxon that is the subject of the proposal.' The information supplied needs to define clearly the entity proposed for listing. This information also allows the Nomenclature Committee and individual Parties to assess the validity of the taxon proposed for listing. Parties can and do reject proposals to list species on CITES if they consider the taxonomic status of the taxon to be unsure or likely to give rise to major problems in implementation. At the 12th meeting of the Conference of the Parties, in 2002, the United States withdrew its proposal to list the cactus *Slerocactus spinosior* subsp. *blainei* on Appendix I due to concerns expressed over its status.

In CITES terms, the taxonomic information supplied in the proposal defines the scope of the listing. If a meeting of the Conference of the Parties adopts a proposal and lists a species for CITES control, the taxon as defined in the original proposal is the plant or animal controlled. If that taxon is later subject to taxonomic review and its status reassessed, the CITES listing will be amended. However, the amended listing cannot go beyond the scope of the original listing, i.e. it cannot be broadened to include an entity not covered in the original proposal. Such a change requires a new listing proposal to be put to a future meeting of the Conference of the Parties.

The CITES Conference of the Parties

The CITES Conference of the Parties meet every two years to consider amendments to the CITES listings and to address challenges to effective implementation of the Convention. The meeting strives for decision making by consensus, but this is often impossible for contentious issues – such as the consideration of the listing of whales, sharks, elephants and timber. If proposals lead to a vote, a two-thirds majority of those present is required, with each Party having one vote. Attendance at the meeting may reach 1500 participants with observers out-numbering the official representatives of the Parties. The mechanisms of decision making at the meeting attempt to reflect the structure of the Convention. All decisions are confirmed in a plenary session of the meeting with the full discussion occurring in two subsidiary groups, Committee I and Committee II. Reflecting the role of the CITES Scientific Authority and Management Authority, Committee I handles the scientific issues, while Committee II covers implementation issues. It is in Committee I where controversial issues, such as listing or delisting species, are covered. It is in Committee I where Parties ensure that their scientific experts, including taxonomists, are represented.

The CITES Plants Committee

The CITES Plants Committee is the scientific advisory group to the CITES Conference of the Parties. Its most important role is to review trade in CITES species, identify trade that is not sustainable and formulate the remedial action required to make that trade sustainable. As such, the Plants Committee is at the core of the Convention – the active management of sustainable trade.

Individual specialists are elected to the Plants Committee to represent the CITES regions of Africa, Asia, Europe, North America, South Central America, and the Caribbean and Oceania. It had always been the view of the Parties that members of the technical committees be elected as individual scientific experts from the CITES regions. This method differs from that used in the CITES Standing Committee.

In that Committee (a management committee), individual countries are elected to represent their region. In the situation where a country is elected, different individuals may represent that country at different meetings of the committee. In the case of the scientific advisory groups (the Animals and Plants Committees), the experts are named individuals whose curricula vitae have been scrutinised by the countries of the region prior to their election and only they can represent the region. At the 12th meeting of the Conference of the Parties, amendments were proposed to amalgamate the two scientific committees into one technical committee and for the representatives to that committee to be elected on a country basis – not as individual experts. This was done in an attempt to streamline the CITES processes and conserve financial resources. The Parties roundly rejected this as they saw the likely result to be erosion of the scientific basis of the Convention and an increased politicisation of debate. The election of individual scientists ensures scientific representation, whereas election on a country basis would not ensure that the country would send a scientist to the meetings – it would be more likely that a member of a government-department policy unit would attend. The CITES Conference of the Parties thus confirmed the requirement that scientific expertise – including taxonomic expertise – be at the core of the Convention.

The regional representatives of the Plants Committee are individuals with the power to make decisions and, when necessary, vote. Representatives of countries and non-governmental organisations (NGOs) may attend the meeting as observers and speak as such but they have no formal role in decision making. The CITES Plants Committee has been in operation since 1988; to date there have been 12 meetings of the Plants Committee. At present there are 10 representatives from the 6 CITES regions of the globe and 7 of those 10 have taxonomic expertise or are from taxonomic institutions. Figure 17.1 looks at the composition of the CITES Plants Committee since its inception and illustrates the significant input of taxonomists and taxonomic institutions in the scientific decision-making processes. Taxonomic expertise is clearly at the core of the decision-making process.

CITES at the national level: the role of the Scientific Authority

Article IX 1(b) of the Convention states that each Party shall designate 'one or more Scientific Authorities'. Resolution Conf.10.3 outlines the role of Scientific Authorities (CITES, 2001). To ensure scientific independence the first recommendation in this resolution is that 'all Parties designate Scientific Authorities independent from Management Authorities'. The aim of this recommendation is to attempt to ensure that institutions with appropriate expertise are designated as Scientific Authorities and that there is an opportunity for independence in the Scientific Authorities'

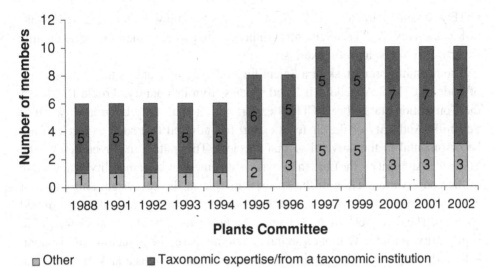

Figure 17.1 Taxonomic expertise in the CITES Plants Committee.

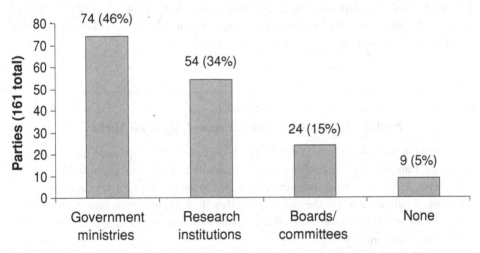

Figure 17.2 Scientific Authorities designated by national governments.

decision-making process. Figure 17.2 gives a breakdown of the types of institutions designated by national governments as their Scientific Authorities. Despite the recommendation of the above resolution, 46% of Scientific Authorities are government ministries. However, 49% of Scientific Authorities are research institutes or some form of board/committee made up of scientific experts. It is within these groups that you find taxonomic expertise. Unfortunately, the information supplied in the

CITES Country Directory (CITES, 2003) does not allow a more detailed break-down. A survey of CITES Scientific Authorities to ascertain their use of taxonomic expertise would be a useful exercise.

The most important task of a Scientific Authority is to assess the sustainability of trade in CITES Appendix II-listed species from its country. Article IV 2(a) of the Convention states that a CITES export permit shall only be granted when 'a Scientific Authority of the State of export has advised that such export will not be detrimental to the survival of that species'. This puts a major burden on the Scientific Authority – the burden to make a statement of sustainability. Knowledge of the conservation status of the species, its role in the ecosystem and a host of other data are required to ensure that the Scientific Authority can make an informed decision. The first question the Scientific Authority must be able to answer is: what is the entity in trade? Without a sound taxonomic base, the remaining information is flawed. This is especially important in cases where there is a lack of other data. Very often there is limited data for a Scientific Authority to make a decision. In these cases, Scientific Authorities may recommend that a conservative export quota be put in place for the species concerned and that exports be closely monitored. For such 'adaptive-management' procedures to be effective taxonomic expertise is required for informed decision making. A good example of CITES at work is the sustainable trade in bulbs from Turkey (see Box 17.1).

Box 17.1
Sustainable trade in bulbs: a case study from Turkey

The trade in bulbs (geophytes) from Turkey is centuries old. Tulips were introduced to Western Europe from Constantinople (now Istanbul) by the Viennese ambassador to Turkey in 1554 and subsequent introductions led to the great speculation in bulbs, known as 'tulipomania', in the 1630s. Since then the trade has waxed and waned and it was only in the 1970s and 1980s that the scale of exports to Western gardeners began to cause concern.

In 1988, Turkey exported some 70 million geophytes to the European Community. The European Community had the facility to restrict or ban imports of the CITES-listed *Cyclamen*, if Member States perceived a problem with the sustainability of the trade. At the time, the only CITES-controlled 'bulb' group was *Cyclamen* and Turkey was not then a Party to CITES. Turkish scientists had been successful in prompting the government to control exports but there was confusion in the importing countries on the levels of propagation and wild collection in Turkey. In 1988, a visit of a mission from the EC CITES Scientific Working Group (SWG) to Turkey proved pivotal in the development of a sustainable trade (interestingly, four out of the five members of that group were from taxonomic institutions).

Turkey used the regulation system of the Convention to control its trade in *Cyclamen* and supported the listing of *Galanthus* (snowdrops) and *Sternbergia* on CITES in 1989. The *Galanthus* listing was strongly opposed by the Dutch government who thought that it would produce needless bureaucracy with no conservation benefit. Today, the Dutch CITES officials acknowledge that the CITES listing has been a notable success and has ensured a sustainable trade.

Turkey became a Party to CITES in 1996 and, for plants, has one of the best implementation structures of CITES for plants within the Convention. At the core is a strong Scientific Authority with a strong taxonomic component. There is an evolving series of national regulations to control export quotas and the exploitation of CITES- and non-CITES-listed species. Collection sites are visited and monitored. Local collectors are trained to use sustainable methods and cultivation trials of selected species are being undertaken. Turkey has led the way in doing this while also ensuring that local villagers retain important income from the trade.

However, the impact of the collection of wild plants requires long-term research and Turkish scientists continue to research the taxonomy, distribution, ecology and conservation status of their geophyte flora to support CITES implementation (Figures 17.4 and 17.5). In 2002, the CITES Scientific Authorities of the United Kingdom and Turkey started a project to investigate the conservation status and sustainable use of all *Galanthus*, *Cyclamen* and *Sternbergia* taxa to help direct decision making about the long-term conservation and the impact of international trade on these taxa.

H. N. McGough and M. J. Mustard

CITES capacity building: taxonomy and the enforcement of CITES

To be successful, as a Convention which manages and controls trade in wildlife, CITES requires trained and properly resourced authorities at every level. CITES is still struggling to find adequate resources. Essential to implementation is taxonomic capacity at key points in the CITES process. CITES Scientific Authorities need at least basic taxonomic skills and access to specialist expertise. Enforcement authorities require an understanding of the species that are controlled and some expertise in the identification of these taxa (Figure 17.3). Again they need access to a centre of expertise on the identification of the species, and the parts and derivatives of those species that are in trade (Royal Botanic Gardens, Kew, 2004). If enforcement authorities undertake prosecutions for smuggling endangered species, these authorities require expert witnesses to appear in court. Very often these are taxonomists coming from institutions that have an international reputation. It is important that the court be certain that the species said to have been smuggled are those that are controlled under the relevant legislation. It is here that clear evidence from an expert witness is vital. The court must also be certain, in simple terms,

Figure 17.3 CITES plants being checked by enforcement official (photo: Royal Botanic Gardens, Kew).

that the species concerned are 'good'. If there is any dispute over the validity of the taxon this will undermine the seriousness of the case in the view of the court and, indeed, the prosecution. If a prosecution case fails when a defence witness undermines the taxonomic validity of a listed species, this has a negative impact on the reputation of the Convention. The prosecuting authorities are unlikely to put scarce resources into pursuing similar cases in the future. Thus taxonomic expertise is vitally important in the effective enforcement of CITES.

The new molecular techniques are also proving vital in the fight against wildlife crime. In the United Kingdom, the government initiative – the Partnership Against Wildlife Crime (PAW) – brings together government departments, the police, HM Customs and Excise and the major conservation NGOs. They combine to be active against a wide range of wildlife crime, both national and international. This initiative includes a working group whose remit is to look at how the new molecular

Figure 17.4 Taxonomic work on *Galanthus* (photo: Royal Botanic Gardens, Kew).

techniques can best be applied to fight wildlife crime. In the United States, the Fish and Wildlife Service forensic laboratory leads the world in using new methods to assist in countering wildlife crime. However, few countries have specialists teams to enforce CITES at national borders and within their state boundaries. National customs officials or plant and animal health inspectors usually carry out this role. Internal controls are often then enforced by the police. It is here that some basic taxonomic training is necessary.

Nomenclature is the naming of groups of organisms and the rules governing the application of these names. In effect, nomenclature is the everyday language of taxonomy. It is not the everyday language of police officers and customs officials. An understanding of some basic taxonomic principles and nomenclature is required for an individual to understand, interpret and apply the CITES controls. In training enforcement officers for CITES it is necessary to give them some basic training in taxonomy. This training is often the first time that these officials have recognised the need for taxonomic capacity in the application of CITES. The species lists, the Latin names, the binomials, the arrangement of the species lists by genus and family, and the use of authorities all form a barrier to their understanding. Taxonomy to

Figure 17.5 Tools for fieldwork on bulb species (*Galanthus*) (photo: Royal Botanic
Gardens, Kew).

them has a poor image, they associate it with dusty museums and dead specimens
of animals and plants – a dead science. However, their introduction to taxonomy as
the international currency of biodiversity – the units by which we measure 'life on
Earth' – enthralls them. The Royal Botanic Gardens, Kew, provides HM Customs
and Excise with regular training for UK customs officials and police officers. All
are individuals with a commitment to fighting wildlife crime and eager to learn new
skills. Their introduction to taxonomy, the role of a modern taxonomist and that
of a working herbarium is what they most frequently describe as an 'eye opener'.
They leave the Kew herbarium seeing it as a vital cog in biodiversity conservation –
without which they could not enforce the conservation controls. The application
of taxonomy to real problems such as CITES enforcement reveals its role at the
core of the Convention. It is unfortunate that in some CITES training programmes
participants are not aware that they are being taught some basic taxonomic skills.
In such cases, the skills are taught as part of an identification module and the term
'taxonomy' is not used as it is thought it would deter students!

Wildlife enforcement also provides an opportunity for displaying some of the most innovative tools in taxonomy. Keys are the hardware of applied taxonomy. Wildlife enforcement agencies need user-friendly keys to the plants and animals controlled by CITES. Critical here is the fact that CITES specimens need to be identified based on the form in which they are traded. Just as field botanists and zoologists struggle to identify specimens in the wild based on a small percentage of the characters presented in a key, field enforcement officers have to identify plants based on the parts that are traded. If enforcement officers are daunted by poorly constructed and jargon-ridden keys, they will not attempt to use them again. If the specimen remains unidentified, the Convention remains unenforced and unenforceable. Plant taxonomists have yet to fully apply themselves to this challenge – there is ample opportunity and indeed funding available for those willing to take up this task. Few taxonomists have sat down with enforcement agencies and worked out the type of tools they need to develop and tailor directly to the needs of day-to-day CITES enforcement. It is an area that needs urgent attention and it can also be used to demonstrate the role of taxonomy in biodiversity conservation.

Taxonomy as a core resource

CITES requires a robust taxonomic base to allow it to function effectively. From the selection of species for listing to front-line enforcement of the Convention, access to taxonomic expertise is vital. This is clearly illustrated by the case study on the sustainable trade in bulbs from Turkey (see Box 17.1). It is the role of taxonomists and taxonomic institutions to ensure that CITES Parties have access to this resource. It is the duty of CITES Parties to acknowledge the debt they owe to taxonomy and to ensure that the taxonomic expertise they require is adequately resourced.

References

CITES (2001). *CITES Handbook*. Geneva, Switzerland: CITES Secretariat.
  (2003). *CITES Country Directory*. Geneva, Switzerland: CITES Secretariat. http://www.cites.org.
Royal Botanic Gardens, Kew (2002). *CITES and the Royal Botanic Gardens, Kew*. Kew Information Sheet K10. Kew, UK: The Royal Botanic Gardens.
  (2004). *CITES and Plants, a Users Guide*, CD ROM. Version 2.0 and its updates. Kew, UK: The Royal Botanic Gardens.
Wijnsteckers, W. (2002 and future updates). *The Evolution of CITES*. Geneva, Switzerland: CITES Secretariat.

18

Gardening the Earth: the contribution of botanic gardens to plant conservation and habitat restoration

Stephen Blackmore and David S. Paterson

Introduction

The idea of the Earth as a garden is scarcely a new one. The image of the pristine Garden of Eden is powerful and compelling with equivalent images occurring in many cultures. Equally widespread today is awareness of the detrimental impact of human activities on the natural world. It is easy to get the impression that, ever since Adam and Eve, it has all been downhill for biodiversity. The aim of this chapter is to explore the garden and gardening as a metaphor (Blackmore, 2001) that might contribute to the debate about how best to conserve biodiversity in general and plants in particular.

Two things in particular appeal to us about the notion of gardening on a planetary scale. First, is the implicit acceptance of our collective responsibility for the outcome of our actions. The gardener knows and accepts that the garden will only flourish with continued care and attention. Secondly, in gardening there is an appreciation that small actions taken now may lead to significant outcomes in the future. This is clearly the case, as the gardener is willing to invest time and energy for future benefits that he may not live to enjoy personally. Thus, the mindset with which people approach their gardens is – we consider – appropriate, timely and urgently needed both in conservation and in achieving the goal of a sustainable future. More passive approaches to conservation, i.e. letting nature run its course, simply do not recognise the extent or rate of biodiversity loss around the world (Pimm et al., 1995) or the scale of the biodiversity crisis (Wilson & Peters, 1988; Heywood & Watson, 1995).

Several things may be considered contentious in the notion of gardening the Earth. First, gardening is an active form of intervention which seeks to shape nature, with varying degrees of strictness, according to some preconceived human design. Conservation, on the other hand, usually aims to preserve nature as it is. Secondly, if we accept that it is appropriate to intervene more actively, altering the chances of long-term survival for some species or even ecosystems, what principles of design or, to put it another way, what outcomes should we work towards? Finally, even if we could agree on the desired outcome, who can actually deliver it? We consider each of these issues in relation to the relevance and potential contribution of botanic gardens and examples have been taken from our work at the Royal Botanic Garden Edinburgh (RBGE). We recognise that botanic gardens in general have been increasingly involved in plant conservation and that the botanic garden community has become increasingly strategic and well coordinated in its approach to the issue (see for example, Touchell and Dixon (1997), and Wyse Jackson and Sutherland (2000)). Furthermore, we acknowledge that success in delivering the objectives of the Convention on Biological Diversity (CBD) (UNEP, 1992) will require collaboration with many essential players in biological conservation. Our intention is that the particular views expressed here will contribute to the wider debate about plant conservation that has received new impetus from the Global Strategy for Plant Conservation (GSPC) (UNEP, 2002).

Should we seek to shape nature?

We consider that the first issue, whether we should try to shape nature or not, requires little discussion. The fact is that human activities have been influencing and modifying the natural environment since the very origin of our species. It is well known that the scale of anthropogenic impact has increased greatly since the industrial revolution and it is scarcely surprising that human population growth has been widely recognised as the principle driver underlying the major causes of biodiversity loss (Meyer & Turner, 1992; Sala *et al.*, 2000). What is less often appreciated is that the relentless growth in global human population may well peak within a few decades before gradually declining (Engelman, 2000). If this is so then, in terms of biodiversity conservation, the next few decades may be the most critical period in human history, a period when our efforts to conserve species are crucial.

The earliest forms of conservation tended to involve restricting access to areas of land in order to protect natural resources, usually for the purposes of hunting by those who owned the land. In Scotland, for example, the wetlands at Wigtown Bay in Galloway and the Eden Estuary in Fife were established as nature reserves at the behest of wildfowling clubs. This pattern has been particularly important in Europe where many species, especially larger mammals, survive only in former hunting reserves. Perhaps the most important example in Europe is the Bielowieza Forest in eastern Poland which, in addition to containing ancient trees, is home to European bison, wolves, wild boar and lynx. Such traditions of land management were never a completely passive process and often predatory species or vermin were exterminated to favour the increased survival of stocks for hunting. This was scarcely a balanced approach to biodiversity conservation but had the effect of protecting many species that are now extinct elsewhere. The protection of nature, or the maintenance of biological resources, lay behind the establishment of many of the earliest national parks and protected areas around the world. Other protected areas have cultural or spiritual meaning to the societies that recognised them, perhaps as burial grounds or sacred groves (Posey, 2000). Today, such different categories of protected area are recognised formally (IUCN, 1994). In recent decades the preservation of wildlife has often been motivated by the benefits that can be derived from ecotourism rather than from hunting. In Scotland, for example, tourism is now one of the largest business sectors (Scottish Executive, 2004). More than 90% of visitors to Scotland cite beautiful scenery as one of the reasons for their visit, with birds and cetaceans also being important specific attractions.

The setting aside of protected areas makes a great deal of sense and since the recognition of biodiversity hotspots (Myers *et al.*, 2000) it has become an increasingly selective process. An entire discipline has arisen within conservation biology that is dedicated to determining conservation priorities by analysing the effectiveness of selected areas to conserve the maximum biodiversity, as measured according to several indices (see Chapter 8). The continued survival of healthy populations of species in their natural environment is universally regarded as the ideal outcome of conservation (except for those unfortunate species that have strong negative impacts on human welfare, such as invasive species, pests and disease-causing organisms). This ideal preserves what Wilson called the 'theaters of biodiversity' (Wilson, 1992), where the processes of natural selection and evolution can continue to operate. The level of intervention required to maintain the protected area may vary, but generally it is as restricted as possible.

It makes sound practical sense for protected areas to be situated in biodiversity hotspots, although overemphasis on this approach has two unfortunate unintended consequences. First, attention is drawn away from regions that do not qualify as biodiversity hotspots but which will tend to include at least some species, ecosystems or genetic varieties that are distinctive. Scotland, for example, is not a biodiversity hotspot although it has 31 endemic species (out of an estimated total of 90 000 species (Usher, 1997)) and several examples of ecosystems, such as blanket bog, that are more varied and extensive than anywhere else in Europe (Curran *et al.*, 2003). Nevertheless, the conservation of Scotland's biodiversity is, or perhaps it is more accurate to say, should be, of paramount concern to its people. This is true for all countries. A second problem is that emphasising protected areas of wilderness, such as national parks or Sites of Special Scientific Interest draws attention away from areas where the impact of human activities is more pronounced. It is often in just such urban or agricultural landscapes that an emphasis on biodiversity conservation is most urgently needed both to preserve the best quality of life for local people and to protect important elements of biodiversity (whether at species level or in terms of genetic diversity). Around the world, most people live in urban communities and migration towards cities continues to be a significant trend. Not surprisingly, therefore, urban biodiversity is a major theme in many Local Biodiversity Action Plans (see, for example, the North East Scotland Local Biodiversity Action Plan (2004) at www.nesbiodiversity.org.uk).

Our perspective is that conservation should aim to achieve the best possible outcome (in terms of preserving biodiversity and its interactions) everywhere. And in all places, except the relatively scarce large tracts of wilderness, careful intervention is at least necessary and often essential. The question remains, what outcomes should we seek and who can take on the challenge of gardening in wild places?

What outcomes do we want?

Given that the vegetation of the Earth has continually changed through geological time as a result of evolution and through the influences of climatic change and anthropogenic activities, an obvious question is: what exactly do we wish to conserve and maintain? This issue is often debated in terms of what moment in time we wish to preserve or which stage in ecological succession (see Miller *et al.* (1995) for a discussion). Clearly, any specific answer will be arbitrary. However, two important principles would be to preserve the 'theaters of biodiversity' where opportunities for interactions and adaptation continue and to preserve habitats as they were before anthropogenic impact. The reality is that in many places and for many taxa, the opportunities for natural evolution have been lost because of the fragmentation and discontinuity of natural ecosystems. In every case, our ability to look back before significant human impact depends on the quality of information we have about the past and present vegetation of an area. Such information exists in various forms as a result of biological recording programmes and in biological reference collections. Biological reference collections such as herbaria and living collections of plants may preserve only a partial record of the past vegetation of any region, but they frequently extend back before the industrial revolution and are the best record we can now obtain except through palaeoecological approaches.

The point we wish to emphasise is the critical role that biological collections such as those held in botanic gardens can play in guiding habitat restoration efforts. Not only do they provide spatially referenced (with varying degrees of precision) examples collected at different points in time, increasingly they can provide DNA to support investigations into genetic diversity. As analytical tools for modelling are developed it is increasingly possible to use spatially referenced biological collections, whether living plants from the garden or preserved specimens from the herbarium, to predict the geographical range of species in terms of their required climatic and other conditions. The Lifemapper project (2004) (see http://www.lifemapper.org/) for example, draws on specimen information in a growing number of reference collections to make predictions about the possible range of species and to relate this to future environmental changes such as global warming. Not only can such methods provide powerful tools for decision makers as they have in Mexico (Blackmore, 1998), they provide the basis for informed habitat restoration. The importance of the major global biological reference collections is that they are a good place to start a regional or national inventory of biodiversity, rather than an immediate excursion into the field (Blackmore, 1996). Botanic gardens, museums and other biological collections are rich in information about past vegetation and can provide the knowledge needed to consider intervention in terms of which taxa should grow and where.

Who are the players?

If botanic gardens, museums and other biological collections – together with biological-records centres – can provide the information we need in order to understand the appropriate plants to grow in habitat-restoration projects, who are the players to carry out such work?

Considerable practical experience in what we think of as gardening the Earth has already been built up. Not surprisingly, the greatest experience exists in relatively wealthy countries that have suffered relatively recent environmental degradation and the loss of indigenous biodiversity. The United States and Australia provide examples of two such countries faced with considerable problems caused by alien and invasive species or by anthropogenic changes in soil salinity. Both countries can provide numerous examples of restoration projects (see, for example, Close and Davidson (2003)). Similarly, there are many examples of programmes aimed at the removal of alien invasive species around the world and this issue is emphasised in the GSPC (UNEP, 2002).

Appropriate expertise in growing the vast diversity of plants necessary to even contemplate 'gardening the Earth' can be found in two main disciplines: forestry and botanic garden horticulture. Some examples from RBGE include: the restoration of wild populations of *Woodsia ilvensis* in the United Kingdom (Lusby *et al.*, 2003); the repatriation of Chinese rhododendrons (Paterson, 2003); and the ongoing work of the Lijiang Botanic Garden and Field Station in Yunnan Province (Li & Paterson, 2003). The latter two projects, carried out as international partnerships with the Kunming Institute of Botany in China, are good examples of benefit sharing in the spirit of the CBD. The first project in China, funded in 1994 by the UK Government's Darwin Initiative, involved the return of material representing almost 100 species of *Rhododendron* to the Hua Xi Botanic Garden in Sichuan and Guizhou Botanic Garden in China. The plant material was originally of known wild origin in China and included many species that are now threatened in the wild and are, therefore, included in the Chinese *Red Data Book* (Li-kuo & Jian-ming, 1992). The collection formed the nucleus in China of the 'Return to China Garden', a demonstration garden of wild-origin rhododendrons. The plants themselves were not subsequently used to re-establish or increase populations growing in the wild but concurrent with their repatriation a programme of training and capacity building allowed the staff of the Hua Xi Botanic Garden to develop the skills needed to develop hands-on *in situ* conservation programmes. Initially these focused on stabilising the populations of two endemic species in the neighbouring Long Xi Reserve, *Rhododendron davidii* Franch. and *R. calophytum* Franch, and locating other species including *R. zheguoense* Ching & H.P. Yang and *R. balangense* W.P. Fang as well as *Davidia involucrata* Baill. Research at the RBGE investigated ways of breaking seed dormancy of the latter species. The know-how gained through the project has been immense, as we saw when visiting the Hua Xi Botanic Garden in 2002. Using wild material from the remaining individuals of several rare and endangered species from the surrounding mountains, thousands of Davidias and Rhododendrons are now available for restoration projects in Sichuan. At the Guizhou Botanic Garden the project was less successful because the staff training programme could not be carried out fully. What the project shows is that living collections can serve as an *ex situ* source of plants for repatriation to the country of origin where, with appropriate comparisons at the genetic level, they could be used as source material for reintroduction to the wild. Such projects have an important capacity-building dimension, targeting the skills needed to bulk up the numbers of plants sufficient for extensive replanting (Paterson, 2003).

The Lijiang Botanic Garden and Field Station was built on earlier work at the Hua Xi and Guizhou Botanic Gardens to create a centre for capacity building, plant conservation and research on the Yulong Xue Shan mountain (Li & Paterson, 2003). In common with best practice for all *in situ* conservation projects, the local community living on and around the mountain are fully engaged with the work. Members of the Naxi ethnic minority will, for example, receive practical training in horticultural techniques in Edinburgh. In addition to species recovery and stabilisation programmes on the mountain, *ex situ* collections at RBGE and other gardens can be used for research projects on the conservation biology of target species. The project has the enthusiastic support of local and provincial governments and receives funding from the Chinese Academy of Sciences through the Kunming Institute of Botany, demonstrating the value that is placed on gaining horticultural expertise for practical conservation purposes and the perceived importance of plant conservation in China (Huang *et al.*, 2002).

Such projects, and others like them at institutions around the world, show the important role that botanic gardens will play in meeting several of the targets of the GSPC.

Conclusions

Recognition of the biodiversity crisis has served as an international call for action. The CBD set out the aspirations of nations around the world. However, achieving these aspirations is likely to depend on specific programmes, such as the GSPC which set out precise targets for attainment. Most of the 16 targets of the GSPC, intended to be achieved by 2010, have relevance to botanic gardens. Many botanic gardens are engaged in taxonomic research that supports Target 1 ('A widely accessible working list of known plant species') and fieldwork which can contribute to the assessment of conservation status (Target 2: 'A preliminary assessment of the conservation status of all known plant species'). As this chapter hopes to demonstrate, botanic gardens can contribute to 'the development of models with protocols for plant conservation and sustainable use' (Target 3) and to the targets directly concerned with conserving plant diversity. Of these, Targets 7 and 8 – relating to *in situ* and *ex situ* plant conservation – are particularly pertinent. Target 8 calls for '60 per cent of threatened plant species in accessible *ex situ* collections, preferably in the country of origin, and 10 per cent of them included in recovery and restoration programmes'.

Achieving the GSPC will require changes of mindset, as well as a scaling up of current efforts, and innovative solutions to many of the challenges involved. It is but one strand in a set of measures needed to secure a sustainable future. Perhaps if we think of the GSPC as 'gardening the Earth' we will see the need for more direct intervention, overcome fears that we do not have a sound knowledge base to guide restoration projects and appreciate the importance of the practical skills embedded in botanic gardens.

References

Blackmore, S. (1996). Knowing the Earth's biodiversity: challenges for the infrastructure of systematic biology. *Science*, **274**, 63–4.

(1998). The life sciences and the information revolution. In *Information Technology in Plant Pathology and Biodiversity*, eds. P. Bridge, D. R. Morse & P. R. Scott. Wallingford, UK: CAB International, pp. 441–50.

(2001). All the world's a garden. *Horticulturist*, **10**, 13–16.

Close, D. G. & Davidson, N. J. (2003). Revegetation to combat tree decline in the Midlands and Derwent Valley Lowlands of Tasmania: practices for improved plant establishment. *Ecological Restoration and Management*, **4**, 29–36.

Curran, J., Fozzard, I., Gibby, M. *et al.* (2003). Scotland's biodiversity resource. In *Towards a Strategy for Scotland's Biodiversity. The Resource and Trends*, ed. M. B. Usher. Edinburgh, UK: HMSO, pp. 1–10.

Engelman, R. (2000). *People in the Balance: Population and Natural Resources at the Turn of the Millennium*. Washington, DC: Population Action International.

Heywood, V. H. & Watson, R. T. (1995). *Global Biodiversity Assessment*. Cambridge, UK: Cambridge University Press.

Huang, H., Han, X., Kang, L. *et al.* (2002). Conserving native plants in China. *Science*, **297**, 935–6.

IUCN (1994). *Guidelines for Protected Area Management Categories*. Gland, Switzerland: IUCN.

Li, D. Z. & Paterson, D. (2003). *Rhododendron* conservation in China. In *Rhododendrons in Horticulture and Science*, eds. G. Argent & M. McFarlane. Edinburgh, UK: Royal Botanic Garden, pp. 171–7.

Li-kuo, F. and Jian-ming, J. (eds.) (1992). *China Plant Red Data Book: Rare and Endangered Plants, I*. Beijing, China: Science Press.

Lifemapper (2004).
http://www.lifemapper.org/.

Lusby, P., Dyer, A. & Lindsay, S. (2003). The role of botanic gardens in species recovery: the Oblong Woodsia as a case study. *Sibbaldia*, **1**, 5–10.

Meyer, W. B. & Turner, B. L. (1992). Human population growth and global land-use/cover change. *Annual Review of Ecology and Systematics*, **23**, 39–62.

Miller, K. Allegretti, M. H., Johnson, N. & Jonsson, B. (1995). Measures for conservation of biodiversity and sustainable use of its components. In *Global Biodiversity Assessment*, eds. V. H. Heywood & R. T. Watson. Cambridge, UK: Cambridge University Press, pp. 915–1061.

Myers, N., Mittermeier, R. A., Da Fonseca, C. G. & Kent, J. (2000). Biodiversity hotspots for conservation priorities. *Nature*, **403**, 853.

North East Scotland Local Biodiversity Action Plan (2004). www.nesbiodiversity.org.uk.

Paterson, D. (2003). Repatriation of *Rhododendron* to China. *Sibbaldia*, **1**, 29–34.

Pimm, S. L., Russell, G. J., Gittleman, J. L. & Brooks, T. M. (1995). The future of biodiversity. *Science*, **269**, 347–50.

Posey, D. A. (ed.) (2000). *Cultural and Spiritual Values of Biodiversity: a Complementary Contribution to the Global Biodiversity Assessment*. London: UNEP and Intermediate Technology Publications.

Sala, O. E, Chapin, F. S., Armesto, J. J. *et al.* (2000). Global biodiversity scenarios for the year 2010. *Science*, **281**, 1770–4.

Scottish Executive (2004).
http://www.scotland.gov.uk.

Touchell, D. & Dixon, K. W. (eds.) (1997). *Conservation into the 21st Century. Proceedings of the Fourth International Botanic Gardens Conservation Congress, Kings Park Botanic Garden, Perth, Western Australia*. London: BGCI.

UNEP (1992). *Convention on Biological Diversity (CBD): Text and Annexes*. Geneva, Switzerland: CBD Interim Secretariat.
www.biodiv.org.

(2002). *Global Strategy for Plant Conservation (GSPC)*. Decision VI/9, English, Spanish and Chinese from the Secretariat.)
www.biodiv.org/programmes/cross-cutting/plant/.

Usher, M. B. (1997). Scotland's biodiversity: an overview. In *Biodiversity in Scotland: Status, Trends and Initiatives*, eds. L. V. Fleming, A. C. Newton, J. A. Vickery & M. B. Usher. Edinburgh, UK: HMSO, pp. 5–20.

Wilson, E. O. (1992). *The Diversity of Life*. London: Allen Lane.

Wilson, E. O. & Peters, F. M. (1988) *Biodiversity*. Washington, DC: National Academy Press.

Wyse Jackson, P. S. & Sutherland, L. A. (2000). *International Agenda for Botanic Gardens in Conservation*. London: Botanic Gardens Conservation International.

19

Taxonomy: the framework for botanic gardens in conservation

Etelka Leadlay, Julia Willison and Peter Wyse Jackson

> A botanic garden is an institution holding documented collections of living plants for the purposes of scientific research, conservation, display and education.
>
> P. S. Wyse Jackson, 1999

This chapter will show that taxonomy and taxonomic literature provide the framework and reference point for plant conservation in botanic gardens. Botanic gardens contribute in many ways to plant conservation, particularly through the maintenance and development of scientific living collections of plants and by providing taxonomic expertise for identifying, monitoring and managing plant diversity. These skills are important for successful ecosystem restoration and sustainable development. The chapter will also show how teaching taxonomy in botanic gardens promotes education and awareness about plant diversity, its importance and conservation.

Botanic gardens are dedicated to the collection, study, development and display of plants and over the last 30 years have become centres for conservation action. This work supports a wide range of national and local priorities, as well as – at the international level – the Convention on Biological Diversity (CBD) (UNEP, 1992); more specifically, the work will help meet the 16 targets of the Global Strategy for Plant Conservation (GSPC) (UNEP, 2002a) which was adopted at the Sixth Conference of the Parties to the CBD in April, 2002.

Taxonomy provides a measure of the diversity of organisms. Taxonomy is the theory and practice of the classification of living organisms and its final product is a system of classification of defined taxa (e.g. families, genera or species). The results of taxonomic research on plants are published in the form of Floras, monographs and checklists. A Flora is an account of the plants of a particular area, a monograph is an account of a group of related plants throughout its geographical range and a checklist is a list of all the plants of a particular area. The taxonomic literature aims to provide a comprehensive overview of the plant diversity of the world and information about each specific plant taxon. Furthermore, these taxa have been identified from the study of plants in the field and in herbaria, and as far as possible the chosen taxonomic system represents the relationship between taxa and how close they are in terms of their evolutionary origins. It provides fundamental baseline information required for conservation. At its simplest level, it is necessary to describe and identify the biodiversity of the world in order to conserve it; but more importantly, it is essential in order that conservation work can be directed towards maintaining the *maximum* diversity.

Taxonomy provides the framework for identifying plants. Floras and monographs generally include descriptions of the diagnostic characters of taxa so that plants can be identified consistently in the field, the herbarium and in cultivation, and so we can be sure that we are talking about the same taxon. The name of a species is a tool for identifying, measuring and monitoring plant diversity.

The name of a species is a basic communication and reference tool to which other information can be attached. The name of a plant is the key to all the information known about that taxon and the classification (taxonomic structure) is a means of organising information in a logical and retrievable way.

Taxonomic research is based on the study of as many characters as possible (e.g. morphological, chemical, cytological) and research on the distribution, ecology, breeding system and population dynamics of a taxon throughout its geographic range. A knowledge of the complete distribution of a species, both past and present, and the variability of the taxon in a population is used to estimate the diversity of a taxon and assess its conservation status – e.g. whether it is a local endemic, a weedy species or one threatened by habitat reduction. The information on ecology, breeding systems and variability of a taxon is essential for designing *ex situ* plant-conservation programmes in botanic gardens. Furthermore, the fact that the taxonomic structure illustrates the relationship between taxa means that it is also predictive. For instance, a plant that has been assigned to a particular taxon will share properties and characteristics with closely related taxa (e.g. members of the Labiatae are frequently aromatic). This predictive feature can be useful for developing procedures that are required in the management of species, such as cultivation or eradication procedures or providing new sources of plant material (e.g. for medicines or food) from a related taxon.

Floras and monographs also often include information on the vernacular name; the local uses of plants; whether they are wild relatives of domesticated or cultivated plants; or whether they are of social, scientific or cultural importance. This information is also very useful for illustrating the relevance of plants to the public and in helping to suggest priorities for conservation.

Living collections

There are over 2200 botanic gardens in 156 countries; they maintain over 6 million accessions of living plants, as well as seed genebanks and tissue-culture facilities (Wyse Jackson, 2001). Botanic gardens are thus a unique resource of plant collections, horticultural skills and genebank facilities developed over many years; a unique resource to support *ex situ* conservation.

Almost 300 gardens have signed up to the International Agenda for Botanic Gardens in Conservation which is a document published by Botanic Gardens Conservation International (BGCI) (Wyse Jackson & Sutherland, 2000; Anon., 2004.). This document provides an international framework for botanic gardens in conservation and guides the management, development and utilisation of relevant botanic gardens collections in support of plant conservation.

By maintaining documented living collections, botanic gardens are conserving plant diversity. Botanic gardens worldwide have approximately 80 000 taxa in cultivation (of an estimated 270 000 vascular plant species in the world (Wyse Jackson, 2001; UNEP, 2002a). This is a direct measure of diversity conserved in cultivation. It provides a method for evaluating the diversity of living plant collections. For instance, there are about 500 plant families and if we are to conserve as much diversity as possible, we should make sure all families are represented in conservation programmes. However, apart from their conservation collections, it must be acknowledged that botanic gardens generally maintain only a few individuals (genotypes) of a species and so the botanic garden collection worldwide must be regarded as representative of the Plant Kingdom, rather than a comprehensive collection.

For conservation, living collections are used to provide: back-up collections of wild plants for population reinforcement and reintroduction; material for reference (e.g. plant identification), for research into plant use (e.g. providing plant material from their collections for pharmaceutical screening) and conservation techniques (e.g. seed-storage behaviour, development of cultivation protocols and studies of genetic variation of populations); and displays of plants (to explain and promote the conservation of species and landscapes, and the sustainable use of plants).

Planning plant collections

Each botanic garden will develop their collections (and accessions policy) in support of their mission statement and strategy. Taxonomic products provide the baseline information for developing plant collections for the purposes of supporting conservation and sustainable use. This taxonomic framework helps set priorities and prevents duplication of effort and the omission of important plant diversity from conservation plans.

CBD Article 9(a) (*Ex situ* Conservation) states the importance of adopting 'measures for the *ex situ* conservation of components of biological diversity, preferably in the country of origin of such components' (UNEP, 1992). CBD Article 7(a) (Identification and Monitoring) highlights the importance of the 'components of biological diversity important for its conservation and sustainable use' identifying more closely the priority areas for conservation in Annex I.2. of the CBD: 'Species and communities which are: threatened; wild relatives of domesticated or cultivated species; of medicinal, agricultural or other economic value; or social, scientific or cultural importance; or importance for research into the conservation and sustainable use of biological diversity, such as indicator species'.

Taxonomic research helps us identify the taxa we need to conserve to reach these objectives. The importance of this role for taxonomy is emphasised by the first two targets of the GSPC in the provision of 'A widely accessible working list of known plant species, as a step towards a complete world flora' (Target 1) and 'A preliminary assessment of the conservation status of all known plant species, at national, regional and international levels' (Target 2; UNEP, 2002a). The refinement of the draft targets of the GSPC explain that 'A working list of known plant species is considered to be a fundamental requirement for plant conservation.'(UNEP, 2002b).

The GSPC has set a target of '60 per cent of threatened plant species in accessible *ex situ* collections, preferably in the country of origin, and 10 per cent of them included in recovery and restoration programmes' (Target 8). Plants conserved *ex situ* are therefore available to contribute to species-recovery programmes and can provide long-term back-up collections.

The conservation status of a species can only be assessed if the taxonomic status is clear and there is a thorough knowledge of the present and previous distribution of the taxon. For instance, the Sweet thorn (*Acacia karoo*) is a common and widespread tree in southern Africa. Taxonomic research indicated that the southern-African representatives of this taxon can be subdivided into six different taxa; one of these taxa is confined to a small locality and is now regarded as rare and needs to be conserved (Botha, 1996). According to the 1997 *IUCN Red List of Threatened Plants* (Walter & Gillett, 1998) nearly 34 000 taxa are considered threatened. Currently, approximately one-third of them (10 000 taxa) are known to be in botanic garden cultivation (BGCI, 2004). Comprehensive threatened plant lists with well-researched taxa are essential if the conservation community is not to waste time and resources conserving entities which are poorly understood and which may not in fact be taxonomically distinct from other non-threatened taxa. Hybrids, subspecies and variants may not warrant the same conservation priority as distinct species. For instance, special conservation efforts (which in this case have been very costly) for the recovery of *Grevillea williamsonii* by the Royal Botanic

Gardens Melbourne, Australia (working with the Department of Natural Resources and Environment; the Trust for Nature, Victoria; and Parks Victoria) were unnecessary as subsequent taxonomic work showed that *G. williamsonii* was only a rare variant of the common *G. aquifolium* and should be subject to general management for that habitat (James, 2004).

Examples of the use of threatened plant lists in developing collections are given below.

- The Center for Plant Conservation (CPC), with its headquarters in St Louis in the United States is dedicated to preventing the extinction of US native plants through the development of a National Collection of Endangered Plants of over 600 taxa in a network of 32 botanic gardens (CPC, 2004). Plants protected by Federal and State legislation provide the focus for its work. Participating institutions range from, for example, the Amy B. H. Greenwell Ethnobotanical Garden in Hawaii (e.g. *Kokia drynarioides*) to the Desert Botanical Garden in Phoenix (e.g. *Agave parviflora*) and are responsible for maintaining the living collections of the taxa in their region and for providing plant material for restoration in the wild.

- The Mexican Association of Botanic Gardens (Asociación Mexicana de Jardines Botánicos) has produced an Action Plan to support the conservation of plants of national importance and encourage sustainable use (Rodríguez-Acosta, 2000). In Mexico, there are 17 gardens in xerophytic habitats, which represents 50% of the area and 20% of the total flora of Mexico. This area is dominated by cacti and xerophytic monocotyledons such as *Agave, Yucca, Dasylirion, Nolina* spp. and Bromeliaceae, many of which are of local economic importance. Institutions such as Cante A. C., Guanajuato and ITESM-Campus Querétaro help to take pressure off the wild species by extensive propagation of cacti and other succulents. The Action Plan uses as a baseline the list of threatened species included in the *Norma Oficial Mexicana* (SEDESOL, 1994).

- Cycads are a taxonomic group of global social, scientific and cultural importance, and now conservation importance (82% of the world's cycad species are listed as threatened compared with 12.5% of the world's vascular plants (Walter & Gillett, 1998)). The Jardín Botánico Francisco Clavijero in Xalapa, Mexico and the Fairy Lake Botanical Garden, Shenzen, China have developed important cycad collections in cultivation (Donaldson, 2003). In addition, three further institutions have developed collections of cycads which represent a significant genetic sample of known populations and are managed as genebanks: the National Botanical Institute, South Africa; the Montgomery Botanical Center in Miami, United States; and the Nong Nooch Tropical Garden, Thailand. The Jardín Botánico Francisco Clavijero in Xalapa, Mexico has developed cycad nurseries of *Dioon edule* for local farmers to alleviate the threat to that species from habitat destruction and illegal collection for leaf crowns. *The Cycads Status Survey and Conservation Action Plan* (Donaldson, 2003) produced by the IUCN/Species Survival Commission (SSC) Cycad Specialist Group has been able to highlight the need for genebanks for two species of *Chigua* and several 'critically endangered' or 'endangered' species for which there are no genebanks. The SSC of the IUCN-The World Conservation Union coordinates a network of volunteer Specialist Groups on a wide range of plants and animals (IUCN/SSC, 2004). The baseline data for the conservation of cycads has been taken from taxonomic treatments (e.g. Stevenson, 1992). An up-to-date list of cycad species is maintained on The Cycad Pages (Royal Botanic Gardens, Sydney, 2004).

Botanic gardens also maintain important seed genebanks of native flora, which include plants of medicinal, agricultural or other economic importance and wild relatives of domesticated or cultivated species. For example, the UK Flora Programme of the Millennium Seed Bank Project (MSBP) of the Royal Botanic Gardens, Kew, United Kingdom has already collected seed from around 90% of the United Kingdom's native higher plants. This was the first time that any country has underpinned efforts in the conservation of its wild flora in this way (MSBP, 2004). The list of taxa to be collected was developed using the current national UK and regional Floras and monographs with taxonomic expertise provided by taxonomists, both amateur and professional (see Chapter 20 for a discussion of the practical aspects of planning and undertaking seed collecting expeditions for the MSBP). Elsewhere, the botanic gardens of the Macaronesian Islands (Azores, Madeira and the Canary Islands) have set up the Macaronesian Seed Bank (BASEMAC) to conserve their endemic plants. Floras and local taxonomic expertise have identified 3900 species of vascular plants existing in the three island groups, of which 925 are endemic and an estimated 418 are threatened (Salinas *et al.*, 2004).

These examples highlight the synergies achieved through networking (Target 16 of the GSPC). The gardens have assumed a responsibility for plants in their areas but benefit from the overall coordination provided by national Floras and monographs of plants throughout their range. Networks enhance communication, provide mechanisms to exchange information and technology, allow coordination to help avoid duplication of effort and optimise the efficient allocation of resources.

Maintenance of the collections

The cultivation of wild species is a core skill in botanic gardens. The information provided by monographs and Floras can help in two ways. First, knowledge of the morphological structure, the ecology and breeding system of a taxon can be used to suggest methods of propagation and cultivation. Secondly, the knowledge of the cultivation of related taxa can help develop successful protocols. In the case of extremely rare taxa, developing germination and propagation protocols can be undertaken using closely related but more common plants. Many conservation programmes are developing protocols for propagating and cultivating plants that have not been cultivated before. For instance, the Eden Project in Cornwall, United

Kingdom is developing protocols for the cultivation of the endemic plants of the Seychelles (supported with a grant from the UK Darwin Initiative).

Plants can be kept *ex situ* for a few generations without severe dysgenic effects (Havens *et al.*, 2004). Gardens can serve as a short- to medium-term shelter where numbers of individuals can be increased before release into appropriate habitats, and can provide the last refugium for the most highly threatened species. A sound scientific basis is needed to improve the success of reintroduction programmes and taxonomic expertise is essential for the initial planning (Vallee *et al.*, 2004). Horticulturists, taxonomists and conservation biologists can work together to cultivate wild plants for reintroduction, reinforcement, habitat restoration and management.

Documentation of the collections

Taxonomy and herbarium techniques are at the heart of documenting plants in cultivation. Documentation will include the name and source of the material, its history in the garden, verification status and location of the voucher specimen. Plants in cultivation, seed banks and tissue culture need to be correctly identified and adequately documented if they are to be of value for research and reference purposes. For instance, for research – such as pharmaceutical screening, developing cultivation protocols or determining seed storage behaviour – verified material is necessary so that the research will be repeatable on other material that has been identified as the same taxon. If there are any anomalies in the research they might be explained by the history of the plant in the garden or the source of the material. In the 1980s, the Arnold Arboretum, United States undertook a methodical verification review of their collections and removed plants that did not meet the exacting criteria for their scientific, horticultural or historical value to the garden (Michener, 1989). For conservation collections, where numbers of individuals are increased for reintroduction, material from a known source is essential so that the wild population is not changed (see Chapter 14).

It is necessary to monitor plants from the moment they arrive in the garden and to keep records even after the plant has died, been lost or given away or discarded because the plant may have been used in a research programme and the records and voucher specimen may need subsequent checking. For instance, seed of *Encephalartos eugene-maraisii* was collected at various localities and germinated by the Transvaal Department of Nature and Environment Conservation, South Africa but a subsequent taxonomic study led to the recognition of two species, two subspecies and an undescribed taxon. The reintroduction programme had to be cancelled because the seed had not been kept separately during germination and the conservation programme could have led to hybridisation (Botha, 1996). Plant material in a garden is known as an accession (seed or vegetative material from one source at one time) to the garden. It is given a reference number (accession number) which will be the key to the information about the plant: name as received, date obtained, propagule type, location of plant in the institution, source (e.g. collector, date, collection locality, location of voucher specimen).

The taxonomic name will be the key to all published information about this plant. The identity of each accession in a garden is (or should be) regularly checked or verified using the garden's library and herbarium resources; this involves two separate procedures. The first procedure is identification, which is the determination of a plant or taxon (such as a species or subspecies) as being identical with or similar to another, already known taxon. This procedure uses taxonomic expertise, taxonomic reference books – such as Floras, keys and monographs – and other reference scientific material such as herbarium specimens. The second procedure is ensuring that the correct nomenclature is used and is concerned with the determination of the correct scientific name of a known plant according to a chosen nomenclatural system. This naming is regulated by internationally agreed and accepted rules laid down in the International Code of Botanical Nomenclature (Greuter & Hawksworth, 2000) (see Chapter 6 in this book). This determination establishes that the name used is the current and preferred one (and correctly spelled) under the rules of nomenclature and also that it is the appropriate one under the system of classification used in the garden.

A collection of voucher specimens is maintained (e.g. herbarium specimens of vegetative material, flowers, fruits, bark and bulbs; or photographs) which can be used as reference material to check the identification at a later date. It is especially important for voucher specimens to be preserved for conservation collections and for when plants are used for research. A later researcher will need to verify the voucher material to be sure that it is what it was thought to be and to ascertain whether any changes (genetic modification) have taken place in the material while in cultivation or storage. If the definition of a taxon changes, or through new research it is discovered that the original population was a local hybrid or should be ascribed to a different subspecies, then it is only by examining a herbarium specimen of the original collection that a new name can be ascribed with certainty and the living collection can still be of value for research.

Botanic gardens have always had record systems, from ledgers through to card indexes and more recently electronic record-keeping systems (for which a wide range of computer applications have been developed, e.g. BG-Recorder (BGCI, 2004) and BG-BASE (2004)). These systems can be used to accumulate and consolidate relevant information on each taxon for future use and provide a secure system for long-term storage and retrieval of data. The classification of plants into families and species (using the binomial of genus and species) provides a sound structure for designing databases. Furthermore, lists of accepted names provide consistency within and between databases; BG-Recorder used Brummitt's *Vascular Plant Families and Genera* (2001).

Many record-keeping systems also include information about the propagation and cultivation of each taxon which can ensure continuity within an organisation, and has the added advantage that the information can be shared easily with other gardens. Keeping records of each trial is important to optimise the success rates for propagation; it avoids wasteful repetition and helps by building on past experience to improve future success rates. The continuity between such trials is provided by the use of verified material.

Very significant efforts have been made by botanic gardens over the last two decades to standardize the ways in which they hold and exchange data about their living collections. The most significant landmark in these efforts was the development and publication of *The International Transfer Format for Botanic Garden Plant Records (ITF)* (BGCS, 1987) with an updated version of this standard, ITF2 prepared in 1998 (BGCI, 2004). It provides a standard and agreed structure for exchange of information on living plant collections between institutions (and botanic gardens have also found the format very helpful in developing databases for plant records). BGCI has also developed an on-line database of plants in cultivation worldwide (Plant Search at www.bgci.org) which will help prioritise further collection and allow the organisations and institutions worldwide that are maintaining plant collections to share information on plants. Botanic gardens can register with BGCI and obtain a remote log-in facility to allow them to add details of their collections on-line and analyse the information on their holdings in relation to all other data sets included. This list of plants can also be compared with available on-line threatened plant lists and will thus help to monitor the achievement of Target 8 of the GSPC.

The availability of a working list of known plant species with the correct and accepted name of a species (Target 1 of GSPC) will be invaluable for measuring progress in meeting the requirements of Target 8. A list of names without synonyms is essential to identify work to be done and to ensure that the resources are better focused. Furthermore, unless such a list exists it is impossible to gain a comprehensive and accurate global overview of the number of species that need to be addressed and considered for conservation action. It is important that the correct name is established and accepted for use as the unique identifier for each species.

Taxonomic expertise

The CBD highlights the importance of the identification and monitoring of biological diversity (Article 7) for its conservation. Many thousands of trained taxonomists are employed by botanic gardens worldwide and their expertise is considerably enhanced by experience and data gained through the management of the living collections and from their extensive libraries and herbaria. Botanic garden herbaria include at least 142 million specimens (Wyse Jackson, 2001). Verified reference collections of herbarium specimens are essential for identifying plants. For example, the Foundation for Revitalization of Local Health Traditions (FRLHT) based in Bangalore, India has a Herbarium–Seed–Raw Drug Centre (HSRD) to authenticate the identity of medicinal plants and to correlate the vernacular and botanical names to support their network of medicinal-plant nurseries designed to improve primary health care for millions of Indian households (FRLHT, 2004).

For historical reasons, some large botanic gardens throughout the world contain extensive collections and comprehensive expertise on the biodiversity of other countries and regions. These gardens often use this expertise in support of bilateral aid agreements. For instance, the Royal Botanic Gardens, Kew has provided taxonomic expertise to the Limbe Botanic Garden in Cameroon, with support from the British Overseas Development Administration (ODA) and later the German Technical Cooperation and the World Bank's Global Environment Facility (GEF), to help conserve the unique biodiversity of the Mount Cameroon area (Ndam & Sunderland, 1997).

Garden staff provide an interface between the taxonomic literature and the garden's constituency to ensure that plants are correctly identified and named so that the relevant information can be accessed and required action taken to manage or conserve particular taxa of concern. These skills are used by botanic garden staff to monitor the populations of local and regional floras, recognise invasive plants and other organisms, advise on land development and the implementation of plant protection laws, and identify plants for conservation and sustainable development.

For example, the Donetsk Botanic Garden in the Ukraine has developed computer databases to monitor and analyse the state of the regional flora in relation to changes due to human interference (Gluckov, 2000). The Tver University Botanic Garden has surveyed local areas so that the flora can be monitored for changes over time (Naumtsev & Notov, 2000). The Garden is also ensuring that endangered plants are brought into cultivation for conservation and is researching methods for their management and reintroduction to the wild. This work supports Target 3 of the GSPC, which identifies the importance of the 'Development of models with protocols for plant conservation and sustainable use, based on research and practical experience', and Target 8.

Botanic gardens work with other agencies to develop and implement management plans for invasive species (Target 10 of the GSPC calls for 'management plans for at least 100 major alien species that threaten plants, plant communities and associated habitats and ecosystems'). Botanic gardens use their skills for identifying invasive species and assessing the invasive potential of a taxon through their knowledge of, for example, the dispersal mechanism, growth habit and ecological requirements of the taxa. They can also use their horticultural skills to suggest methodologies for the management and control of invasive plant taxa and other pathogens. Gardens have developed invasive plant policies to assess the potential risk of a plant becoming invasive and to ensure that the gardens are not themselves responsible for dispersing invasive plants (Havens, 2003).

Botanic gardens are often consulted over planning permission to develop land. A key role for some botanic gardens has been to ensure that populations of wild plants are taken into cultivation for conservation as part of rescue operations when natural vegetation is being cleared (e.g. during road widening, forest clearance, expansions of settlements, etc.) The Aburi Botanic Garden, Ghana plays a regular role in surveying sites where planning permission is being sought. The Garden can then make recommendations, and if required take wild plants into cultivation for conservation purposes. The Kings Park Botanic Garden, Perth, Australia has removed native plants from local areas cleared for urban development under a CALM (Department of Conservation and Land Management, Government of Western Australia) licence system.

Botanic gardens promote the sustainable use of plants through implementing plant protection laws and such measures as the Convention on International Trade in Endangered Species (CITES). The GSPC sets a target of 'No species of wild flora endangered by international trade' by 2010 (Target 11) (see Chapter 17 in this book). The GSPC also sets a target of '30 per cent of plant-based products derived from sources that are sustainably managed' (Target 12). A major cause of biodiversity loss is the collection of wild plants for horticulture, medicine and food. Botanic gardens can be the licence-granting body for the collection of wild plants. For example, hundreds of tonnes of wild mushrooms are collected and exported each year from Russia. The Mycological Department of the Komarov Botanical Institute, St Petersburg is authorised to issue licences in the region to regulate the unsustainable harvesting of wild mushrooms for trade (Kovalenko, 1998).

Ecosystem restoration

Botanic gardens worldwide are increasingly involved in ecosystem management and restoration. In that regard botanic gardens can play an important role in supporting the thematic work programmes of the CBD (e.g. the biodiversity of islands, dry and sub-humid lands, inland waters, forests, mountains and coastal systems).

The GSPC has also adopted a target of '60 per cent of the world's threatened species conserved *in situ*' (Target 7) to be achieved by 2010.

Surveys and herbarium research can be directed towards producing checklists of plants in protected areas; these are then used for monitoring and identifying species which need management. Botanic garden expertise and skills are often valuable to help restore native areas both within and outside the protected-area system. For instance, in Hawaii, the National Tropical Botanical Gardens have an integrated approach to the conservation of their island biodiversity which uses taxonomic studies, monitoring, cultivation and reintroduction, and management of these areas (Chapin *et al.*, 2001). The purpose of the Turpan Eremophyte Botanic Garden in China is to collect a wide range of desert plants (over 400 species) and study their taxonomy, cultivation and use, particularly in sand stabilisation (Pan, 1996). The conservation of wetlands, including mires and bogs, is well acknowledged as an international priority, recognised both by a special CBD thematic work programme and by the Ramsar Convention on Wetlands. Botanic gardens can support urgent actions in wetland conservation. For example, the Atlanta Botanical Garden (ABG) in Georgia, United States is working with endangered seepage bogs, home to many endangered carnivorous plant species. The ABG has assessed their simple but effective restoration techniques by monitoring pitcher plants (Determan & Groves, 1998). The Singapore Botanic Garden is undertaking a long-term monitoring programme of a 4-ha remnant of primary rain forest in the Gardens (Kiew & Chan, 2001). It still contains a remarkable diversity of 314 plant species, including a range of Dipterocarpaceae species and 10 species that were thought to have become extinct in Singapore. The programme is following the flowering, fruiting and regeneration patterns of the trees; in addition there are active rehabilitation programmes, which include the removal of exotics, control of climbers, and collection and germination of seed for enrichment planting and to restore the canopy gaps. At the Mount Annan Botanic Garden, Australia (a satellite garden of the Royal Botanic Garden, Sydney) the plant growth and life cycles in the remnant of Cumberland Plain Woodland is being monitored to find out more about this community and how to manage it in order to ensure that all native species it contains are conserved for the long term – not just flowering plants, but also fungi and lichens, as well as animals (Botanic Gardens Trust, Sydney, 2004).

Promoting sustainable development

Target 13 of the GSPC seeks to halt 'The decline of plant resources . . . innovations and practices that support sustainable livelihoods, local food security and health care'. The taxonomic literature often includes information on the local use of plants – e.g. for medicine, tools and food (vegetables, fruits, tubers) – or whether they are wild relatives of domesticated or cultivated plants or of social, scientific or cultural importance. Botanic gardens in many countries are involved in the development of plant resources for the local economy and to take pressure off wild plants, which helps deliver this Target.

In northern Vietnam, as part of a UK Government Darwin Initiative-supported project undertaken by BGCI with its partners in Vietnam, anthropologists, taxonomists and conservationists worked with local herbalists and local villagers to develop a list of medicinal and food plants that are collected in the wild from the Tam Dao National Park. These plants were then brought into cultivation in the Garden of Useful Plants especially created for the project near the headquarters of the National Park; they were also propagated for use by local villagers for cultivation in village and community gardens (Figure 19.1). The identification and naming of the plants was the key to the information about their use, methods of cultivation, harvesting, drying and preparation (Khanh *et al.*, 2000). The Botanic Gardens of Adelaide, Australia is working on the development of the bushfood industry and also provides important advice on the sustainability of wild harvesting and the potential environmental impacts (Christensen & Beal, 2001).

Environmental education

Displays of plants in botanic gardens are widely used to explain and promote conservation of species and landscapes, and the sustainable use of plants. Teaching taxonomy helps people appreciate the diversity of plants and the fact that taxonomy is fundamental for identification, communication and information about plants in many disciplines – from horticulture to medicine. The awareness of the importance of diversity can be used to illustrate the need for its conservation. This approach helps to implement Article 13(a) of the CBD (Public Education and Awareness), 'Promote and encourage understanding of the importance of, and the measures required for, the conservation of biological diversity, as well as its propagation through media, and the inclusion of these topics in educational programmes', and to achieve Target 14 of the GSPC, 'The importance of plant diversity and the need for its conservation incorporated into communication, education and public awareness programmes'.

At present, the importance of taxonomy is little understood by decision makers, the public, students and children; this is a result of the current trend of teaching broadly based biological sciences (which do not include significant taxonomic studies) rather than single subjects such as botany.

Botanic gardens are particularly good resources for teaching taxonomy as they hold documented collections of living plants and often have practising taxonomists

Figure 19.1 The Garden of Useful Plants, Tam Dao National Park, Vietnam (photo: Peter Wyse Jackson)

on the staff. The Cambridge University Botanic Garden, United Kingdom has systematic flower beds, representing nearly 100 plant families, which provide a valuable teaching and research resource. It is utilised for University teaching and for school parties, and it is the focus of a week-long residential course on plant systematics aimed at students and at professionals in environmental subjects (Upson, 2003). Botanic garden staff not only identify plants but can describe the diagnostic characters and the distribution and evolution of the taxa in relation to other members of the genus. They can give additional information showing how the plants are locally relevant – the local use of the plants, vernacular names, the conservation status of plants – and how the ecology and breeding system affect their conservation.

At the Royal Botanic Gardens, Kew, explanatory panels have been installed next to the Order Beds where plants are grown in family groups, e.g. Compositae, Cruciferae, Polygonaceae, based on the Bentham and Hooker system of classification (Bentham & Hooker, 1862–83). These panels describe 'Taxonomy in Action', 'Identification' and 'Naming'. By displaying plants in this way the similarities and differences in the form and flower structure between members of the same family can be seen and the aim is to show that taxonomy provides a method of organising our knowledge about the Plant Kingdom so that it becomes more manageable (Griggs, 2001). The Mairie de Paris, the Parks and Recreation Department of the City of Paris has developed a series of panels to help people discover the individual stories of plants, e.g. one panel describes the classification and history of the *Dahlia* (Jossien, 1996).

Identifying, naming and classifying objects are all activities we undertake everyday. For instance, in the office we might sort envelopes into different sizes, colours, design, types of paper depending on our needs and this classification system would help us quickly find a suitable envelope for our purpose. Plants and animals are in fact classified basically in the same way as non-living objects on the basis of possessing various characters that they have in common. In its most simple form it is the identification of differences in characters. Objects can be classified according to their shape, size, colour, smell, age, location and so on depending on the purpose of the classification. Education staff in the Royal Tasmanian Botanical Gardens use games to help children recognize differences: 'Hug a tree', to recognise that trees are different in texture, size and smell; 'Rainbow splats', to appreciate different colours in nature, through matching colour cards with plants in the garden; 'Tree ID', to introduce the concept of species names by looking at the leaves from three different trees (Smith, 1996).

The Gurukula Botanical Sanctuary in Kerala, India run a 'School in the Forest' programme to educate children and adults, through observation and participation in nature, about their region's biodiversity and the urgent need for conservation.

Figure 19.2 In response to a request from the state's traditional healers association, in collaboration with the Santo Domingo Historical Ethnobotanical Garden and the Mexican People's Institute, traditional healers from Oaxaca completed a basic plant taxonomy course conducted by the Instituto de Biología Universidad Nacional de Mexico (Project: Bioactive Agents from Dryland Biodiversity of Latin America SU01/TW00316-09 (Linares *et al.*, 1999) (photo: Robert Bye).

This experience includes classification activities where the students compare their own made-up taxonomic systems with scientific taxonomic systems (Seshan, 2000; S. Seshan, personal communication, 2004).

The Botanic Garden of the Instituto de Biologia, Universidad Nacional de Mexico (UNAM) has a well-developed education programme on the documentation and use of medicinal plants in Mexico. In response to a request from the state's traditional healers association, in collaboration with the Santo Domingo Historical Ethnobotanical Garden and the Mexican People's Institute, traditional healers from Oaxaca completed a basic plant taxonomy course held at the garden, conducted by the Institute de Biologia, UNAM (Linares *et al.*, 1999) (Figure 19.2). The UNAM programme includes training teachers (regarded as 'botanical information multiplicators') on themes such as: the diversity in the Plant Kingdom and the garden, developing dichotomous keys, and observation and description of the cycad collection (Linares, 2001). Edelmira Linares firmly believes that 'The first step to appreciate plants is the development of knowledge . . . which will promote the conservation of natural resources.'

Over 200 million people visit botanic gardens each year. This number of visitors provides botanic gardens with an excellent and perhaps unique opportunity to demonstrate the relevance of taxonomy to everyone and its importance in conservation. With better understanding of the need for taxonomy, it is hoped that there will be greater support from the public which in turn will lead to a call for increased teaching of taxonomy as a contribution to the understanding of biodiversity conservation.

Conclusion

This case study of botanic gardens shows that the taxonomic framework is an essential component for the successful conservation and sustainable use of plants. It provides baseline information which prevents duplication and omissions, and identifies priorities for conservation – such as threatened and endemic plants and medicinal or minor crop plants that can be developed to improve the livelihoods of local people. It also gives a framework for measuring and monitoring.

Taxonomic expertise is essential for identifying plants correctly so that relevant information can be accessed and required action taken to manage or conserve particular taxa of concern. Teaching taxonomy in botanic gardens promotes education and awareness about plant diversity, its importance and conservation.

In a world where resources to support taxonomy for plant conservation continue to decline, botanic gardens have become among the last bastions where the science of plant taxonomy continues to be practised to the highest scientific standards. Continued efforts need to be made to ensure that taxonomy is comprehensively supported and continues to thrive through botanic gardens.

Acknowledgements

We would like to thank Chris Hobson and Caspar Davey for help and advice on the manuscript.

References

Anon. (2004). International Agenda for Botanic Gardens in Conservation: Registration update. *BGJournal*, **1**(1), 24–5.

Bentham, G. & Hooker, J. D. (1862–83). *Genera plantarum ad exemplaria imprimis in Herbariis Kewensibus servata definita/auctoribus G. Bentham et J. D. Hooker.* London: Reeve.

BG-BASE (2004).
http://rbg-web2.rbge.org.uk/BG-BASE/.

BGCI (Botanic Gardens Conservation International) (2004). The International Transfer Format for botanic garden plant records (ITF2).
http://www.bgci.org.

BGCS (1987). *The International Transfer Format for Botanic Garden Plant Records.* Pittsburgh, PA: Hunt Institute for Botanical Documentation, Carnegie Mellon University.
    www.bgci.org.

Botanic Gardens Trust, Sydney (2004).
    http://www.rbgsyd.gov.au.

Botha, D. J. (1996). Developing horticulture in botanic gardens: the role of research for the benefit of plant conservation. In *Botanic Gardens in a Changing World,* ed. C. Hobson. Richmond, UK: BGCI, pp. 119–22.
    www.bgci.org.

Brummitt, R. K. (2001). *Vascular Plant Families and Genera.* Kew, UK: The Royal Botanic Gardens.

Chapin, M., Lorence, D. H., Perlman, S. & Wood, K. R. (2001). Support for the conservation of endemic Pacific palms though *ex situ* collections at the National Tropical Botanical Garden, Hawaii, U.S.A. *Botanic Gardens Conservation News,* **3**(6), 46–8.

Christensen, T. J. & Beal, A. (2001). Australian bush tucker: new crops, new industry. In *Plants, People and Planet Earth: the Role of Botanic Gardens in Sustainable Living,* eds. G. Davis & A. Scott, CD ROM. Cape Town, South Africa: National Botanic Institute, Kirstenbosch.
    www.bgci.org.

CPC (Center for Plant Conservation) (2004).
    http://www.centerfor plantconservation.org.

Determan, R. & Groves, M. (1998). Conservation of carnivorous plants and bog communities at Atlanta Botanical Garden, Georgia, U.S.A. *Botanic Gardens Conservation News,* **2**(10), 48–51.

Donaldson, J. S. (2003). *The Cycads Status Survey and Conservation Action Plan.* Gland, Switzerland and Cambridge, UK: IUCN.

FRLHT (Foundation for Revitalization of Local Health Traditions) (2004).
    http://www.frlht-india.org/.

Gluckov, A. Z. (2000). Conserving the plants of the south-eastern Ukraine: the role of the Donetsk Botanical Garden. *Botanic Gardens Conservation News,* **3**(4), 40–1.

Greuter, W. & Hawksworth, D. L. (2000). International Code of Botanical Nomenclature (St Louis Code). *Regnum Vegetabile,* **138**, vii–xviii.

Griggs, P. (2001). Order, Order: taxonomy in action: interpreting the order beds at the Royal Botanic Gardens, Kew. In *Teaching for the 21st Century: Botanic Garden Education for a New Millennium,* eds. C. Hobson & J. Willison. Brooklyn, NY: BGCI (US) Inc., pp. 126–7.

Havens, K. 2003. Developing an invasive plant policy: the Chicago Botanic Garden's experience. *The Public Garden,* **17**(4), 16–17.
    www.chicagobotanic.org (accessed 2004).

Havens, K., Guerrant, E. O., Jr, Maunder, M. & Vitt, P. (2004). Guidelines for *ex-situ* conservation collection management minimizing risks. In *Ex-situ Plant Conservation Supporting Species Survival in the Wild,* eds. E. O. Guerrant, Jr, K. Havens & M. Maunder. Washington, London: Island Press, pp. 454–73.

IUCN/SSC (IUCN Species Survival Commission) (2004).
    http://www.iucn.org/themes/ssc.

James, E. A. (2004). Conserving *Grevillea williamsonii:* the importance of taxonomic research for appropriate action. *BGJournal,* **1**(1), 17–19.

Jossien, C. (1996) Exhibitions in Paris. *Roots,* **12**, 9–10.

Khanh, T. C., On, T. V., Tien, D. D. *et al.* (2001). Establishment of a botanic garden for the protection of the plant genetic resources of the Tam Dao National Park, Vietnam. In *Plants, People and Planet Earth: the Role of Botanic Gardens in Sustainable Living*, eds. G. Davis & A. Scott, CD ROM. Cape Town, South Africa: National Botanic Institute, Kirstenbosch. www.bgci.org.

Kiew, R. & Chan, L. (2001). Habitat management: the strategy to preserve the biodiversity of the Singapore Botanic Gardens rain forest. In *Plants, People and Planet Earth: the Role of Botanic Gardens in Sustainable Living*, eds. G. Davis & A. Scott, CD ROM. Cape Town, South Africa: National Botanic Institute, Kirstenbosch. www.bgci.org.

Kovalenko, A. (1998). The present state of the conservation of fungi in Russia. In *Conservation of Fungi in Europe*, ed. C. Perini. Siena, Italy: Centro Stampa dell'Università, pp. 65–8.

Linares, E. (2001). Traditions and medicinal plants: a valuable field of knowledge and a great challenge for science. In *The Power for Change*, eds. L. A. Sutherland, T. K. Abraham & J. Thomas. Kerala, India: Tropical Botanic Garden and Research Institute, pp. 171–6.

Linares, E., Balcázar, T., Herrera, E. *et al.* (1999). Ethnobotanical education: beyond the garden. *Roots*, **19**, 23–5.

Michener, D. C. (1989). To each a name: verifying the living collections. *Arnoldia*, **49**(1), 36–41.

MSBP (Millennium Seed Bank Project) (2004). http://www.kew.org/msbp/.

Naumtsev, Y. V. & Notov, A. A. (2000). Regional ex situ biodiversity conservation programmes at the Botanic Garden of the Tver State University, Russia. *Botanic Gardens Conservation News*, **3**(5), 47–8.

Ndam, N. & Sunderland, T. (1997). The role of Limbe Botanic Garden in the conservation and development of the Mount Cameroon region. In *Conservation into the 21st Century, Proceedings of the Fourth International Botanic Gardens Conservation Congress, Kings Park Botanic Garden, Perth, Western Australia*, eds. D. H. Touchell & K. W. Dixon. London: BGCI, pp. 277–86.

Pan, B. (1996). Turpan Eremophyte Botanic Garden, Academia Sinica, China. *Botanic Gardens Conservation News*, **2**(7), 20–2.

Rodríguez-Acosta, M. (ed.) (2000). *Estrategia de Convservación para los Jardins Botánicos Mexicanos*. Mexico: Asociación Mexicana de Jardins Botánicos A.C.

Royal Botanic Gardens, Sydney (2004). The Cycad Pages. (2004). http://PlantNet.rbgsyd.gov.au/PlantNet/cycad.

Salinas, A. R., Vilches Navarrete, B., Jardim, R. *et al.* (2004). Banco de semillas de macaronesia (BASEMAC). Un programa para La conservación racional de la diversidad vegetal. http://www.bcn.es/medciencies/botanicgardens2004/abstracts/pdf_abstracts/ Alicia_Roca/final.pdf.

SEDESOL (Secretaría de Desarrollo Social) (1994). Norma Oficial Mexicana NOM-059-ECOL-1994. *Diario Oficial de la Federación, México, D.F.*, **CDLXXXVIII** (10), 1–25.

Seshan, S. (2000). School in the forest: educating the young at the Gurukula Botanical Sanctuary. *Roots*, **20**, 29–32.

Smith, A. (1996). Classification by senses. In *Cultivating Green Awareness*, eds. J. D. Rodrigo Pérez & N. González Henríquez. Las Palmas de Gran Canaria, Spain: Jardín Botánico Canario 'Viera y Clavijo', pp. 115–17.

Stevenson, D. W. (1992). A formal classification of the extant Cycads. *Brittonia*, **44**, 220–3.

UNEP (1992). *Convention on Biological Diversity (CBD): Text and Annexes*. Geneva, Switzerland: CBD Interim Secretariat. www.biodiv.org.

(2002a). *Global Strategy for Plant Conservation (GSPC)*. Decision VI/9, UNEP/CBD/ COP/6/20. Montreal, Canada: CBD Secretariat. (Available in hard copy in English, Spanish and Chinese from the Secretariat.) www.biodiv.org/programmes/cross-cutting/plant/.

(2002a). *Global Strategy for Plant Conservation: follow-up on SBSTTA Recommendation VII/8 paragraph 2 on the refinement of the 16 draft targets included in the proposed strategy*. UNEP/CBD/GSPC/1/2. Montreal, Canada: CBD Secretariat.

Upson, T. (2003). University botanic gardens and systematic research. In *European Botanic Gardens – Studies in Conservation and Education: Papers from the EuroGard 2000 Congress*, ed. C. Hobson. Richmond, UK: BGCI, pp. 119–22.

Vallee, L., Hogbin, T., Monks, L. *et al.* (2004). *Guidelines for the Translocation of Threatened Plants in Australia*, 2nd edn. Canberra, Australia: Australian Network for Plant Conservation.

Walter, K. S. & Gillett, H. J. (eds.) (1998). *1997 IUCN Red List of Threatened Plants*. Cambridge, UK: World Conservation Monitoring Centre and Gland, Switzerland: IUCN.

Wyse Jackson, P. S. (1999). Experimentation on a large-scale: an analysis of the holdings and resources of Botanic gardens. *Botanic Gardens Conservation News*, **3**(3), 27–32.

(ed.) (2001). *An International Review of the Ex situ Plant Collections of the Botanic Gardens of the World: Reviewing the Plant Genetic Resource Collections of Botanic Gardens Worldwide, as a Contribution to Decision V/26 on Access to Genetic Resources of the Conference of the Parties to the Convention on Biological Diversity*. London: UK Government, Department of the Environment Transport and the Regions (DETR) and Montreal, Canada: Secretariat of the CBD and *Botanic Gardens Conservation News*, **3**(6), 22–33. http://www.biodiv.org/programmes/socio-eco/benefit/bot-gards.asp.

Wyse Jackson, P. S. & Sutherland, L. (2000). *International Agenda for Botanic Gardens in Conservation*. Richmond, UK: BGCI, p. 56.

20

Wild-seed banks and taxonomy

Paul P. Smith

Introduction

Seed banking is óne in a series of tools that can be employed in the conservation of plant species.[1] Seed banking cannot directly protect the biological diversity of ecosystems, but it can ensure the protection of diversity between, and within, plant species. In particular, banking seeds provides a last resort for the protection of plant species that are condemned to extinction. In doing so, it balances the greatly increased certainty of short-term survival against the risk of genetic stasis and reduced adaptation. Seed banks also provide many further benefits that directly support the wider range of plant-conservation activities. This chapter describes the rationale for seed banking in the context of plant conservation as a whole, then examines the role of taxonomy in seed banking. Problem areas are identified, and recommendations are made as to how taxonomists and their products might become more relevant to the conservation effort. Examples of wild-seed banking in practice are taken from the Millennium Seed Bank Project (MSBP), Royal Botanic Gardens, Kew, United Kingdom.

The applications of taxonomy to seed banking

Taxonomy unlocks everything that is known about a species and is therefore crucial to all seed-banking efforts, but particularly when conservation of wild species is the main focus of effort. It is in providing plant-diversity information, identification tools and expertise that taxonomy makes its greatest contribution to seed banking for conservation purposes. Plant-diversity data are necessary for efficient targeting of species during the planning phase, and plant identification tools and expertise are essential for accurate identification of species in the field and after collection.

---

[1] Throughout this chapter, the term 'species' will be employed in lieu of 'taxon' or 'taxa'.

Table 20.1. *Criteria for selecting wild species for seed banking and*
ex situ *conservation*

Orthodox seeds	Seeds that retain their viability after drying, and which are therefore likely to be bankable
Indigenous or endemic species	Species native to an area, and neither introduced nor a pan-tropical weed
Endangered, threatened or vulnerable species	Species of restricted distribution or threatened on a local, national or global scale
Economically important species	Species valued/used by local people
Species suitable for research	Species targeted by research projects
Seed not widely available	Seed not already in the bank or available through the List of Seeds or from commercial sources.
Rare seeds	Often difficult to find high-quality or -quantity seed of this species in this region

Species names are the most important measure of success in any wild-species seed-banking programme because they indicate collection quality.

Targeting species for collection

The most frequent outcome of a targeting and prioritisation exercise is a list of target species that have been selected as priorities for collecting, together with information about their biology, distribution, identification, utility, etc. Identifying priority species before a field trip will usually involve a great deal of research. Species will be prioritised according to various criteria, depending on the focus of the seed-banking programme (Table 20.1) but, once selected, further information will be needed about those species. A comprehensive review of botanical data sources available to assist in this task can be found in Prendergast (1995) but it is worth looking more closely at some examples of taxonomic products here.

Regional, national or local Floras (including annotated checklists) are a useful source of information about species endemism (through distribution data) or rarity (through number of collections made). They may also provide information about phenology (i.e. when a species is likely to be in seed) and location, although the level of detail is often insufficient to pin down a species precisely in space and time. Monographs are useful where the planner is interested in a particular systematic group. In the unusual cases where monographs or Floras carry conservation assessments, they are helpful for targeting threatened species. Of most use to conservation planners, however, are the raw data collected by taxonomists in the form of herbarium specimens, and the information that accompanies them. Herbarium sheets often have detailed locality data and, with the exception of sterile specimens, they always provide information about species phenology. Additional information about aute-cology, rarity, habitats, local uses, vernacular names, etc. may also be valuable to the expedition planner. Herbarium specimen data are most useful in electronic format because they can then be integrated into a geographical information system (GIS). Using GIS, specimen databases can be combined with geophysical data to provide detailed planning information, e.g. accessibility, predictive distributions, intactness of habitat, etc. (Sawkins *et al.*, 1999; Kolberg, 2003; Moat & Smith, 2003).

Identification of plants in the field

Targeting priority species for seed collection and *ex situ* conservation is a pointless exercise unless the field team is able to recognise the target plants in the field. For this reason it is essential that the seed collector is equipped with the means to identify the plants that he/she encounters. The best way to achieve this is to include in the expedition a person or persons with the requisite plant identification skills. This will often be a taxonomist with generalist or specialist skills. For example, a recent expedition made by the MSBP in the Northern Cape Province of South Africa included a taxonomist specialising in the family Mesembryanthemaceae. There are an estimated 1800 members of this family in South Africa, and they make up a significant proportion of the Karoo flora in the Northern Cape. With the help of this particular taxonomist, 46 rare and threatened species in the family Mesembryanthemaceae were collected on this one expedition (Smith *et al.*, 2001).

Where taxonomic expertise is not available in the field, the seed collector may employ a range of identification aids. Floras are useful for their species descrip-tions and illustrations, and for their keys when microscopic characters are not used. Monographs are less useful in the field unless the focus of the expedition is on collecting from one systematic group. Of greater utility than either Floras or mono-graphs are field identification guides of local circumscription with field-friendly keys, descriptions, illustrations and vernacular names. Where published material is not available, digital images, cibachromes or photocopies of herbarium sheets can be useful in field identification, especially if they include additional information such as vernacular names.

Identification of collections in the herbarium

Even with meticulous planning and the best plant-identification tools and skills available, it is often not possible to identify plants in the field. For example, species in the petaloid monocotyledon families such as Iridaceae and Orchidaceae are notoriously difficult to identify when they are in seed. Likewise, deciduous plants often set seed when they are leafless, making field identification almost impossible. In addition, because of the vagaries of seed-set timing, collectors will often collect seed opportunistically even if the identity of the plant is unknown. Regardless of whether the collector is sure of the plant's identity or not, it is accepted good practice to collect a herbarium voucher specimen with the seed. In cases where this is not possible, or where voucher specimens are lost or destroyed, a voucher is normally grown from the seed to ensure accurate identification at a later date.

Accurate naming of herbarium voucher specimens and hence the plant from which the seed was collected is very much the province of the taxonomist. In the case of the MSBP, a permanent Herbarium Liaison Officer (HLO), with generalist knowledge across the taxonomic spectrum, is employed by the Project in the Kew herbarium to manage the voucher identification process. Essentially, the HLO sorts the incoming vouchers into their relevant taxonomic families and genera, and passes on the vouchers from difficult groups to the relevant taxonomic expert (who may be in a different institution in a different part of the world). The remaining vouchers are identified by the HLO, using the reference collections in the herbarium and the Floras and monographs produced by the taxonomic experts. This identification process, which provides a measure of the success of any seed-banking project, requires both generalist and specialist taxonomic skills as well as the vast repository of taxonomic data held in the world's herbaria.

Evaluation

By naming collections, taxonomists also have an important role to play in evaluating projects with species-based targets. For example, the MSBP has the central aim of collecting and conserving 24 000 species by the year 2010. This means that the Millennium Commission releases funds for the project according to the quantity (number of new species) and quality (number of target species) of the collections coming in to the bank. The taxonomists employed by the project provide both these measures.

Discussion

The role of taxonomy in conservation is currently a matter of great debate, and various aspects of this topic are discussed in this book. The overall view, however, from both conservationists and taxonomists, is that there is currently insufficient synergy between the two disciplines (Lowry & Smith, 2003). The reasons for this are explored elsewhere (e.g. Schatz, 2002a; Golding & Timberlake, 2003), and will not be re-iterated here. Instead, it would probably be more useful to look critically at taxonomy and the products of systematic research from the perspective of the *ex situ* conservationist.

Targeting species for collection – discussion

The particular plant species targeted for conservation will invariably depend on institutional, local, national or regional strategies and plans. For example, threatened species will be defined by Red Lists or similar publications, and conservation workers will plan their activities around these. However, these lists have certain limitations: in particular, they do not tell the field worker either where the plant occurs, or what it looks like. This makes it very difficult for *ex situ* conservationists to go out and collect seed from a particular threatened species. Taxonomists invariably have the answers to these problems – they know from herbarium records where a species is likely to be found, and they know from herbarium specimens what the plant looks like. However, the fundamental problem with taxonomic products such as Floras, monographs and specimen databases is that they are designed for use by taxonomists, not conservationists. The point, which has been made by conservationists, is that with little extra effort these products could be made useful for all. For example, Golding and Smith (2001) have published a 13-point strategy for Flora writers that would make Floras a more useful tool for assessing the conservation status of plant species. The strategy was mainly about adding a little more of the data available from the herbarium specimens consulted by the authors, and being consistent in presenting that information. This point can be taken even further by convincing authors of taxonomic treatments to carry out conservation assessments themselves, and publish them with their treatments. For rare or obscure species, herbarium specimens are often the only source of information available, and the taxonomists who study those specimens are better placed than anyone to assess their conservation status. For *ex situ* conservationists trying to target severely threatened species, the current lack of conservation assessments is a serious impediment to any collecting programme.

A perhaps more difficult trend to reverse among taxonomists is the movement away from Floras towards monographs. Monographs are generally of little use to conservationists, who invariably are interested in conserving the flora of a particular geographical region. The same goes for specimen databases. Taxonomists are interested in creating and using specimen databases with a systematic focus, while conservationists need specimen information from a particular locality. Of course, all specimen information is valuable and adapting systematic databases for use in geographical areas of interest is not technically difficult. For example, the Royal Botanic Gardens, Kew, United Kingdom has specimen databases for the

Leguminosae, Rubiaceae and Orchidaceae in Madagascar; and the MSBP uses these databases in its targeting programme. Taken together these represent perhaps 30% of the flora, which is better than nothing but still far short of what is needed.

The trend towards monographs, driven by the need for systematists to understand the worldwide relationships between taxa, has also reduced the number of taxonomists with generalist skills, i.e. the ability to recognise species across the taxonomic range. This has had a negative impact on the amount of botanical survey and inventory work being carried out by taxonomists. Local checklists, particularly those that include conservation assessments are an invaluable tool for conservation planners.

The lack of useful plant-diversity information emanating from taxonomic institutions has forced conservation planners to develop rapid biodiversity assessment techniques (Beattie & Oliver, 1994), which use surrogates for species diversity (e.g. land systems). These have some utility for *in situ* conservationists but are of little use to *ex situ* conservation planners, who need to target particular species for collection. Participation in inventory work by taxonomists (Vollesen *et al.*, 1999; Cheek *et al.*, 2000), and the inclusion of Red List assessments in revisions and monographs (Davis, 2001; Lowry *et al.*, 2002) is happening on a small scale, but these activities need to be given far greater priority by taxonomic institutions.

Identification of plants in the field – discussion

As indicated above, the plant-identification products of taxonomy are generally not designed for use by non-taxonomists, and certainly not by field botanists. The species descriptions in Floras and monographs are often intellectually inaccessible and full of taxonomic jargon, and the keys are frequently based around microscopic characters or features that may be absent when the seed collector is in the field. Clearly, Floras and monographs do what they are designed to do, and that is to provide the herbarium-based taxonomist with the information they require to identify a plant specimen. From this perspective, it makes no sense to fundamentally redesign Floras and monographs themselves. Instead, the development of derivatives should be encouraged. For the field botanist the most useful by-products of Floras and monographs are field identification guides. Of course field identification guides should not necessarily be written by taxonomists – ecologists, conservationists and amateur naturalists have all traditionally contributed to the genre. However, the fact remains that the taxonomic institutions have both the data and expertise at hand to produce plant identification guides useful to a wide range of users. Useful field guides are already being produced by taxonomic institutions (e.g. Evans *et al.*, 2001; Schatz, 2002b), often in partnership with conservation organisations, but more effort is needed in this respect.

Perhaps more worrying than the lack of field identification guides currently available is the trend away from generalist expertise among botanists in taxonomic institutions. The MSBP currently has partnerships in 16 countries, and in every single one of these the collection programme centres around threatened, rare or economically important species. In other words, nowhere do we have a collecting programme with a purely systematic focus. This means that botanists with a broad knowledge of plant taxonomy, who are able to recognise species across the taxonomic range, are essential to our collecting programmes. These are increasingly difficult to find among the younger generation of taxonomists. Some institutions have recognised this, and begun to implement measures to ensure that this kind of knowledge is not lost, e.g. through providing training opportunities for studying specimens across the taxonomic range both in the herbarium and in the field. Examples of this include 'family sorts' in which incoming herbarium material from a particular geographical region is sorted into families and genera by a team of taxonomists with a range of skills and experience, so that everyone learns from each other and generalist skills are developed in this way. Fieldwork opportunities and extended visits to local herbaria (often on an exchange basis) should also be made available to young botanists in order for them to develop their generalist skills. The MSBP routinely works with generalists on seed-collecting expeditions, however it is essential that the collaboration is mutually beneficial: the seed collectors gaining plant identification knowledge, and the taxonomist gaining field experience and the opportunity to observe and collect a wide range of taxa in the field.

Identification of collections in the herbarium – discussion

The identification of herbarium voucher specimens in the herbarium is currently the most valuable service provided by taxonomists to the seed-banking effort. The perennial goal of achieving taxonomic stability/consistency and reaching consensus on species concepts so that names and classifications of taxa are not constantly changing, is still worth pursuing from the perspective of the user, but naming specimens is something that the taxonomic community generally does very well.

Conclusions

It is hoped that this chapter has demonstrated that taxonomic products and expertise are essential to wild-seed banking and the *ex situ* conservation effort. The reason for this is that if the strategy of banking wild seed for long-term conservation and utilisation is to be successful, we must concentrate our effort where it is needed most, and that means knowing what we should be collecting. It is perhaps worth reinforcing this point by citing the recently agreed Global Strategy for Plant Conservation (Smith, 2002) which calls for 60% of the world's threatened species to be conserved *in situ* and 60% of threatened plant species to be conserved in accessible *ex situ* collections by the year 2010. In practice this means that someone must provide reliable information about what species these are, where they occur and what threats they face today. The information and expertise held in the world of plant taxonomy has never been more relevant.

Box 20.1
The seed-banking rationale

The history of seed banking has been driven by the need to conserve and utilise crop species, and the technologies developed for this purpose have more recently been adapted towards conserving and utilising wild species. A useful review of seed-banking history and methodology, and its role in plant conservation is provided by Linington and Pritchard (2001). Crop seed banks primarily conserve infraspecific diversity among domesticated species, and intensively sample allellic-diversity-conferring traits such as disease resistance, high yields, etc. Wild-seed banks, on the other hand, seek to conserve the interspecific diversity of non-domesticated species, focusing mainly on utilitarian, rare and threatened species. For a variety of historical reasons, the world's wild-seed-bank network is not as well developed as the international crop seed bank network (Laliberté, 1997). However, wild-seed banks are playing an increasingly important role in conservation science today (Smith *et al.*, 2002; Heywood & Iriondo, 2003). In particular:

- seed banks provide insurance against threats to plants *in situ*;
- seed banks provide options for the future conservation and utilisation of plants;
- seed banks provide a controlled source of plant material for research;
- skills, knowledge and data from seed banks support wider plant conservation aims;
- the benefits of seed banking are long term and achievable at relatively low cost;
- seed banks contribute to education and raising public awareness about plant conservation.

The wild species selected for seed banking will depend on institutional, local, national and regional conservation strategies, as well as on the seed characteristics of particular species (Table 20.1). Seed banking is not an end in itself. Long-term conservation in seed banks, where seed may remain viable for hundreds of years, is one option open to conservation planners, but seed-bank accessions are also available for research as part of conservation and sustainable-utilisation strategies designed to take pressure off wild plant populations. These strategies rely on the development of germination and propagation protocols, and this work is routinely carried out by seed banks (Smith *et al.*, 2002; Linington *et al.*, 2003).

Box 20.2
The Millennium Seed Bank Project

The Millennium Seed Bank Project (MSBP), funded by the UK Millennium Commission, is the largest wild-seed banking project ever conceived (Smith *et al.*, 1998). The International Programme of the MSBP is a 10-year global conservation programme (2000–10), conceived, developed and managed by the Seed Conservation Department at the Royal Botanic Gardens, Kew. The two principal aims of the Project are to:

• collect and conserve 10% of the world's seed-bearing flora, principally from the drylands, by the year 2010;
• develop bilateral research, training and capacity-building relationships worldwide in order to support and to advance the seed conservation effort.

The Millennium Seed Bank is based on about 30 years of work by its forerunner the Kew Seed Bank and currently holds the largest and most diverse collection of wild species in the world, including 90% of the flora of the United Kingdom. To date, it has developed partnerships in 16 countries, with many further collaborations on the way (Linington *et al.*, 2003). By 2010 it will have attempted to collect and conserve 10% of the world's flora, some 24 000 species, as defined by governments and institutions in partner countries.

One of the most important aspects of the MSBP is that, through its bilateral access and benefit-sharing agreements, the MSBP ensures duplication of conserved seed collections at facilities all over the world; at the same time providing capital input, training and technical expertise. Where agreements allow and quantities are sufficient, the germplasm stored in the Millennium Seed Bank and the other banks worldwide is made available to the world scientific community; in this way the Project has become a world focal point for *ex situ* conservation research.

At the same time, taxonomists and the institutions where they work still have much to do if they are to meet the challenges ahead. The above examples of how this might be achieved are provided from the perspective of the *ex situ* conservationist, but will have equal resonance to all conservationists who use plant-diversity information. Ultimately, the responsibility to conserve the world's plant species belongs to us all, because if we don't act in concert now, our own species will end up on the Red List.

References

Beattie, A. J. & Oliver, I. (1994). Taxonomic minimalism. *Trends in Ecology and Evolution*, **9**, 488–90.

Cheek, M., Odana, J.-M. & Pollard, B. J. (2000). *The Plants of Mount Oku and the Ijim Ridge, Cameroon: a Conservation Checklist*. Kew, UK: The Royal Botanic Gardens.

Davis, A. P. (2001). Two new species of *Coffea* from eastern Madagascar. *Kew Bulletin*, **56**, 479–89.

Evans, T. D., Senghala, K., Viengham, O. V. & Thammavong, B. (2001). *A Field Guide to the Rattans of Lao PDR*. Kew, UK: Royal Botanic Gardens.

Golding, J. S. & Smith, P. P. (2001). A 13-point flora strategy to meet conservation challenges. *Taxon*, **50**, 475–7.

Golding, J. S. & Timberlake, J. (2003). How taxonomists can bridge the gap between taxonomy and conservation science. *Conservation Biology*, **17**(4), 1177–8.

Heywood, V. H. & Iriondo, J. M. (2003). Plant conservation: old problems, new perspectives. *Biological Conservation*, **113**, 321–35.

Kolberg, H. (2003). Targeting collecting for conservation: an example from Namibia. In *Seed Conservation: Turning Science into Practice*, eds. R. D. Smith, J. Dickie, S. H. Linington, H. W. Pritchard & R. J. Probert. Kew, UK: The Royal Botanic Gardens, pp. 209–17.

Laliberté, B. (1997). Botanic garden seed banks/genebanks worldwide, their facilities, collections and network. *Botanic Gardens Conservation News*, **2**(9), 18–23.

Linington, S. H. & Pritchard, H. W. (2001). Gene Banks. In *Encyclopedia of Biodiversity*, vol. 3, ed. S. Levin. London: Academic Press, pp. 165–81.

Linington, S. H., Tenner, C. & Smith, R. D. (2003). Seed banks: *ex situ* is not out of place. *Biologist*, **50**(5), 163–8.

Lowry, P. P., II & Smith, P. P. (2003). Closing the gulf between botanists and conservationists. *Conservation Biology*, **17**(4), 1175–6.

Lowry, P. P., II, Schatz, G. E. & Wolf, A.-E. (2002). Endemic families of Madagascar. VIII. A synoptic revision of *Xylolaena* Baill. (*Sarcolaenaceae*). *Adansonia*, **24**, 7–19.

Moat, J. & Smith, P. P. (2003). Applications of geographical information systems in seed conservation. In *Seed Conservation: Turning Science into Practice*, eds. R. D. Smith, J. Dickie, S. H. Linington, H. W. Pritchard & R. J. Probert. Kew, UK: The Royal Botanic Gardens, pp. 79–87.

Prendergast, H. D. V. (1995). Published sources of information on wild plant species. In *Collecting Plant Genetic Diversity*, eds. L. Guarino, V. Ramanatha Rao & R. Reid. Wallingford, UK: CAB International, pp. 153–79.

Sawkins, M. C., Maxted, N., Jones, P. G., Smith, R. D. & Guarino, L. (1999). Predicting species distributions using environmental data: case studies using *Stylosanthes* Sw. In *Linking Genetic Resources and Geography;: Emerging Strategies for Conserving and Using Crop Biodiversity*, eds. S. L. Green & L. Guarino. CSSA Special Publication 27 Madison, WI: ASA and CSSA, pp. 87–99.

Schatz, G. E. (2002a). Taxonomy and herbaria in service of plant conservation: lessons from Madagascar's endemic families. *Annals of the Missouri Botanical Garden*, **89**, 145–52.

(2002b). *Generic Tree Flora of Madagascar*. Kew, UK: The Royal Botanic Gardens and Missouri: Missouri Botanical Garden.

Smith, P. P. (2002). The Global Strategy for Plant Conservation. *Oryx*, **36**(4), 325.

Smith, P. P., Burgoyne, P. & van Wyk, E. (2001). Rare plants rediscovered in the Northern Cape. *SABONET News*, **6**(1), 51–52.

Smith, P. P., Smith, R. D. & Wolfson, M. (2002). The Millennium Seed Bank Project in South Africa. In *Rebirth of Science in Africa: a Shared Vision for Life and Environmental Sciences*, eds. H. Baijnath & Y. Singh. Hatfield, South Africa: Umdaus Press, pp. 87–97.

Smith, R. D., Linington, S. H. & Wechsberg, G. E. (1998). The Millennium Seed Bank, the Convention on Biological Diversity and the dry tropics. In *Plants for Food and Medicine*, eds. H. D. V. Prendergast, N. L. Etkin, D. R. Harris & P. J. Houghton. Kew, UK: The Royal Botanic Gardens, pp. 251–61.

Vollesen, K., Abdallah, R., Coe, M. & Mboya, E. (1999). Checklist: vascular plants and pteridophytes of Mkomazi. In *Mkomazi: the Ecology, Biodiversity and Conservation of a Tanzanian Savanna*, eds. M. Coe, N. McWilliam, G. Stone & M. Packer. London: Royal Geographical Society, pp. 81–116.

21

Good networks: supporting the infrastructure for taxonomy and conservation

Stephen L. Jury

Network collaboration has been the way Vernon Heywood has conducted much of his research. This chapter gives an insight into some of the research projects undertaken at The University of Reading – projects for which Vernon Heywood has played a leading, if not the leading, role since his appointment in 1968 to the Chair of Botany. They continue apace even now, well after his retirement from the University. As Sir Peter Crane and Laura Pleasants have noted in Chapter 1, science has become highly internationalised and 'ever-more dependent on the coor-dinated activities of multiple practitioners working together in teams'. Vernon Heywood recognised this long ago and his scientific successes stem from it – and it is something British Airways ought also to recognise!

The Umbelliferae

Following a symposium, *The Biology and Chemistry of the Umbelliferae*, held in Reading in September 1970 (Heywood, 1971), a major research network (RCP, a French acronym) was formed with French research groups, funded by the UK Science Research Council and the French equivalent, CNRS. This international group, which soon came to involve Spain, examined in detail the members of the Umbelliferae, tribe Caucalideae with the different laboratories contributing towards a better understanding and taxonomy of the tribe by their own research specialties: Reading, fruit morphology and anatomy, and phytochemistry; Paris, palynology and phytochemistry; Dijon, phytodermology; Perpignan, cytology; Madrid, fruit morphology and anatomy, etc. A considerable number of research papers resulted, many as the result of a second international symposium held in Perpignan in 1977 (Cauwet-Marc & Carbonnier, 1977). The obvious advantages of such international collaborative networks is something the Commission of the European Community now fosters by awarding its scientific research grants to two or more institutions working in two or more EC countries. This disparate multilingual group 'gelled', developing international friendships and came to share not only scientific data but gastronomic experiences in restaurants near the various laboratories!

Flora Europaea

A *Flora Europaea* project was discussed during a formal session of the International Botanical Congress in Paris in 1954, but no feasible organisation could be envisaged and the project was thrown out. However, a group of British delegates – together with David Webb from Trinity College, Dublin – had adjourned to a small Paris café opposite the Sorbonne where they agreed over a bottle of Calvados that a Flora of Europe *could* be written and that *they* would do it. A brief history of the project is given in the printed programme of the final symposium held in Cambridge in 1977 at which the manuscript of the final fifth volume was handed over (Anon, 1977). Further details are given by David Webb in his '*Flora Europaea*: a retrospective' (Webb, 1978) and Max Walters in his papers on the relation between the British and Irish Floras (Walters, 1984) and the influence of the *Flora Europaea* project on the taxonomy of European vascular plants (Walters, 1995). It is clear that it was realised at the outset that if *Flora Europaea* was to succeed, then a large army of committed taxonomists was needed, and that there would be no money to pay them. It was also understood that an author should prepare an account of a genus which, after editing by one of the Editors, should be sent out to 'Regional Advisors' in as many countries and territories as possible for vetting. The work would have a European stamp of approval, not just one from the United Kingdom and Ireland. This needed a strong Editorial Committee which met, under the chairmanship of T. G. Tutin, to set the plan and rules in Leicester in January 1956. These were duly compiled and published in 1958 (with a supplement in 1960 by the Secretary, Vernon Heywood) as: *The Presentation of Taxonomic Information: a Short Guide to Contributors of Flora Europaea* (commonly known as 'the Green Book' (Heywood, 1958, 1960)).

Flora Europaea not only had a large number of good Regional Advisors making it international, but an incredible network of 187 contributing authors – only half from the United Kingdom and Ireland (Webb, 1978, p. 9). In the days before computers, Xerox machines and the internet, the *Flora Europaea* Secretariat in Reading was a hive of activity in the 1970s with a very large postage bill.

The Editorial Committee based in the United Kingdom and Ireland deliberately tried to prevent an 'insular view' developing by arranging a series of symposia, which could also function as networking events with opportunities for personal friendships to develop. The great success of these and the project overall can be seen from the first symposium held in Vienna in 1959, when few delegates knew each other, to the last in Cambridge where the Programme reports: 'There must be very few instances of such successful international cooperation involving so many countries over so many years. This alone we believe would justify *Flora Europaea*.' Some of the symposia were held in places where it was important to gain further information about the local botanical scene and obtain further collaborators, especially true of the one held in Cluj, Romania in 1963 (Pop, 1965).

Although the main success of the *Flora Europaea* organisation must be the publication of a complete Flora of Europe (Tutin *et al.*, 1964–80), the way it stimulated European plant taxonomy (despite the fact that at the time it was said the opposite would happen) and brought European botanists together had a significant effect on the development of European botanical science, something which should not be underestimated.

In 1981, John Akeroyd started work on a revision of volume one of *Flora Europaea*, some 20 years after its publication (Akeroyd & Jury, 1991; Jury, 1991). Although originally intended only to correct major errors and to bring the work up to the standard of the later volumes (by refining the geographical phrases and providing descriptions of the subspecies), the revision was far more extensive than the Editorial Committee had planned. Over 10% more species were added, almost all the un-numbered species mentioned in observations were given a full treatment or sunk in synonymy, as appropriate. The new volume (Tutin *et al.*, 1993) appeared 12 years after Akeroyd first started work on it, and has been much used, but with not all genera thoroughly revised and insufficient financial resources and time available to revive a full team of Regional Advisors, the work lacks the European authority of the earlier volumes. With the royalties from previous volumes spent on this revision and no further funding forthcoming, it was not possible to continue revisions in this or the *Flora Europaea* manner. (It was a great credit to Vernon Heywood that he had the foresight at the start of the project to set up a trust fund to be held by the Linnean Society of London to receive royalties from publishing the work. Recently, the *Flora Europaea* Trust Fund supported the follow-on Euro+Med PlantBase project, but has now been wound up.)

ESFEDS

Also in the autumn of 1981, a second project, ESFEDS (European Science Foundation European Documentation System), was started to create a database of the European flora (Heywood *et al.*, 1984; Packer & Kiger, 1989). This was part of an Additional Activity in Taxonomy that was approved by the European Science Foundation in November 1979. All names and the geographical territories were among the data included. The project published an updated checklist of European pteridophytes (Derrick *et al.*, 1987), data on the Leguminosae were transferred electronically to the International Legume Database and Information Service (ILDIS) and subsequently incorporated in *Legumes of Africa* (Lock, 1989) and a list of the non-native plants of Italy provided. A synonymic checklist of European plants was planned, but was never published. After funding had ended in 1985, the project ended and the database was given to BIOSIS (a Thompson Corporation information service for life sciences) in York (only to be later returned to the Euro+Med PlantBase Project Secretariat in Reading in 2003). However, a copy was also given to Richard Pankhurst, working first at The Natural History Museum and later at the Royal Botanic Garden Edinburgh. Pankhurst upgraded ESFEDS to include the revised volume one of *Flora Europaea* and also undertook an enormous amount of work, including that on the bibliographic references. He changed the database platform from ORACLE to his Pandora system with Advanced Revelation (Pankhurst, 1991). His significant contribution to the development of European plant databases should not be underestimated.

ESFEDS could be criticised for having spent too much time struggling to solve computer issues (such as using colour to display the various classes of data at a time when this was not readily available), rather than concentrating on the botanical aspects of the project. ESFEDS was a project rather ahead of its time.

Euro+Med PlantBase

In 1988 the Linnean Society of London celebrated its bicentenary and initiated discussion groups on a number of topics; number 3 was called the 'Pan-European Flora and Database Initiative'. A paper was prepared and circulated with the heading 'Possibilities for co-operation amongst European Flora projects' and an international group met in February and October 1991 in London. A steering group was set up which met again in November 1991 in Geneva and March 1992 before submitting a 'Concerted Action' to the European Union. This was not successful, but a new group was convened, 'Pan-European Initiative in Plant Systematics' in Reading on 24th and 25th October 1996, sponsored by the *Flora Europaea* Trust Fund (Figure 21.1). A small Steering Group of Vernon Heywood, Stephen Jury, Franco-Maria Raimondo and Benito Valdés met in the library of Palermo University's Dipartimento di Scienze Botaniche in December 1996 and began the formulation of a new project, EMIPS (Euro-Mediterranean Initiative in Plant Systematics (EMIPS); later to become Project Sisyphus and finally, Euro+Med PlantBase). (This group was supplied with a bottle of Sardinian Mirto by Professor Andrea Di Martino, Euro+Med's Calvados.) An international meeting was held at Castello Utveggio, Palermo in June 1997 with 40 delegates from 14 countries represented. A further meeting was held in Sevilla in June 1998 which resulted in an application being submitted to the EU Commission under Framework V of their Research Infrastructures programme in June 1999. This was successful, although seriously cut back, resulting in three-years of funding for Euro+Med PlantBase. The key role

Figure 21.1 Members of the 'Pan-European Initiative in Plant Systematics' fore-runner of the Euro+Med Organisation in Reading in 1996 (from left to right: front row, Pertti Uotila, Ulla-Mai Hultgard, Dominique Richard, Rupert Wilson; second row, Tomas Raus, Mohamed Rejdali, Richard Pankhurst, Santiago Castroviejo; third row, Hennig Heupler, Stephen Jury, Dmitry Geltman, John Edmondson; back row, Stephen Blackmore, Franco-Maria Raimondo, Alan Burges, Benito Valdés and Vernon Heywood.

played by Vernon Heywood throughout as an Honorary Advisor – from chairing meetings and editing reports, to planning and help with writing the application – cannot be overemphasised. His great interest and support continues.

The ESFEDS database provided the starting point for Euro+Med PlantBase. Pankhurst developed software to merge the data from the three published (of a proposed six) volumes of *Med-Checklist* (Greuter *et al.*, 1984–9) and the fourth edition of the *Flora of Macaronesia: Checklist of Vascular Plants* (Hansen & Sunding, 1993). The project also added in the extra names accepted in over 100 standard Floras for the Euro+Med area, published after *Flora Europaea* or *Med-Checklist* (or from the corresponding families not yet covered by *Med-Checklist*). This 'appending' process was carried out by Euro+Med partners with regional responsibilities (the 'Input Centres') in Sevilla (Western Europe and northwest Africa),

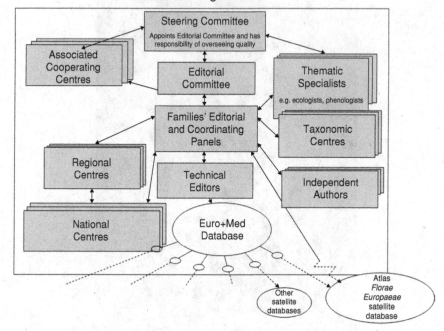

Figure 21.2 The Euro+Med PlantBase organisation. Shortage of time prevented the Families' Editorial and Coordinating Panels operating to any great extent. Regional Centres, such as Madrid (already coordinating the Iberian region for *Flora Iberica*) provided important advice and help with National Centres, local authors, etc.

Bratislava (central and northern Europe), Palermo (southern Europe, Egypt, Libya and the Middle East) and Sofia (Eastern Europe), using more software written by Pankhurst. All this data was merged and family reports produced for first editing by a team of editors. This process has not yet been completed. It is then proposed that the equivalent of the *Flora Europaea* Regional Advisors scrutinise the data for their territories, now increased to over 70. (It is no longer acceptable to maintain Sicily and Malta as a combined territory as *Flora Europaea* did, nor to maintain the old countries of the former Yugoslavia and USSR, etc.) The Euro+Med model was based on the successful *Flora Europaea* organisation and is shown in Figure 21.2.

It was recognised at the outset that the editing would have to be done by a truly international team, not now confined as in the past to one or two countries (perhaps only Spain could claim enough trained taxonomists, but they are heavily engaged on their own *Flora Iberica* project (Castroviejo *et al.*, 1986–2003)). This resulted in the proposal for Euro+Med partner Walter Berendsohn, and his team in Berlin, to develop software for an Internet editor that can access the database through the Web and allow editorial changes to the database (Geoffroy *et al.*, 2004). This novel facility has just been installed and debugged and should allow dispersed taxonomists to take responsibility for their areas of expertise. It should also reduce costs significantly.

Euro+Med PlantBase also very significantly transferred its database from Pandora to the Berlin Taxonomic Information Model which was already being used by a number of other large plant-based projects (see http://www.bgbm.org/biodivinf/docs/bgbm-model/software.htm).

Apart from providing a database with the accepted names and synonyms for the flora of the Euro+Med area, the project had a second main objective to produce new taxonomic revisions for the whole Euro+Med area. Groups were chosen to represent different areas of interest (e.g. plants important for conservation and listed by CITES, crop wild relatives, critical taxa, etc.). These included: *Acis, Galanthus, Leucojum* (Amaryllidaceae); *Anthriscus* (Apiaceae); *Launaea* (Asteraceae); *Brassica, Cardamine* (Brassicaceae); *Erodium, Geranium* (Geraniaceae); and *Asphodelus* (Liliaceae). Sixteen authors in six countries were involved in the production of these revisions. Note that lack of time and resources in the project prevented any changes to the family delimitations used in *Flora Europaea*, although this is desirable following recent systematic studies and is planned when finances allow. (This will be seen when the new edition of Heywood's well-known *Flowering Plants of the World* (Heywood, 1978) is published under the new editorial team of Messrs Heywood, Brummitt, Culham and Humphries.) The Euro+Med revisions were only available as downloadable PDF files from the Euro+Med website (www.euromed.org.uk), and will be incorporated fully into the database in time (most of the nomenclature already has).

However, Euro+Med had also been concerned to link its verified synonymic core data to other databases with additional information. So far, Euro+Med has worked with Pertti Uotila and the *Atlas Florae Europaeae* project in Helsinki, Finland (http://www.euromed.helsinki.fi/euromed/); the PhytoKaryon database of cytological information in Patras, Greece (http://www.karyoplant.biology.upatras.gr/); and the University of Bern, Switzerland and their contractors, Verlag für interactive Medien, Gaggenau (http://www.s2you.com/euromed/) for information on conservation. It is presently looking for additional funding to develop these links further and to add other data sources. These 'beads' (or databases that can share information) using Euro+Med-verified data link with other websites (see Figure 21.3). Beads with summary data for distribution, caryology and conservation have been

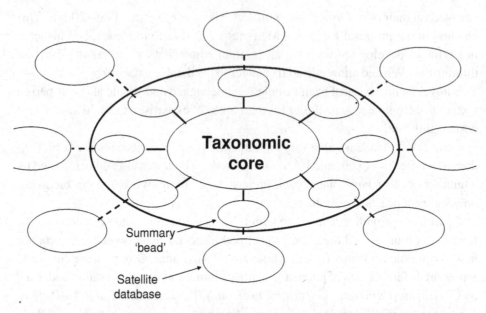

Figure 21.3 Taxonomic core consisting of taxon name, author citation, place of publication, basionym, other synonyms, geographical distribution, status (native/naturalised/cultivated), world-distribution completeness (is it endemic to the Euro+Med region?), bibliographic details (data source) and input centre. The surrounding Summary beads within the circle are linked to satellite databases outside the Euro+Med organisation. These are summary databases (not full databases) which use the Euro+Med-verified data. Beads with summary data for distribution, caryology and conservation have been developed.

developed ('Summary beads'). This was an idea developed by Vernon Heywood to obtain maximum usefulness for the greatest number of people.

The Euro+Med PlantBase checklist can be queried over the internet using software produced by partners in Palermo, but is present being upgraded in Berlin.

There is also a project called Species 2000 Europa led by Frank Bisby at Reading to develop a coordinated hub as part of Species 2000 (www.sp2000.org) for all organisms covered by Euro+Med, *Fauna Europaea* (a sister EU project developing an on-line catalogue of the animals of Europe and coordinated in Amsterdam, The Netherlands – see www.faunaeur.org) and ERMS (the European Register of Marine Species; originally funded by the European Community and now based at the Flanders Marine Research Institute in Oostende, Belgium – see www.vliz.be/vmdcdata/erms). These three databases provide a very significant checklist of species' names (with a wealth of further information attached in most cases) for the region.

At the global level, the Global Biodiversity Information Facility (GBIF), based in Copenhagen, Denmark and funded by governments, is taking the lead in making biodiversity data available via the Web (www.gbif.org). Target 1 of the Global Strategy for Plant Conservation (adopted by the Convention for Biological Diversity at the last Conference of the Parties) is to produce a list of all plants.

Vernon Heywood was also much involved in the setting up and running of another database in 1996, the Medusa network (see Chapter 10). This concerned the identification, conservation and use of wild plants in the Mediterranean region, funded partly by the European Union and the Centre International des Hautes Etudes Agronomiques Méditerranéennes (CIHEAM) and its constituent organ the Mediterranean Agronomic Institute of Chania, Crete (MAICh) (Heywood & Skoula, 1999; Skoula *et al.*, 2003). As the website says:

It is imperative that as much accurate information as possible is collected and disseminated concerning both the plants themselves and the use-related knowledge pertaining to them, both traditional and modern. The collection of this information is an essential prerequisite for the development of programmes for increasing the utilisation of these plants in a sustainable manner. It is to facilitate the achievement of this task that the MEDUSA network was conceived, and the MEDUSA database is being developed. (http://medusa.maich.gr/index.html.)

Despite a recent editorial in *Nature* (Anon. 2004), the taxonomic community has been galvanised into action, although much more remains to be done, since it is aware – more than any other scientific community – of the urgency of the task of maintaining the planet's biodiversity.

It is relatively easy to put taxonomic data on the Web but the gathering, analysis and evaluation of such information needs more skilled taxonomists with field experience without which web-based systems alone will be of little value.

References

Akeroyd, J. R. & Jury, S. L. (1991). Updating '*Flora Europaea*'. *Botanika Chronika*, **10**, 49–54.

Anon. (1977). Brief history of the *Flora Europaea* project. In *Flora Europaea Final Symposium Cambridge 31 August–4 September 1977. Programme*. Cambridge, UK: Cambridge University Press.

(2004). Ignorance is not bliss. Nature, **430**, 385.

Castroviejo, S. *et al.* (eds.) 1986–2003. *Flora Iberica*, Plantas vasculares de la Península Ibérica e Islas Baleares. Vol. I (1986); vol. II (1990); vol. III (1993); vol. IV (1993); vol. V (1997); vol. VI (1998); vol. VII(1) (1999); vol. VII(2) (2000); vol. VIII (1997); vol. X (2003); vol. XIV (2001). Madrid, Spain: CSIC.

Cauwet-Marc, A.-M. & Carbonnier, J. (eds.) (1977). *Les Ombellifères: contributions pluridisciplinaires à le systématique*. Perpignan, France: Centre National de la Recherche Scientifique and Centre Universitaire de Perpignan.

Derrick, L. N., Jermy, A. C. & Paul, A. M. (1987). *Checklist of European pteridophytes.* *Sommerfeltia* **6**, i–xx, 1–94. Oslo; Norway: Sommerfeltia.

Geoffroy, M., Güntsch, A. & Berendsohn, W. G. (2004). Teleworking for taxonomists: The Berlin Model Internet Editor. In *Abstracts and Programme, International Scientific Symposium 'Botanic Gardens: Awareness for Biodiversity'*, eds. E. Zippel, W. Greuter & A.-D. Stevens, pp. 54–5.

Greuter, W., Burdet, H. M. & Long, G. (eds.) (1984–9) *Med-Checklist.* Vol. 1 (1984); vol. 3 (1986); vol. 4 (1989). Genève: Conservatoire et Jardin Botanique de la Ville du Genève and Berlin: Botanic Garden and Botanical Museum Berlin-Dahlem.

Hansen, A. & Sunding, P. (1993). *Flora of Macaronesia: Checklist of Vascular Plants*, 4th edn. *Sommerfeltia* **17**. Oslo, Norway; Sommerfeltia.

Heywood, V. H. (ed.) (1958). *The Presentation of Taxonomic Information: a Short Guide to Contributors of Flora Europaea.* Leicester, UK: Leicester University Press.

    (1960). *The Presentation of Taxonomic Information: a Short Guide to Contributors of Flora Europaea, Supplement.* Leicester, UK: Leicester University Press.

    (1971). *The Biology and Chemistry of the Umbelliferae.* London: Linnaen Society of London.

    (1978) *Flowering Plants of the World.* Oxford, UK: Oxford University Press.

Heywood, V. H. & Skoula, M. (1999). The Medusa Network: conservation and sustainable use of wild plants of the Mediterranean region. In *Perspectives on New Crops and New Uses*, ed. J. Janick, pp. 148–51. Alexandria, VA: ASHS Press.

Heywood, V. H., Moore, D. M., Derrick, L. N., Mitchell, K. A. & Scheepen, J. van (1984). The European Taxonomic, Floristic and Biosystematic Documentation System. In *Databases in Systematics*, eds. R. Allkin & F. A. Bisby, pp. 79–89. London: Academic Press.

Jury, S. L. (1991). Some recent computer-based developments in plant taxonomy. *Botanical Journal of the Linnean Society*, **106**, 121–8.

Lock, J. M. (1989). *Legumes of Africa: a Checklist.* Kew, UK: The Royal Botanic Gardens.

Packer, J. G. & Kiger, R. W. (1989). *Flora Europaea* and the European Documentation System. In *Floristics for the 21st Century*, eds. N. R. Morin, R. D. Whetstone, D. Wilken & K. L. Tomlinson, pp. 8–10. St Louis, MO: Missouri Botanical Garden.

Pankhurst, R. (1991). *Flora Europaea* database status report: April 1991. Unpublished research report, Natural History Museum, London.

Pop, E. (ed.) (1965). Symposium '*Flora Europaea*'. *Revue Roumaine de Biologie*, **10**, 1–182.

Skoula, M., Rakic, Z., Boretos, N., Johnson, C. B. & Heywood, V. H. (2003). The Medusa Information System: a tool for the identification, conservation and sustainable use of Mediterranean plant diversity. *Acta Horticulturae*, **598**: 219–25.

Tutin, T. G., Heywood, V. H., Burges, N. A. *et al.* (eds.) (1964–80). *Flora Europaea*, vols. 1 to 5. Cambridge, UK: Cambridge University Press.

Tutin, T. G., Burges, N. A., Chater, A. O. *et al.* (eds.) (1993). *Flora Europaea*, 2nd edn., vol. 1. Cambridge, UK: Cambridge University Press.

Walters, S. M. (1984). The relation between the British and Irish Floras. *New Phytologist*, **98**, 3–13.

    (1995). The taxonomy of European vascular plants: a review of the past half-century and the influence of the *Flora Europaea* project. *Biological Reviews*, **70**, 361–74.

Webb, D. A. (1978). *Flora Europaea*: a retrospective. *Taxon*, **27**, 3–14.

Index

315

Printed in the United States
By Bookmasters